Intelligent Systems, Control and Automation: Science and Engineering

Volume 73

More information about this series at http://www.springer.com/series/6259

Chien Chern Cheah · Xiang Li

Task-Space Sensory Feedback Control of Robot Manipulators

Springer

Chien Chern Cheah
Electrical and Electronic Engineering
Nanyang Technological University
Singapore
Singapore

Xiang Li
Electrical and Electronic Engineering
Nanyang Technological University
Singapore
Singapore

ISSN 2213-8986 ISSN 2213-8994 (electronic)
Intelligent Systems, Control and Automation: Science and Engineering
ISBN 978-981-10-1352-2 ISBN 978-981-287-062-9 (eBook)
DOI 10.1007/978-981-287-062-9

Springer Singapore Heidelberg New York Dordrecht London

Softcover reprint of the hardcover 1st edition 2015

Printed on acid-free paper

Springer Science+Business Media Singapore Pte Ltd. is part of Springer Science+Business Media (www.springer.com)

Contents

Chapter 1
Introduction

A robotic manipulator consists of rigid links connected by joints, as illustrated in Fig. 1.1. One end of the manipulator is attached to a fixed base, and an end effector or a tool is connected to the other end. Each joint of the manipulator is connected to an actuator such as a motor. The actuators apply torques or forces to the joints, resulting in movement of the joints. In robotics, the term *dynamics* describes the relationship between the joint motion and the input forces and torques, as illustrated in Fig. 1.2.

The movements of the joints result in motions of the links and thus the end effector is able to perform a task. The motion of the end effector is dependent on the joint motion and mechanical structure of the manipulator, and *robot kinematics* describes the relationship between them. A forward kinematics model is used to determine the motion of the end effector given the joint motion, while an inverse kinematics model is used to calculate the joint motion given the motion of the end effector. Block diagrams illustrating the concepts of forward kinematics and inverse kinematics are

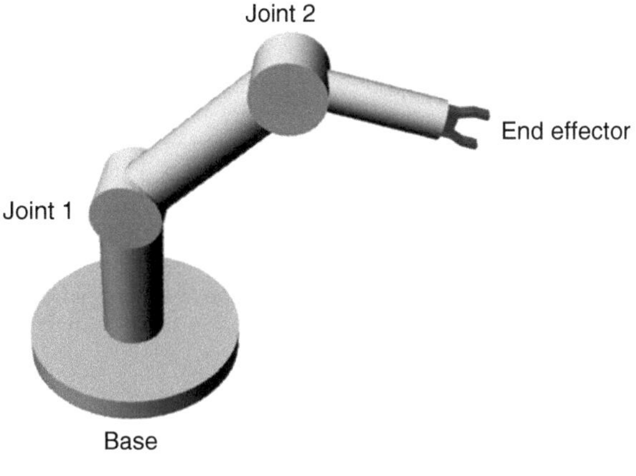

Fig. 1.1 A robotic manipulator

C.C. Cheah and X. Li, *Task-Space Sensory Feedback Control of Robot Manipulators*, Intelligent Systems, Control and Automation: Science and Engineering 73, DOI 10.1007/978-981-287-062-9_1

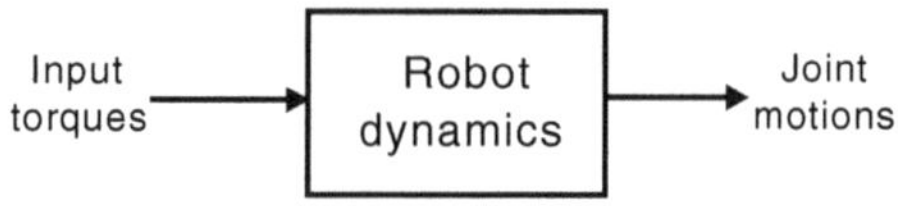

Fig. 1.2 The relationship between input torques and joint motions

shown in Fig. 1.3. Therefore, the motion of the robot end effector is related to the input torques based on the robot dynamics and kinematics, as shown in Fig. 1.4.

In order for the robot manipulator to perform a task, a desired motion of the end effector is first planned or specified. The coordinates in which the manipulator task is specified are referred to the *task space*, and the vector space in which the joint displacements are defined is referred to as the *joint space*. To execute a task, a robot controller is designed to provide suitable input torques to the robot joints so that the output motion of the end effector follows the desired or specified motion in task space.

Given a desired motion of an end effector in task space, such as Cartesian space, the inverse kinematics model can be used to compute the corresponding desired motion in joint space. In the joint-space control method, the desired motion in joint space is set as a reference input to the controller and compared with the actual joint motion to form a feedback error in joint space. A block diagram of a typical joint-space control method is shown in Fig. 1.5.

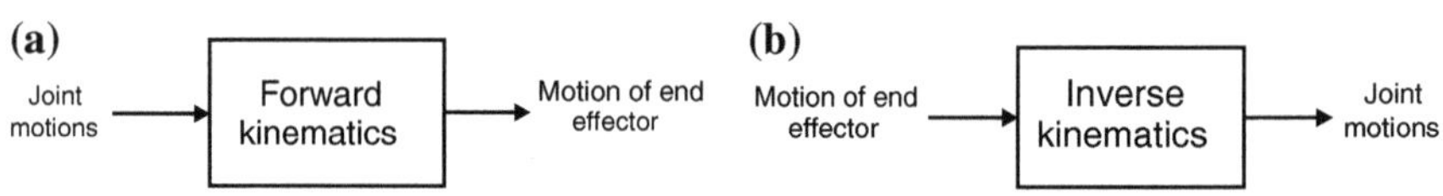

Fig. 1.3 Robot kinematics. **a** Forward kinematics. **b** Inverse kinematics

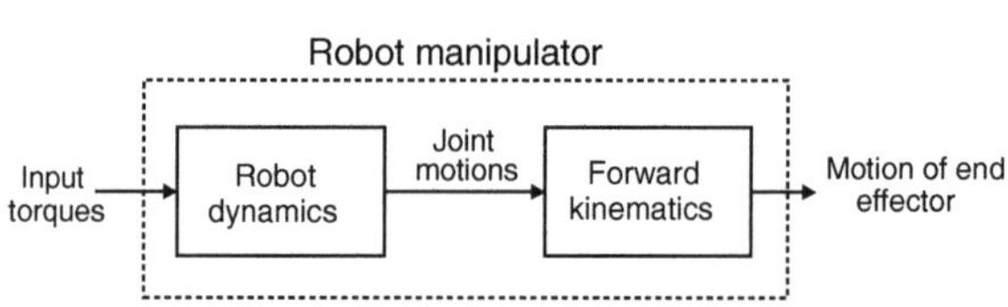

Fig. 1.4 The relationship between input torques and motion of a robot end effector

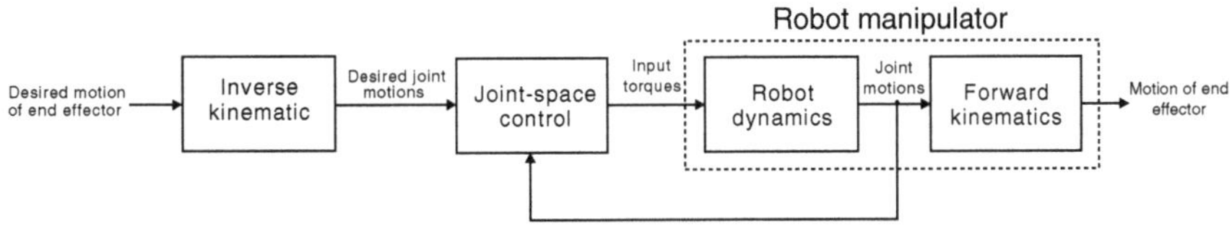

Fig. 1.5 Joint-space control

In the Cartesian-space control method, the desired motion of the end effector in Cartesian space is used directly as the reference input of the control system, as illustrated in Fig. 1.6. The motion of the end effector can be computed from the forward kinematics equation and compared with the reference input to form a feedback error in Cartesian space. The exact kinematics model is required in the Cartesian-space control, and thus this method is sensitive to modeling errors in kinematics.

When sensors are used to monitor the motion of a robot end effector, the task coordinates are defined in sensory task space. When sensory feedback is used, the robot control systems are robust to modeling and calibration errors. A block diagram of the sensory task space robot control system is illustrated in Fig. 1.7.

This book presents recent advances in robot control theory on task space sensory feedback control of robot manipulators. Various sensory task-space control methods that do not require exact knowledge of the kinematics or dynamics of robots are

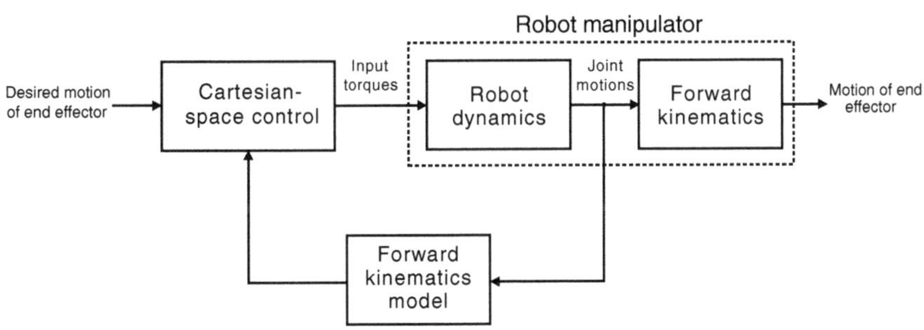

Fig. 1.6 Cartesian-space control

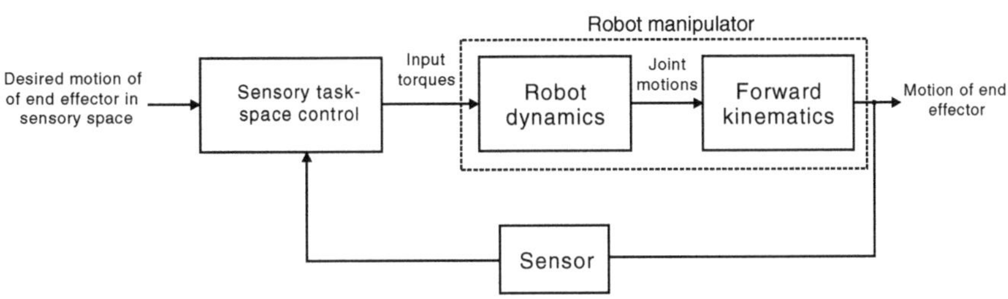

Fig. 1.7 Sensory task-space control

presented. Though identifications of kinematics and dynamics models are not required in sensory task-space control methods, a good understanding of the structure and properties of the models is essential for the design of simple and effective controllers. The dynamics and kinematics of robot manipulators and their basic properties are introduced in the remainder of this chapter. This chapter serves an introductory purpose, and more detail on kinematics and dynamics modeling can be found in most undergraduate textbooks on robotics.

1.1 Robot Dynamics

Consider a robot manipulator with n joints with the joint displacements denoted by $q_1, q_2, \ldots, q_n$. An illustration of a robot manipulator with three revolute joints q_1, q_2, q_3 is shown in Fig. 1.8.

Let $\boldsymbol{q}$ be a vector of joint displacements defined as

$$\boldsymbol{q} = [q_1, q_2, \ldots, q_n]^T \in \Re^n. \tag{1.1}$$

A corresponding vector of joint velocities $\dot{\boldsymbol{q}}$ is obtained by differentiating the vector of joint angles, so that

$$\dot{\boldsymbol{q}} = [\dot{q}_1, \dot{q}_2, \ldots, \dot{q}_n]^T \in \Re^n. \tag{1.2}$$

The vector space in which the joint motion is defined is often referred as the joint space. Each joint of the manipulator is driven by an actuator that generates the torque or force to move the link that is connected to the joint. Let $\tau_1, \tau_2, \ldots, \tau_n$ be the corresponding torques or forces that apply to the joint angles $q_1, q_2, \ldots, q_n$

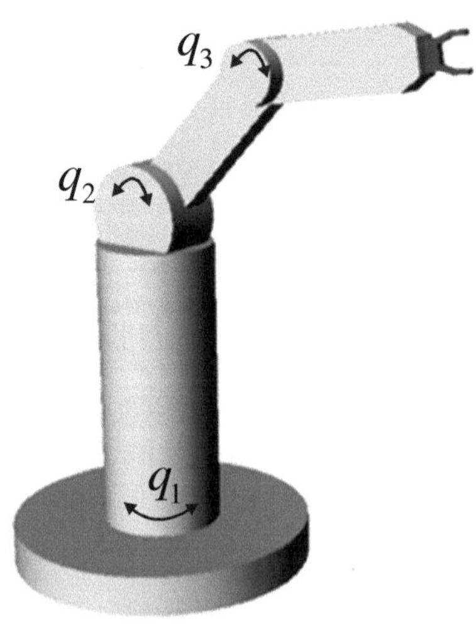

Fig. 1.8 A robot manipulator with three degrees of freedom

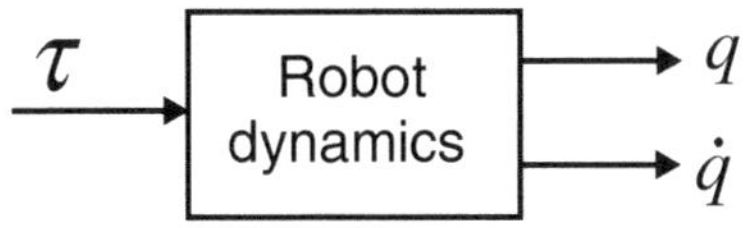

Fig. 1.9 Robot dynamics

respectively. The vector of input torque $\boldsymbol{\tau}$ is defined as follows:

$$\boldsymbol{\tau} = [\tau_1, \tau_1, \ldots, \tau_n]^T \in \Re^n. \quad (1.3)$$

The dynamics model describes the relationship between the joint variables and input forces and torques, as illustrated in Fig. 1.9.

Here, a Lagrange formulation is adopted for the derivation of the robot dynamics model. Using such a Lagrange formulation, the equations of motion can be derived in a systematic way independently of the reference coordinate frame. The Lagrangian L is defined as the difference between the kinetic energy K and potential energy P, and the Lagrangian of the manipulator system can be written as

$$L(\boldsymbol{q}, \dot{\boldsymbol{q}}) = K(\boldsymbol{q}, \dot{\boldsymbol{q}}) - P(\boldsymbol{q}) = \tfrac{1}{2}\dot{\boldsymbol{q}}^T \boldsymbol{M}(\boldsymbol{q})\dot{\boldsymbol{q}} - P(\boldsymbol{q}), \quad (1.4)$$

where $\boldsymbol{M}(\boldsymbol{q}) \in \Re^{n\times n}$ is the manipulator inertial matrix. Using the definition of $L(\boldsymbol{q}, \dot{\boldsymbol{q}})$, the dynamics equation of the robot manipulator with n degrees of freedom is given in joint space as

$$\boldsymbol{M}(\boldsymbol{q})\ddot{\boldsymbol{q}} + \left(\tfrac{1}{2}\dot{\boldsymbol{M}}(\boldsymbol{q}) + \boldsymbol{S}(\boldsymbol{q}, \dot{\boldsymbol{q}})\right)\dot{\boldsymbol{q}} + \boldsymbol{g}(\boldsymbol{q}) = \boldsymbol{\tau}, \quad (1.5)$$

where $\boldsymbol{S}(\boldsymbol{q}, \dot{\boldsymbol{q}})\dot{\boldsymbol{q}} = \frac{1}{2}\left(\dot{\boldsymbol{M}}(\boldsymbol{q})\dot{\boldsymbol{q}} - \left\{\frac{\partial}{\partial \boldsymbol{q}}\dot{\boldsymbol{q}}^T \boldsymbol{M}(\boldsymbol{q})\dot{\boldsymbol{q}}\right\}^T\right), \boldsymbol{C}(\boldsymbol{q}, \dot{\boldsymbol{q}})\dot{\boldsymbol{q}} = \left(\frac{1}{2}\dot{\boldsymbol{M}}(\boldsymbol{q}) + \boldsymbol{S}(\boldsymbol{q}, \dot{\boldsymbol{q}})\right)\dot{\boldsymbol{q}}$ represents the Coriolis/centripetal forces, and $\boldsymbol{g}(\boldsymbol{q}) = \frac{P(\boldsymbol{q})}{\partial \boldsymbol{q}} \in \Re^n$ denotes the vector of gravitational forces.

For a robot manipulator with all revolute joints, some important properties of the dynamic Eq. (1.5) are listed that will be used in the design of the controllers throughout this book.

Property 1.1 *The inertia matrix $\boldsymbol{M}(\boldsymbol{q})$ is symmetric and positive definite. Since each entry of $\boldsymbol{M}(\boldsymbol{q})$ is constant or a trigonometric function of $\boldsymbol{q}$, there are positive constants γ_m and γ_M ($\gamma_m < \gamma_M$) such that*

$$\gamma_m \boldsymbol{I}_n \leq \boldsymbol{M}(\boldsymbol{q}) \leq \gamma_M \boldsymbol{I}_n, \quad (1.6)$$

where $\boldsymbol{I}_n \in \Re^{n\times n}$ is the $n \times n$ identity matrix. □

Property 1.2 *The matrix $\boldsymbol{S}(\boldsymbol{q}, \dot{\boldsymbol{q}}) \in \Re^{n\times n}$ is skew-symmetric and satisfies*

$$\boldsymbol{y}^T \boldsymbol{S}(\boldsymbol{q}, \dot{\boldsymbol{q}})\boldsymbol{y} = 0 \;\forall\; \boldsymbol{q}, \dot{\boldsymbol{q}}, \boldsymbol{y} \in \Re^n. \quad (1.7)$$

In addition, $\boldsymbol{C}(\boldsymbol{q},\dot{\boldsymbol{q}})\dot{\boldsymbol{q}} = \left(\frac{1}{2}\dot{\boldsymbol{M}}(\boldsymbol{q})+\boldsymbol{S}(\boldsymbol{q},\dot{\boldsymbol{q}})\right)\dot{\boldsymbol{q}}$ *is quadratic in* $\dot{\boldsymbol{q}}$ *and periodic in each component of* $\boldsymbol{q}$*, and*

$$||\boldsymbol{C}(\boldsymbol{q},\dot{\boldsymbol{q}})\dot{\boldsymbol{q}}|| \leq c||\dot{\boldsymbol{q}}||^2, \tag{1.8}$$

where c *is a positive constant and* $||\cdot||$ *denotes the* Euclidean norm. □

Throughout this book, the Euclidean norm is defined as follows.

Definition 1.1 The Euclidean norm $||\boldsymbol{y}||$ of a vector $\boldsymbol{y}$ is defined as

$$||\boldsymbol{y}|| = \sqrt{\boldsymbol{y}^T\boldsymbol{y}}. \tag{1.9}$$

□

Property 1.3 *The dynamic model as described by Eq. (1.5) is linear in a set of physical parameters* $\boldsymbol{\theta}_d = [\theta_{d1}, \ldots, \theta_{dn_d}]^T \in \Re^{n_d}$ *as follows:*

$$\boldsymbol{M}(\boldsymbol{q})\ddot{\boldsymbol{q}} + \left(\tfrac{1}{2}\dot{\boldsymbol{M}}(\boldsymbol{q}) + \boldsymbol{S}(\boldsymbol{q},\dot{\boldsymbol{q}})\right)\dot{\boldsymbol{q}} + \boldsymbol{g}(\boldsymbol{q}) = \boldsymbol{Y}_d(\boldsymbol{q},\dot{\boldsymbol{q}},\dot{\boldsymbol{q}},\ddot{\boldsymbol{q}})\boldsymbol{\theta}_d, \tag{1.10}$$

where $\boldsymbol{Y}_d(\boldsymbol{q},\dot{\boldsymbol{q}},\dot{\boldsymbol{q}},\ddot{\boldsymbol{q}}) \in \Re^{n\times n_d}$ *is a dynamic regressor matrix.* □

A special case of Property 1.3 is the following, which is usually used in setpoint control problems of robot manipulators.

Property 1.4 *The gravitational force vector* $\boldsymbol{g}(\boldsymbol{q})$ *is bounded and can be characterized by a set of physical parameters* $\boldsymbol{\theta}_g = (\theta_1, \ldots, \theta_{n_g})^T \in \Re^{n_g}$ *as follows:*

$$\boldsymbol{g}(\boldsymbol{q}) = \boldsymbol{Y}_g(\boldsymbol{q})\boldsymbol{\theta}_g, \tag{1.11}$$

where $\boldsymbol{Y}_g(\boldsymbol{q}) \in \Re^{n\times n_g}$ *is a gravity regressor matrix.* □

For prismatic joints, the joint variables represent the translation displacements of joints, which are bounded by the mechanical limits. Thus the above properties also hold if the prismatic joint variables are bounded.

Example 1.1 Consider a single-link robot as shown in Fig. 1.10, where the joint angle is q and the input torque is τ. The robot dynamics that describes the relationship between input torque and joint motion is described as

$$ml^2\ddot{q} + mlg\cos(q) = \tau, \tag{1.12}$$

where m denotes the mass of the link, l is the length from the base to the centroid of the link, and g denotes the acceleration due to gravity.

The inertia $M = ml^2$ is a positive scalar, we have $C = \left(\frac{1}{2}\dot{M} + S\right)\dot{q} = 0$, the dynamic model in Eq. (1.12) can be parameterized as $ml^2\ddot{q} + mlg\cos(q) = [\ddot{q}, \cos(q)] \times [ml^2, mlg]^T = Y_d(q,\ddot{q})\boldsymbol{\theta}_d$, and the gravitational force $g(q) =$

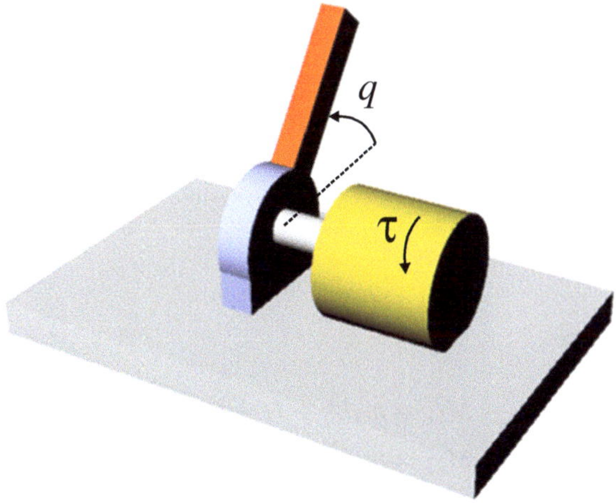

Fig. 1.10 A single-link robot

$mlg\cos(q)$ is bounded and can be parameterized as $g(q) = Y_g(q)\theta_g$, where $Y_g(q) = \cos(q)$, and $\theta_g = mlg$. □

Example 1.2 Consider a planar robot manipulator with two degrees of freedom, as shown in Fig. 1.11.

The vector of joint angles is

$$\boldsymbol{q} = [q_1, q_2]^T, \tag{1.13}$$

and the vector of joint velocities is

$$\dot{\boldsymbol{q}} = [\dot{q}_1, \dot{q}_2]^T. \tag{1.14}$$

Each joint is driven by an actuator, and the vector of input torques is

$$\boldsymbol{\tau} = [\tau_1, \tau_2]^T. \tag{1.15}$$

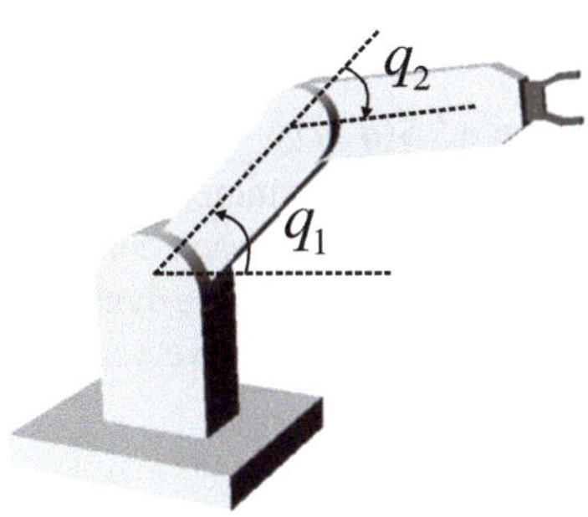

Fig. 1.11 A 2-degree-of-freedom (DOF) planar manipulator

The dynamic equation that describes the relationship between joint motion and input torques can be derived as

$$\begin{bmatrix} M_{11} & M_{12} \\ M_{21} & M_{22} \end{bmatrix}\begin{bmatrix} \ddot{q}_1 \\ \ddot{q}_2 \end{bmatrix} + \left(\frac{1}{2}\begin{bmatrix} \dot{M}_{11} & \dot{M}_{12} \\ \dot{M}_{21} & \dot{M}_{22} \end{bmatrix}\right.$$
$$\left. + \begin{bmatrix} 0 & -m_2\sin(q_2)l_1l_{c2}\left(\dot{q}_1 + \frac{1}{2}\dot{q}_2\right) \\ m_2\sin(q_2)l_1l_{c2}\left(\dot{q}_1 + \frac{1}{2}\dot{q}_2\right) & 0 \end{bmatrix}\right)\begin{bmatrix} \dot{q}_1 \\ \dot{q}_2 \end{bmatrix} + \begin{bmatrix} g_1 \\ g_2 \end{bmatrix} = \begin{bmatrix} \tau_1 \\ \tau_2 \end{bmatrix}, \tag{1.16}$$

where

$$\begin{aligned} M_{11} &= m_1l_{c1}^2 + I_1 + m_2\left[l_1^2 + l_{c2}^2 + 2\cos(q_2)l_1l_{c2}\right] + I_2, \\ M_{22} &= m_2l_{c2}^2 + I_2, \\ M_{12} &= M_{21} = m_2\cos(q_2)l_1l_{c2} + m_2l_{c2}^2 + I_2, \\ g_1 &= m_1g\cos(q_1)l_{c1} + m_2g[\cos(q_1 + q_2)l_{c2} + \cos(q_1)l_1], \\ g_2 &= m_2g\cos(q_1 + q_2)l_{c2}, \end{aligned} \tag{1.17}$$

and m_1, m_2 are the masses of the first and second links, l_1, l_2 are the lengths of the links, I_1, I_2 are the inertias, l_{c1}, l_{c2} denote the respective lengths from the centroid of each joint to the center of gravity.

Therefore, the inertia matrix $\boldsymbol{M}(\boldsymbol{q})$ is obtained as

$$\boldsymbol{M}(\boldsymbol{q}) = \begin{bmatrix} M_{11} & M_{12} \\ M_{21} & M_{22} \end{bmatrix} \in \Re^{2\times 2}. \tag{1.18}$$

Since $M_{11} > 0, \det(\boldsymbol{M}(\boldsymbol{q})) = I_1I_2 + l_1^2l_{c2}^2m_2^2\left(1-\cos^2(q_2)\right) + I_2\left(l_1^2m_2 + l_{c1}^2m_1\right) + I_1l_{c2}^2m_2 + l_{c1}^2l_{c2}^2m_1m_2 > 0$, and $M_{12} = M_{21}$, $\boldsymbol{M}(\boldsymbol{q})$ is symmetric and positive definite. Each entry of $\boldsymbol{M}(\boldsymbol{q})$ is also bounded as in Property 1.1.

From Eqs. (1.5) and (1.16), the matrix $\boldsymbol{S}(\boldsymbol{q}, \dot{\boldsymbol{q}})$ is described as

$$\boldsymbol{S}(\boldsymbol{q}, \dot{\boldsymbol{q}}) = \begin{bmatrix} 0 & -m_2\sin(q_2)l_1l_{c2}\left(\dot{q}_1 + \frac{1}{2}\dot{q}_2\right) \\ m_2\sin\left(q_2\right)l_1l_{c2}\left(\dot{q}_1 + \frac{1}{2}\dot{q}_2\right) & 0 \end{bmatrix} \in \Re^{2\times 2}, \tag{1.19}$$

where $\boldsymbol{y}^T\boldsymbol{S}(\boldsymbol{q}, \dot{\boldsymbol{q}})\boldsymbol{y} = (y_1y_2 - y_1y_2)m_2\sin(q_2)l_1l_{c2}\left(\dot{q}_1 + \frac{1}{2}\dot{q}_2\right) = 0$. Therefore, $\boldsymbol{S}(\boldsymbol{q}, \dot{\boldsymbol{q}})$ is skew-symmetric, and $\boldsymbol{C}(\boldsymbol{q}, \dot{\boldsymbol{q}})\dot{\boldsymbol{q}} = \left(\frac{1}{2}\dot{\boldsymbol{M}}(\boldsymbol{q}) + \boldsymbol{S}(\boldsymbol{q}, \dot{\boldsymbol{q}})\right)\dot{\boldsymbol{q}}$ is also quadratic in $\dot{\boldsymbol{q}}$ and periodic in each component of $\boldsymbol{q}$, as described in Property 1.2.

As in Property 1.3, the dynamic model can be written as the product of the dynamic regressor matrix and a vector of physical parameters as

$$M(q)\ddot{q} + \left(\tfrac{1}{2}\dot{M}(q) + S(q,\dot{q})\right)\dot{q} + g(q) = Y_d(q,\dot{q},\dot{q},\ddot{q})\theta_d$$

$$= \begin{bmatrix} \ddot{q}_1 & \ddot{q}_1+\ddot{q}_2 & 2c_2\ddot{q}_1 + c_2\ddot{q}_2 - s_2\dot{q}_1\dot{q}_2 - s_2\left(\dot{q}_1\dot{q}_2+\dot{q}_2^2\right) & c_1 & c_{12} \\ 0 & \ddot{q}_1+\ddot{q}_2 & c_2\ddot{q}_1 + s_2\dot{q}_1^2 & 0 & c_{12} \end{bmatrix}$$

$$\times \begin{bmatrix} m_1 l_{c1}^2 + m_2 l_1^2 + I_1 \\ m_2 l_{c2}^2 + I_2 \\ m_2 l_1 l_{c2} \\ m_1 g l_{c1} + m_2 g l_1 \\ m_2 g l_{c2} \end{bmatrix}, \tag{1.20}$$

where $c_1 = \cos(q_1)$, $c_{12} = \cos(q_1+q_2)$, $c_2 = \cos(q_2)$, and $s_2 = \sin(q_2)$.

Each entry of the gravity vector is bounded, and the gravity vector $g(q) = [g_1, g_2]^T \in \Re^2$ can also be written as the product of the gravity regressor matrix and a vector of physical parameters as follows (see Property 1.4):

$$g(q) = Y_g(q)\theta_g = \begin{bmatrix} c_1 & c_{12} & c_1 \\ 0 & c_{12} & 0 \end{bmatrix} \begin{bmatrix} m_1 g l_{c1} \\ m_2 g l_{c2} \\ m_2 g l_1 \end{bmatrix}. \tag{1.21}$$

□

1.2 Forward Kinematics and the Jacobian Matrix

When input torques or forces are applied to the joints by the actuators, the joints and links move and result in motion of the end effector. The coordinate space in which the manipulation task of the end effector is performed is referred to as the task space, for example Cartesian space. Let r be a task-space vector defined in Cartesian space as

$$r = [r_1, r_2, \ldots, r_p]^T \in \Re^p. \tag{1.22}$$

The Cartesian-space vector r can be defined as either the position of the end effector or the orientation of the end effector or both the position and orientation of the end effector, depending on the task requirements. A manipulation task is considered realizable if there are sufficient degrees of freedom to perform the task, that is, if the number of joints n or the dimension of the vector of joint space variables is more than or equal to p.

The Cartesian-space vector r is dependent on the joint configuration and can be expressed as a function of joint-space displacements as

$$r = h(q), \tag{1.23}$$

where $h(\cdot) \in \Re^n \rightarrow \Re^p$ represents a nonlinear function that computes the Cartesian-space position variables from the joint-space position variables. Let $\dot{r}$

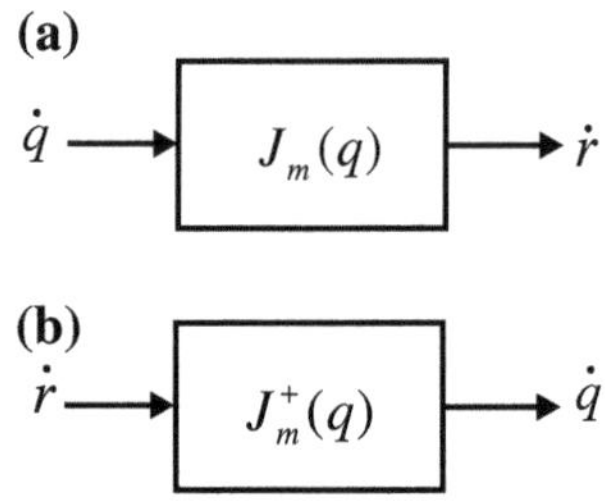

Fig. 1.12 **a** Jacobian matrix $\boldsymbol{J}_m(\boldsymbol{q})$; **b** Pseudo-inverse matrix $\boldsymbol{J}_m^+(\boldsymbol{q})$

denote a velocity vector in Cartesian space. Then $\dot{\boldsymbol{r}}$ is related to the vector of joint-space velocity $\dot{\boldsymbol{q}}$ by

$$\dot{\boldsymbol{r}} = \boldsymbol{J}_m(\boldsymbol{q})\dot{\boldsymbol{q}}, \tag{1.24}$$

where $\boldsymbol{J}_m(\boldsymbol{q}) \in \Re^{p\times n}$ is the manipulator Jacobian matrix.

Robot kinematics describes the relationship between task space and joint space. The forward kinematics model allows the task-space variables to be expressed in terms of the joint-space variables, while the inverse kinematics give the inverse relationship. The Jacobian matrix $\boldsymbol{J}_m(\boldsymbol{q})$ relates the joint-space velocity to the velocity of the robot end effector in Cartesian space, while the pseudo-inverse matrix $\boldsymbol{J}_m^+(\boldsymbol{q})$ (or the inverse matrix $\boldsymbol{J}_m^{-1}(\boldsymbol{q})$ if $p = n$) represents the inverse relationship, as illustrated in Fig. 1.12.

The Jacobian matrix $\boldsymbol{J}_m(\boldsymbol{q})$ is said to be singular when the joint-space velocity becomes infinite to maintain a finite Cartesian-space velocity of the robot end effector. The manipulation task is therefore not realizable if the Jacobian matrix is singular.

Singularities occur when the Jacobian matrix $\boldsymbol{J}_m(\boldsymbol{q})$ is not of full rank. The singularity issue limits the workspace of the robot and impedes the robot in performing its task. It is therefore necessary to handle the singularity problem carefully or avoid it entirely so that the robot can be operated properly and stably within the workspace. The singular configurations of the manipulator can be obtained by investigating the Jacobian matrix, and a singularity occurs whenever the Jacobian matrix is not of full rank.

In general, the manipulator singularities can be divided into external singularities and internal singularities. An external singularity occurs at the external workspace boundary, and the internal singularities include the internal boundary singularities and interior singularities.

Example 1.3 Consider the 2-DOF planar manipulator in Fig. 1.11. Let $\boldsymbol{r} = [r_1, r_2]^T$ denote the position of the end effector in Cartesian space, and the forward kinematic equation is given as

$$\boldsymbol{r} = \begin{bmatrix} r_1 \\ r_2 \end{bmatrix} = \begin{bmatrix} l_1\cos(q_1) + l_2\cos(q_1 + q_2) \\ l_1\sin(q_1) + l_2\sin(q_1 + q_2) \end{bmatrix} \in \Re^2, \tag{1.25}$$

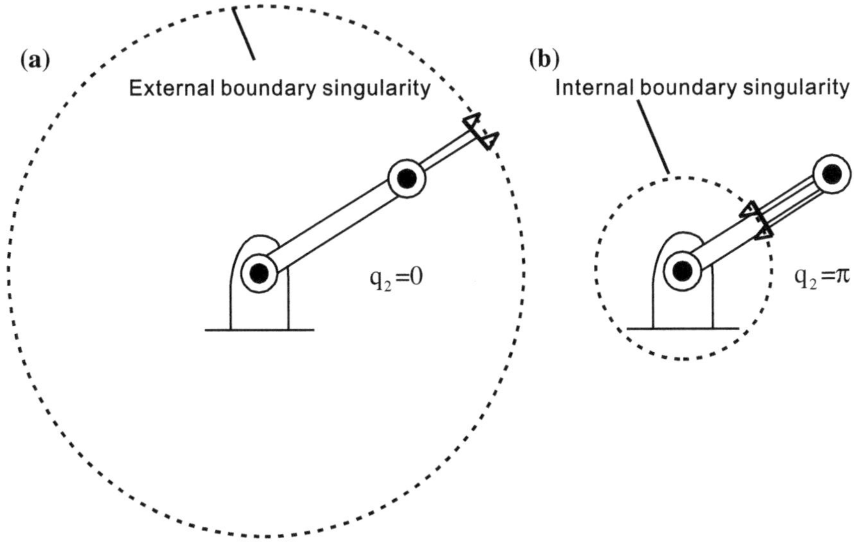

Fig. 1.13 Singularities of a 2-link planar robot manipulator occur **a** when the manipulator is fully stretched out, or **b** when the manipulator is fully folded back

where $\boldsymbol{q} = [q_1, q_2]^T$, q_1, q_2 denote the first and second joint angles, and l_1 and l_2 are the lengths of the first and the second links.

The Jacobian matrix of the manipulator is described as follows:

$$\boldsymbol{J}_m(\boldsymbol{q}) = \begin{bmatrix} -l_1\sin(q_1) - l_2\sin(q_1+q_2) & -l_2\sin(q_1+q_2) \\ l_1\cos(q_1) + l_2\cos(q_1+q_2) & l_2\cos(q_1+q_2) \end{bmatrix} \in \Re^{2\times 2}. \quad (1.26)$$

The Jacobian matrix is singular when

$$\det[\boldsymbol{J}_m(\boldsymbol{q})] = l_1 l_2 \sin(q_2) = 0. \quad (1.27)$$

Therefore, singularities occur where $q_2 = 0$ (external boundary singularity) and $q_2 = \pi$ (internal boundary singularity), as illustrated in Fig. 1.13. □

Example 1.4 Consider the 3-DOF manipulator illustrated in Fig. 1.14. Let $\boldsymbol{r} = [r_1, r_2, r_3]^T$ denote the position of the end effector in Cartesian space, and the forward kinematic equation is given as

$$\boldsymbol{r} = \begin{bmatrix} r_1 \\ r_2 \\ r_3 \end{bmatrix} = \begin{bmatrix} \cos(q_1)[l_2\cos(q_2) + l_3\cos(q_2+q_3)] \\ \sin(q_1)[l_2\cos(q_2) + l_3\cos(q_2+q_3)] \\ l_1 - l_2\sin(q_2) - l_3\sin(q_2+q_3) \end{bmatrix} \in \Re^3, \quad (1.28)$$

where $\boldsymbol{q} = [q_1, q_2, q_3]^T$, q_1, q_2, q_3 denote the joint angles, and l_1, l_2, l_3 are the lengths of the links. Therefore, the Jacobian matrix is obtained as

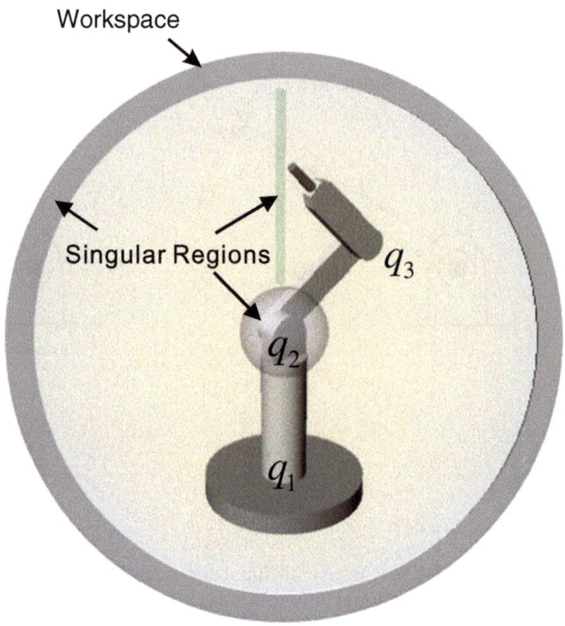

Fig. 1.14 Singular regions which enclose the singular configurations of a 3-DOF robot

$$J_m(q) = \begin{bmatrix} -s_1(l_2c_2 + l_3c_{23}) & -s_2c_1l_2 - s_{23}l_3c_1 & -s_{23}l_3c_1 \\ -c_1(l_2c_2 + l_3c_{23}) & -s_2s_1l_2 - s_{23}l_3s_1 & -s_{23}l_3s_1 \\ 0 & -c_2l_2 - c_{23}l_3 & -c_{23}l_3 \end{bmatrix} \in \Re^{3\times 3}, \quad (1.29)$$

where $s_1 = \sin(q_1)$, $s_{23} = \sin(q_2+q_3)$, $c_{23} = \cos(q_2+q_3)$, and hence the Jacobian matrix of the manipulator is singular when

$$\det[J_m(q)] = [l_2\cos(q_2) + l_3\cos(q_2+q_3)]l_3l_2\sin(q_3) = 0. \quad (1.30)$$

Therefore, singularities occur at the joint configurations where $q_3 = 0$ (external boundary singularity), $q_3 = \pi$ (internal boundary singularity), and $[l_2\cos(q_2) + l_3\cos(q_2+q_3)] = 0$ (interior singularity). □

Example 1.5 Consider the 4-DOF redundant planar manipulator illustrated in Fig. 1.15. The Cartesian-space position of the robot end effector can be derived in terms of the joint variables $q = [q_1, q_2, q_3, q_4]^T$ as

$$r = \begin{bmatrix} r_1 \\ r_2 \end{bmatrix} = \begin{bmatrix} l_1c_1 + l_2c_{12} + l_3c_{123} + l_4c_{1234} \\ l_1s_1 + l_2s_{12} + l_3s_{123} + l_4s_{1234} \end{bmatrix} \in \Re^2, \quad (1.31)$$

where $c_{123} = \cos(q_1+q_2+q_3)$, $s_{123} = \sin(q_1+q_2+q_3)$, $c_{1234} = \cos(q_1+q_2+q_3+q_4)$, $s_{1234} = \sin(q_1+q_2+q_3+q_4)$, l_1, l_2, l_3, l_4 are the lengths of the links.

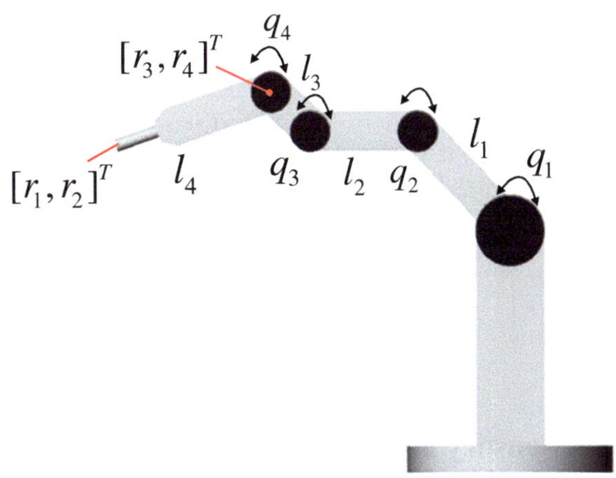

Fig. 1.15 A 4-DOF redundant planar robot

On differentiating Eq. (1.31), the Jacobian matrix $\boldsymbol{J}_m(\boldsymbol{q})$ relating the velocity of end effector in Cartesian space to joint-space velocity is obtained as

$$\boldsymbol{J}_m(\boldsymbol{q}) = \begin{bmatrix} -(l_1 s_1 + l_2 s_{12} + l_3 s_{123} + l_4 s_{1234}) & -(l_2 s_{12} + l_3 s_{123} + l_4 s_{1234}) & -(l_3 s_{123} + l_4 s_{1234}) & -l_4 s_{1234} \\ l_1 c_1 + l_2 c_{12} + l_3 c_{123} + l_4 c_{1234} & l_2 c_{12} + l_3 c_{123} + l_4 c_{1234} & l_3 c_{123} + l_4 c_{1234} & l_4 c_{1234} \end{bmatrix}. \tag{1.32}$$

If the end point of the third link of the redundant robot is specified as a second feature point as shown in Fig. 1.15, the Cartesian-space position of the additional feature is derived as

$$\begin{bmatrix} r_3 \\ r_4 \end{bmatrix} = \begin{bmatrix} l_1 c_1 + l_2 c_{12} + l_3 c_{123} \\ l_1 s_1 + l_2 s_{12} + l_3 s_{123} \end{bmatrix}. \tag{1.33}$$

The following Cartesian-space vector $\boldsymbol{r}$ can be used to describe the positions of two features:

$$\boldsymbol{r} = \begin{bmatrix} r_1 \\ r_2 \\ r_3 \\ r_4 \end{bmatrix}. \tag{1.34}$$

Similarly, the velocities of the features in Cartesian space are related to joint-space velocities by the Jacobian matrix as follows:

$$\dot{\boldsymbol{r}} = \begin{bmatrix} \dot{r}_1 \\ \dot{r}_2 \\ \dot{r}_3 \\ \dot{r}_4 \end{bmatrix} = \boldsymbol{J}_m(\boldsymbol{q})\dot{\boldsymbol{q}}$$

$$= \begin{bmatrix} -(l_1 s_1 + l_2 s_{12} + l_3 s_{123} + l_4 s_{1234}) & -(l_2 s_{12} + l_3 s_{123} + l_4 s_{1234}) & -(l_3 s_{123} + l_4 s_{1234}) & -l_4 s_{1234} \\ l_1 c_1 + l_2 c_{12} + l_3 c_{123} + l_4 c_{1234} & l_2 c_{12} + l_3 c_{123} + l_4 c_{1234} & l_3 c_{123} + l_4 c_{1234} & l_4 c_{1234} \\ -(l_1 s_1 + l_2 s_{12} + l_3 s_{123}) & -(l_2 s_{12} + l_3 s_{123}) & -l_3 s_{123} & 0 \\ l_1 c_1 + l_2 c_{12} + l_3 c_{123} & l_2 c_{12} + l_3 c_{123} & l_3 c_{123} & 0 \end{bmatrix}$$

$$\times \begin{bmatrix} \dot{q}_1 \\ \dot{q}_2 \\ \dot{q}_3 \\ \dot{q}_4 \end{bmatrix}. \qquad (1.35)$$

□

1.3 Sensory Jacobian Matrix

When sensors are employed to measure the motion of an end effector, a vector of feature points in the sensory space can be specified to describe the position of the robot end effector. The feature points in the sensory space are represented as a vector of sensory-space variables as

$$\boldsymbol{x} = [x_1, x_2, \ldots, x_m]^T \in \Re^m. \qquad (1.36)$$

The relationship between the velocities of sensory task-space variables and the velocities of the end effector in Cartesian space is given by

$$\dot{\boldsymbol{x}} = \boldsymbol{J}_s(\boldsymbol{r})\dot{\boldsymbol{r}}, \qquad (1.37)$$

where $\boldsymbol{J}_s(\boldsymbol{r}) \in \Re^{m \times p}$ is the sensory Jacobian matrix. A manipulation task specified in sensory task space is considered realizable if $m \leq p \leq n$.

In general, the Jacobian matrix $\boldsymbol{J}_s(\boldsymbol{r})$ represents the mapping from Cartesian space to sensory task space. From Eqs. (1.24) and (1.37), the velocities of features are related to the joint velocities by

$$\dot{\boldsymbol{x}} = \boldsymbol{J}_s(\boldsymbol{r})\boldsymbol{J}_m(\boldsymbol{q})\dot{\boldsymbol{q}} = \boldsymbol{J}(\boldsymbol{q})\dot{\boldsymbol{q}}, \qquad (1.38)$$

where $\boldsymbol{J}(\boldsymbol{q}) = \boldsymbol{J}_s(\boldsymbol{r})\boldsymbol{J}_m(\boldsymbol{q})$ represents the overall Jacobian matrix from joint space to sensory task space, as illustrated in Fig. 1.16.

A property of the Jacobian matrix is the following.

Property 1.5 *The right-hand side of equation* $\dot{\boldsymbol{x}} = \boldsymbol{J}(\boldsymbol{q})\dot{\boldsymbol{q}}$ *is linear in a set of constant kinematic parameters* $\boldsymbol{\theta}_k = [\theta_{k1}, \ldots, \theta_{kn_k}]^T \in \Re^{n_k}$, *such as link lengths and link twist angles, which can be expressed as*

$$\dot{\boldsymbol{x}} = \boldsymbol{J}(\boldsymbol{q})\dot{\boldsymbol{q}} = \boldsymbol{Y}_k(\boldsymbol{q}, \dot{\boldsymbol{q}})\boldsymbol{\theta}_k, \qquad (1.39)$$

where $\boldsymbol{Y}_k(\boldsymbol{q}, \dot{\boldsymbol{q}}) \in \Re^{n \times n_k}$ *is called the* kinematic regressor matrix. □

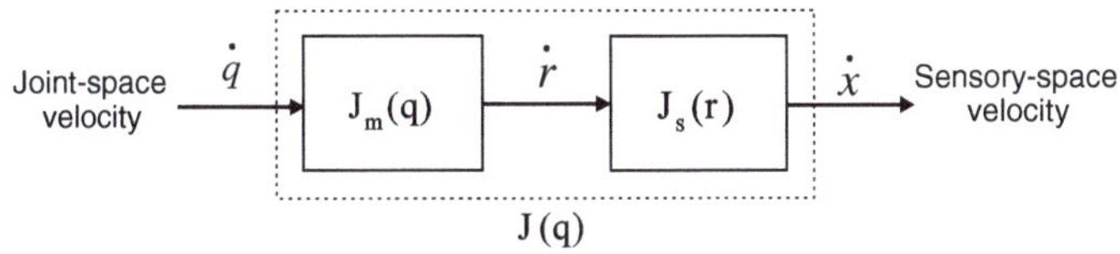

Fig. 1.16 The Jacobian matrix $\boldsymbol{J}(\boldsymbol{q})$ represents the transformation of joint-space velocity to sensory task-space velocity

When a camera is used to measure the position of a robot end effector, the vector $\boldsymbol{x}$ represents the image-space position of the end effector, which is specified as

$$\boldsymbol{x} = [x_h, x_v]^T \in \Re^2, \tag{1.40}$$

where x_h represents the horizontal coordinate, while x_v denotes the vertical coordinate in image space.

Using the visual information from a camera to control the movement of a robotic manipulator is referred to as *visual servoing*. In visual servoing, the relationship between the robot frame and the camera is to be derived. In general, the relationship is defined by a projection between the robot workspace and the camera image plane via the imaging geometry of the camera.

The pinhole camera model is widely used to represent the mapping from Cartesian space to image space, and hence the projection of a point onto the image plane is modeled as a central projection through the center of the lens, which is illustrated in Fig. 1.17.

Based on the pinhole camera model, the image-space velocity of the feature is related to the Cartesian-space velocity of the end effector by

$$\dot{\boldsymbol{x}} = \underbrace{\boldsymbol{Z}^{-1}(\boldsymbol{q})\boldsymbol{L}(\boldsymbol{x})}_{\boldsymbol{J}_s(\boldsymbol{r})}\dot{\boldsymbol{r}}, \tag{1.41}$$

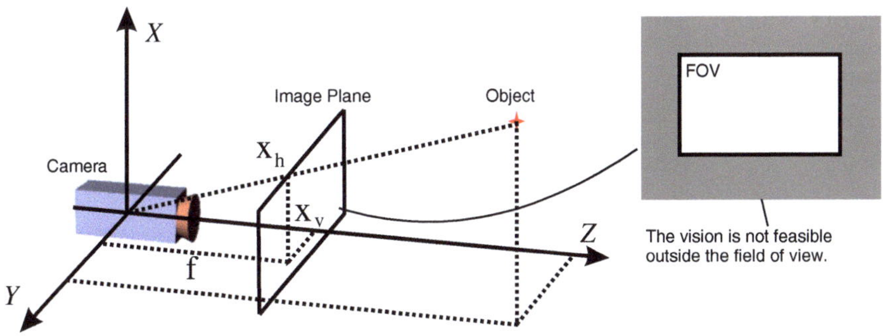

Fig. 1.17 Pinhole camera model

where $\boldsymbol{Z}^{-1}(\boldsymbol{q}) \in \Re^{2\times 2}$ is a diagonal matrix that contains the depth information of the feature point with respect to the camera image frame, and $\boldsymbol{L}(\boldsymbol{x}) \in \Re^{2\times p}$ is a Jacobian matrix. The overall matrix $\boldsymbol{Z}^{-1}(\boldsymbol{q})\boldsymbol{L}(\boldsymbol{x})$ is the image Jacobian matrix or sensory Jacobian matrix.

The dimensions of the image Jacobian matrix $\boldsymbol{J}_s(\boldsymbol{r})$ may vary according to the degrees of freedom of the robotic manipulator and the specific robot tasks. Consider a 6-DOF robot for which both the position and orientation of the end effector are specified. Then the image Jacobian matrix $\boldsymbol{J}_s(\boldsymbol{r})$ is described as

$$\begin{aligned} \boldsymbol{J}_s(\boldsymbol{r})\dot{\boldsymbol{r}} &= \begin{bmatrix} \frac{f}{z} & 0 & -\frac{x_h}{z} & -\frac{x_h x_v}{f} & \frac{f^2+x_h^2}{f} & -x_v \\ 0 & \frac{f}{z} & -\frac{x_v}{z} & -\frac{f^2+x_v^2}{f} & \frac{x_h x_v}{f} & x_h \end{bmatrix} \\ &= \underbrace{\begin{bmatrix} \frac{1}{z} & 0 \\ 0 & \frac{1}{z} \end{bmatrix}}_{\boldsymbol{Z}^{-1}(\boldsymbol{q})} \underbrace{\begin{bmatrix} f & 0 & -x_h & -\frac{z x_h x_v}{f} & \frac{z(f^2+x_h^2)}{f} & -z x_v \\ 0 & f & -x_v & -\frac{z(f^2+x_v^2)}{f} & \frac{z x_h x_v}{f} & z x_h \end{bmatrix}}_{\boldsymbol{L}(\boldsymbol{x})}, \end{aligned} \tag{1.42}$$

where z is the depth information of the feature point with respect to the camera image frame, and f is the focal length of the camera.

If the orientation of the end effector is not specified, the image Jacobian matrix reduces to

$$\boldsymbol{J}_s(\boldsymbol{r}) = \begin{bmatrix} \frac{f}{z} & 0 & -\frac{x_h}{z} \\ 0 & \frac{f}{z} & -\frac{x_v}{z} \end{bmatrix} = \underbrace{\begin{bmatrix} \frac{1}{z} & 0 \\ 0 & \frac{1}{z} \end{bmatrix}}_{\boldsymbol{Z}^{-1}(\boldsymbol{q})} \underbrace{\begin{bmatrix} f & 0 & -x_h \\ 0 & f & -x_v \end{bmatrix}}_{\boldsymbol{L}(\boldsymbol{x})}. \tag{1.43}$$

If the end effector evolves in a two-dimensional plane, while the camera is perpendicular to the plane as shown in Fig. 1.18, then the image Jacobian matrix is described as

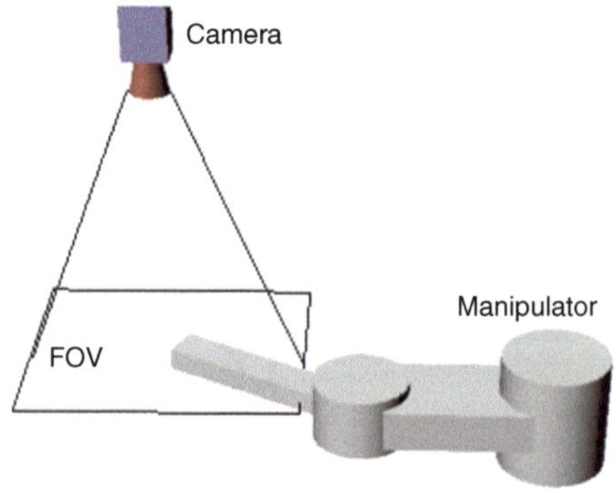

Fig. 1.18 A 2-DOF planar manipulator with a fixed camera configuration

$$J_s(r) = \begin{bmatrix} \frac{f}{z} & 0 \\ 0 & \frac{f}{z} \end{bmatrix} = \underbrace{\begin{bmatrix} \frac{1}{z} & 0 \\ 0 & \frac{1}{z} \end{bmatrix}}_{Z^{-1}} \underbrace{\begin{bmatrix} f & 0 \\ 0 & f \end{bmatrix}}_{L}. \tag{1.44}$$

In the presence of multiple feature points or cameras, the sensory task-space variable $\boldsymbol{x}$ can be specified as

$$\boldsymbol{x} = [\boldsymbol{x}_1^T, \ldots, \boldsymbol{x}_M^T]^T \in \Re^{2M}, \tag{1.45}$$

where M is the number of feature points, and $\boldsymbol{x}_i = [x_{ih}, x_{iv}]^T \in \Re^2$ is the position of the ith feature point in image space. Then the sensory Jacobian matrix is obtained as

$$\boldsymbol{J}_s(\boldsymbol{r}) = \begin{bmatrix} \boldsymbol{J}_{s1}(\boldsymbol{r}) \\ \boldsymbol{J}_{s2}(\boldsymbol{r}) \\ \vdots \\ \boldsymbol{J}_{sM}(\boldsymbol{r}) \end{bmatrix} = \boldsymbol{Z}^{-1}(\boldsymbol{q})\boldsymbol{L}(\boldsymbol{x})\dot{\boldsymbol{r}}, \tag{1.46}$$

where $\boldsymbol{J}_{si}(\boldsymbol{r})$ denotes the sensory Jacobian matrix for the ith feature point, and

$$\boldsymbol{Z}^{-1}(\boldsymbol{q}) = \begin{bmatrix} \frac{1}{z_1(\boldsymbol{q})}\boldsymbol{I}_2 & \boldsymbol{0} & \cdots & 0 \\ \boldsymbol{0} & \frac{1}{z_2(\boldsymbol{q})}\boldsymbol{I}_2 & \cdots & \boldsymbol{0} \\ \vdots & \vdots & \ddots & \vdots \\ \boldsymbol{0} & \boldsymbol{0} & \cdots & \frac{1}{z_M(\boldsymbol{q})}\boldsymbol{I}_2 \end{bmatrix} \in \Re^{2M \times 2M} \tag{1.47}$$

contains the depth information of the multiple feature points, where $\boldsymbol{I}_2 \in \Re^{2\times 2}$ is the 2×2 identity matrix.

Other types of sensors can also be used to measure the position of a robot end effector in sensory task space, such as a position-sensitive detector (PSD). For example, a robot system with a PSD is shown in Fig. 1.19. An LED is installed on the robot end effector, and the PSD camera captures the light emitted from the LED. When there is incident light on the light-sensitive sensor in the PSD camera, photocurrents are generated and collected from four electrodes placed along each side of the sensor near the boundary, and the photocurrents are then converted to voltage signals. The position of the light spot represents the position of the end effector, and the voltages vary with the position of the light spot, as illustrated in Fig. 1.19.

Therefore, the position of the robot end effector is represented by voltages $\boldsymbol{v} = [v_1, v_2]^T$ such that

$$\begin{cases} r_1 = \beta_1 v_1 + c_1, \\ r_2 = \beta_2 v_2 + c_2, \end{cases} \tag{1.48}$$

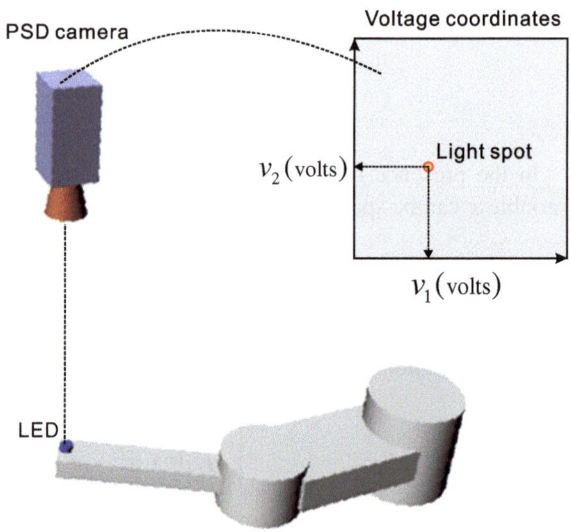

Fig. 1.19 A robot system with a position-sensitive detector (PSD)

where $[r_1, r_2]$ denotes the position of the end effector in Cartesian space, β_1, β_2 are scaling factors, and c_1, c_2 are constant offsets.

If β_1, β_2, c_1, c_2 are known or the relationship between the voltages and the Cartesian space is perfectly obtained by calibration, then the sensory-space variable can be directly specified in Cartesian space as $\boldsymbol{x} = \boldsymbol{r}$. Therefore, the sensory-space velocity of the end effector is related to the joint-space velocity as

$$\dot{\boldsymbol{x}} = \boldsymbol{J}_m(\boldsymbol{q})\dot{\boldsymbol{q}}, \tag{1.49}$$

where the sensory Jacobian matrix $\boldsymbol{J}_s(\boldsymbol{r})$ reduces to the identity matrix.

If the relationship between the voltages and Cartesian space is not known, the sensory-space variable can be specified only by voltages: $\boldsymbol{x} = \boldsymbol{v}$. From Eq. (1.48), we have

$$\begin{cases} v_1 = \frac{1}{\beta_1} r_1 - c_1, \\ v_2 = \frac{1}{\beta_2} r_2 - c_2. \end{cases} \tag{1.50}$$

Differentiating Eq. (1.50) with respect to time yields

$$\begin{cases} \dot{v}_1 = \frac{1}{\beta_1} \dot{r}_1, \\ \dot{v}_2 = \frac{1}{\beta_2} \dot{r}_2. \end{cases} \tag{1.51}$$

Then the sensory space velocity of the end effector is related to the joint-space velocity as

$$\dot{\boldsymbol{x}} = \dot{\boldsymbol{v}} = \begin{bmatrix} \frac{1}{\beta_1} & 0 \\ 0 & \frac{1}{\beta_2} \end{bmatrix} \begin{bmatrix} \dot{r}_1 \\ \dot{r}_2 \end{bmatrix} = \boldsymbol{J}_s(\boldsymbol{r})\boldsymbol{J}_m(\boldsymbol{q})\dot{\boldsymbol{q}}, \tag{1.52}$$

where $\boldsymbol{J}_s(\boldsymbol{r}) = \mathrm{diag}\{\frac{1}{\beta_1}, \frac{1}{\beta_2}\}$ denotes the mapping from Cartesian space to voltage or sensor coordinates. In this case, an additional LED is required to obtain the desired position for the end effector.

Note that the sensory feedback information is valid only within the limited sensing zone. In sensory task-space control, it is usually assumed that the robot end effector stays inside the sensing zone so that sensory feedback information is available throughout the course of manipulation. In addition, the Jacobian matrix is assumed to be nonsingular, so that the inverse Jacobian matrix or the pseudoinverse always exists. Some techniques to avoid singularities or limit the operations inside the sensing zone or singularity-free zone are introduced in Chap. 4. In Chap. 5, some techniques are also introduced to allow the robot to start or enter the regions where singularities occur or sensory feedback is not available.

Example 1.6 For the 2-DOF planar manipulator with a fixed camera configuration illustrated in Fig. 1.18, since the camera is placed perpendicular to the plane where the end effector evolves, the image Jacobian matrix is described as

$$\boldsymbol{J}_s = \begin{bmatrix} \frac{f}{z} & 0 \\ 0 & \frac{f}{z} \end{bmatrix} \in \Re^{2\times 2}. \tag{1.53}$$

From Eqs. (1.26) and (1.53), the composite Jacobian matrix from joint space to image space is given as

$$\boldsymbol{J}(\boldsymbol{q}) = \boldsymbol{J}_s\boldsymbol{J}_m(\boldsymbol{q}) = \begin{bmatrix} -\frac{f}{z}(l_1 s_1 + l_2 s_{12}) & -\frac{f}{z} l_2 s_{12} \\ \frac{f}{z}(l_1 c_1 + l_2 c_{12}) & \frac{f}{z} l_2 c_{12} \end{bmatrix} \in \Re^{2\times 2}. \tag{1.54}$$

The image space velocity $\dot{\boldsymbol{x}}$ can be derived using the Jacobian matrix in Eq. (1.54) as

$$\begin{aligned} \dot{\boldsymbol{x}} = \boldsymbol{J}_s\boldsymbol{J}_m(\boldsymbol{q})\dot{\boldsymbol{q}} &= \begin{bmatrix} \frac{f}{z} & 0 \\ 0 & \frac{f}{z} \end{bmatrix} \begin{bmatrix} -l_1 s_1 - l_2 s_{12} & -l_2 s_{12} \\ l_1 c_1 + l_2 c_{12} & l_2 c_{12} \end{bmatrix} \begin{bmatrix} \dot{q}_1 \\ \dot{q}_2 \end{bmatrix} \\ &= \begin{bmatrix} -\frac{f}{z} l_1 s_1 \dot{q}_1 - \frac{f}{z} l_2 s_{12}(\dot{q}_1 + \dot{q}_2) \\ \frac{f}{z} l_1 c_1 \dot{q}_1 + \frac{f}{z} l_2 c_{12}(\dot{q}_1 + \dot{q}_2) \end{bmatrix}. \end{aligned} \tag{1.55}$$

Hence $\dot{\boldsymbol{x}} = \boldsymbol{J}_s\boldsymbol{J}_m(\boldsymbol{q})\dot{\boldsymbol{q}}$ can be written as the product of a known regressor matrix $\boldsymbol{Y}_k(\boldsymbol{q}, \dot{\boldsymbol{q}})$ and an unknown constant vector $\boldsymbol{\theta}_k$ such that

$$\dot{\boldsymbol{x}} = \begin{bmatrix} -s_1\dot{q}_1 & -s_{12}(\dot{q}_1 + \dot{q}_2) \\ c_1\dot{q}_1 & c_{12}(\dot{q}_1 + \dot{q}_2) \end{bmatrix} \begin{bmatrix} \frac{f}{z} l_1 \\ \frac{f}{z} l_2 \end{bmatrix} = \boldsymbol{Y}_k(\boldsymbol{q}, \dot{\boldsymbol{q}})\boldsymbol{\theta}_k. \tag{1.56}$$

□

1.4 Summary and Notes

This chapter serves a brief introductory purpose, and we assume on the part of the reader some familiarity with robot dynamics and kinematics. More details on the derivation and formulation of robot dynamics and kinematics can be found in various robotics textbooks such as Craig (1989), Schilling (1990), Spong et al. (2006). In Spong et al. (2006), the derivation of the image Jacobian matrix is also given. In the remaining chapters, it is also assumed that the reader is familiar with Lyapunov stability theory. Some basic theorems of the Lyapunov method are listed in the appendix for ease of reference, and more details can be found in various textbooks on nonlinear control, such as Slotine and Li (1991), Khalil and Dombre (2004).

Chapter 2
Sensory Task-Space Setpoint Control

Increasing demand for robotic systems operating in unstructured environments has led to the development of sensory feedback control methods. This chapter introduces the fundamental concepts for design and analysis of task-space sensory feedback control of robotic systems with uncertainty. The joint-space method for control of robots is first reviewed, and several task-space sensory feedback control schemes for setpoint control or point-to-point control of a robot manipulator are introduced. Such control schemes were inspired by human visually guided reaching movements, which do not require an accurate knowledge of kinematics and dynamics of the arm. Using visual sensory feedback information, a human being is able to interact with the environment by picking up a new or unknown tool and manipulating it easily and skillfully to accomplish a task. Such basic ability of sensing and responding to uncertainty and changes without an accurate knowledge of sensory-to-motor transformation gives human beings a high degree of flexibility in dealing with unforeseen changes in the real world.

2.1 Joint-Space Setpoint Control

In this section, a standard joint-space setpoint control design problem based on the Lyapunov method is first reviewed. In the joint-space control methodology, a desired position of the end effector is first specified in Cartesian space and then converted to a corresponding desired joint configuration by solving an inverse kinematic problem, as illustrated in Fig. 2.1. The robot kinematics model is assumed to be known exactly in the joint-space control method. The setpoint control law is then designed so that the robot joints follow the desired joint positions.

To illustrate the use of Lyapunov-based techniques for robot joint-space control design, a single-link robot, as shown in Fig. 2.2, is first considered. The equation of motion of the robot is described by

$$m\ddot{q} + b\dot{q} = \tau, \tag{2.1}$$

C.C. Cheah and X. Li, *Task-Space Sensory Feedback Control of Robot Manipulators*,
Intelligent Systems, Control and Automation: Science and Engineering 73,
DOI 10.1007/978-981-287-062-9_2

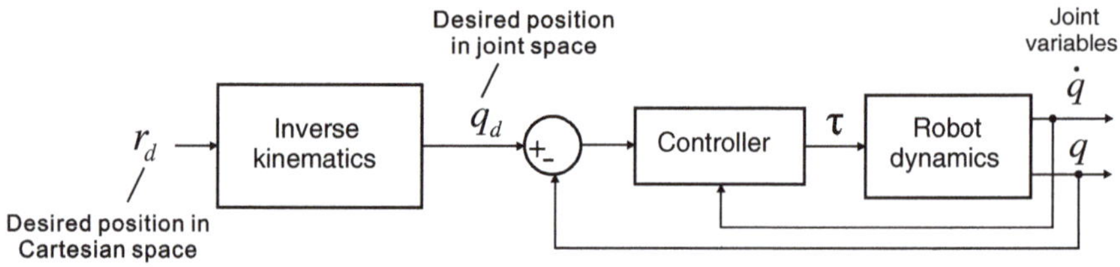

Fig. 2.1 Joint-space control

Fig. 2.2 A single-link robot

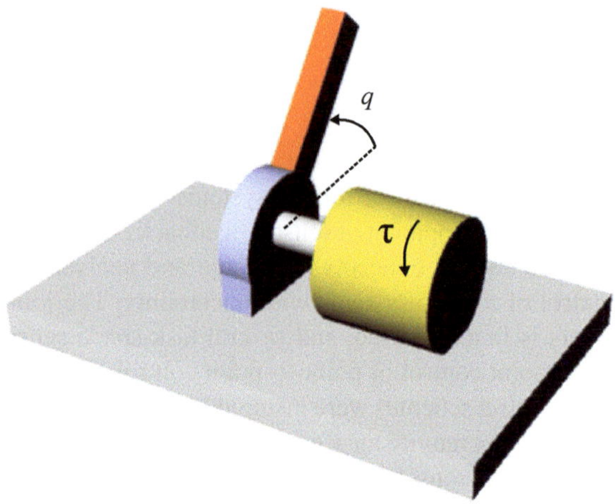

where m is the inertia, b represents the damping coefficient, q represents the joint angle, and τ is the control input torque exerted on the joint.

Consider the following proportional-derivative (PD) controller:

$$\tau = -k_v \dot{q} - k_p \Delta q, \tag{2.2}$$

where q_d is a constant desired position of the joint, $\Delta q = q - q_d$ denotes the position error, and k_v and k_p are positive control gains for the proportional and derivative errors respectively. The PD controller in Eq. (2.2) ensures the stability of the closed-loop system. To prove this, substituting the PD controller into the dynamic model described by Eq. (2.1) yields the following closed-loop equation:

$$m\ddot{q} + (b + k_v)\dot{q} + k_p \Delta q = 0. \tag{2.3}$$

Next, multiplying Eq. (2.3) by an output specified as $y = \dot{q}$ yields

$$m\ddot{q}\dot{q} + (b + k_v)\dot{q}^2 + k_p \Delta q \dot{q} = 0. \tag{2.4}$$

The above equation can be written as

$$\frac{\mathrm{d}}{\mathrm{d}t}\left(\tfrac{1}{2}m\dot{q}^2 + \tfrac{1}{2}k_p\Delta q^2\right) + (b + k_v)\dot{q}^2 = 0, \tag{2.5}$$

where $\frac{1}{2}m\dot{q}^2$ is the kinetic energy of the robot system, $\frac{1}{2}k_p\Delta q^2$ is the artificial potential energy associated with the springlike proportional control term, $b\dot{q}^2$ is the amount of energy dissipated due to joint friction, and $k_v\dot{q}^2$ is the amount of energy loss due to the damping or friction injected into the system by the frictionlike derivative control term.

Let

$$V(\Delta q, \dot{q}) = \tfrac{1}{2}m\dot{q}^2 + \tfrac{1}{2}k_p\Delta q^2, \tag{2.6}$$

and

$$W(\dot{q}) = (b + k_v)\dot{q}^2, \tag{2.7}$$

and represent Eq. (2.5) as

$$\tfrac{\mathrm{d}}{\mathrm{d}t}V(\Delta q, \dot{q}) = -W(\dot{q}). \tag{2.8}$$

The function $V(\Delta q, \dot{q})$ represents the sum of the kinetic energy and the artificial potential energy, and its time derivative is equal to $-W(\dot{q})$. Since $-W(\dot{q})$ is negative unless $\dot{q} = 0$, energy is dissipated by the frictional terms, and the system eventually stops at an equilibrium point.

In control theory, $V(\Delta q, \dot{q})$ is called the *Lyapunov function*, which is positive definite in the state variables Δq and $\dot{q}$, and $-W(\dot{q})$ denotes its time derivative, which is negative definite in $\dot{q}$ only. Thus $V(\Delta q, \dot{q})$ is positive definite, and its derivative is negative semidefinite, and the stability of the closed-loop system can be concluded using the Lyapunov method. That is, both Δq and $\dot{q}$ are bounded. To show the asymptotic stability of the equilibrium state, *LaSalle's Invariance Theorem* is used. Since $W(\dot{q}) = 0$ implies $\dot{q} = 0$, Eq. (2.3) reduces to $k_p\Delta q = 0$, and thus the position error Δq is zero. Therefore, the maximum invariant set is the origin, and the equilibrium at the origin is asymptotic stable.

We now consider the PD control problem of a robotic manipulator with n degrees of freedom. From Eq. (1.5) in Chap. 1, the nonlinear robot dynamics is described as

$$\boldsymbol{M}(\boldsymbol{q})\ddot{\boldsymbol{q}} + \left(\tfrac{1}{2}\dot{\boldsymbol{M}}(\boldsymbol{q}) + \boldsymbol{S}(\boldsymbol{q}, \dot{\boldsymbol{q}})\right)\dot{\boldsymbol{q}} + \boldsymbol{g}(\boldsymbol{q}) = \boldsymbol{\tau}. \tag{2.9}$$

Note that when the manipulator stops at an equilibrium point, the only torque exerted on the system is the gravitational term $\boldsymbol{g}(\boldsymbol{q})$. Consider the following PD controller with gravity compensation:

$$\boldsymbol{\tau} = -\boldsymbol{K}_p\Delta\boldsymbol{q} - \boldsymbol{K}_v\dot{\boldsymbol{q}} + \boldsymbol{g}(\boldsymbol{q}), \tag{2.10}$$

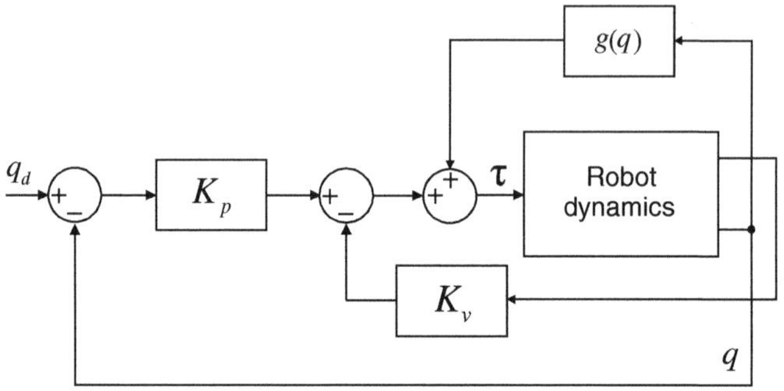

Fig. 2.3 Block diagram of a PD controller with gravity compensation

where $\dot{\boldsymbol{q}} \in \Re^n$ is the joint-space velocity, $\Delta\boldsymbol{q} = \boldsymbol{q} - \boldsymbol{q}_d \in \Re^n$ denotes the joint-space error, where $\boldsymbol{q}_d$ is the desired joint configuration, and $\boldsymbol{K}_p \in \Re^{n\times n}$ and $\boldsymbol{K}_v \in \Re^{n\times n}$ are diagonal and positive definite matrices. A block diagram of the controller described by Eq. (2.10) is illustrated in Fig. 2.3.

The closed-loop equation is obtained by substituting Eq. (2.10) into Eq. (2.9):

$$\boldsymbol{M}(\boldsymbol{q})\ddot{\boldsymbol{q}} + \left(\tfrac{1}{2}\dot{\boldsymbol{M}}(\boldsymbol{q}) + \boldsymbol{S}(\boldsymbol{q},\dot{\boldsymbol{q}})\right)\dot{\boldsymbol{q}} + \boldsymbol{K}_p\Delta\boldsymbol{q} + \boldsymbol{K}_v\dot{\boldsymbol{q}} = \boldsymbol{0}, \tag{2.11}$$

where we note that with the compensation of the gravitational term, $\Delta\boldsymbol{q}$ is zero when the manipulator stops.

Next, the inner product of Eq. (2.11) with an output vector $\boldsymbol{y} = \dot{\boldsymbol{q}}$ yields

$$\dot{\boldsymbol{q}}^T\boldsymbol{M}(\boldsymbol{q})\ddot{\boldsymbol{q}} + \dot{\boldsymbol{q}}^T\left(\tfrac{1}{2}\dot{\boldsymbol{M}}(\boldsymbol{q}) + \boldsymbol{S}(\boldsymbol{q},\dot{\boldsymbol{q}})\right)\dot{\boldsymbol{q}} + \dot{\boldsymbol{q}}^T\boldsymbol{K}_p\Delta\boldsymbol{q} + \dot{\boldsymbol{q}}^T\boldsymbol{K}_v\dot{\boldsymbol{q}} = 0, \tag{2.12}$$

which can be written as

$$\tfrac{\mathrm{d}}{\mathrm{d}t}V(\Delta\boldsymbol{q},\dot{\boldsymbol{q}}) = -W(\dot{\boldsymbol{q}}), \tag{2.13}$$

where

$$V(\Delta\boldsymbol{q},\dot{\boldsymbol{q}}) = \tfrac{1}{2}\dot{\boldsymbol{q}}^T\boldsymbol{M}(\boldsymbol{q})\dot{\boldsymbol{q}} + \tfrac{1}{2}\Delta\boldsymbol{q}^T\boldsymbol{K}_p\Delta\boldsymbol{q}, \tag{2.14}$$

and

$$W(\dot{\boldsymbol{q}}) = \dot{\boldsymbol{q}}^T\boldsymbol{S}(\boldsymbol{q},\dot{\boldsymbol{q}})\dot{\boldsymbol{q}} + \dot{\boldsymbol{q}}^T\boldsymbol{K}_v\dot{\boldsymbol{q}}. \tag{2.15}$$

From *Property* 1.1 in Chap. 1, the inertia matrix $\boldsymbol{M}(\boldsymbol{q})$ is positive definite, and hence the function $V(\Delta\boldsymbol{q},\dot{\boldsymbol{q}})$ is a Lyapunov function that is positive definite in $\dot{\boldsymbol{q}}$ and $\Delta\boldsymbol{q}$. From *Property* 1.2, the term $\dot{\boldsymbol{q}}^T\boldsymbol{S}(\boldsymbol{q},\dot{\boldsymbol{q}})\dot{\boldsymbol{q}}$ in Eq. (2.15) reduces to zero, and hence $W(\dot{\boldsymbol{q}})$ is positive definite in $\dot{\boldsymbol{q}}$. Therefore, the system is stable in the sense that $\dot{\boldsymbol{q}}$ and $\Delta\boldsymbol{q}$ are bounded. To show the global asymptotic stability of the origin, LaSalle's

Invariance Theorem is used. Since $W(\dot{q}) = 0$ implies $\dot{\boldsymbol{q}} = \boldsymbol{0}$, and both $\boldsymbol{M}(\boldsymbol{q})$ and $\boldsymbol{K}_p$ are positive definite matrices, Eq. (2.11) satisfies $\Delta\boldsymbol{q} = \boldsymbol{0}$, and the maximum invariant set is the origin. Therefore, $\dot{\boldsymbol{q}} \to \boldsymbol{0}$ as $t \to \infty$, which also implies $\boldsymbol{q} \to \boldsymbol{q}_d$ from Eq. (2.11).

2.2 Cartesian-Space Setpoint Control

To describe a task for a robotic manipulator, it is convenient to specify the position and orientation of the end effector in Cartesian space. In joint-space control methods, the inverse kinematics problem is used to obtain a desired joint configuration that corresponds to the desired position and orientation of the end effector. To eliminate the requirement of solving the inverse kinematics problem, Cartesian-space control methods can be formulated using feedback information directly in Cartesian space. A coordination transformation in the form of a Jacobian matrix is used in the feedback control law to transform the Cartesian-space feedback information into joint space.

The Cartesian-space setpoint controller is designed as

$$\boldsymbol{\tau} = -\boldsymbol{J}_m^T(\boldsymbol{q})\boldsymbol{K}_p\Delta\boldsymbol{r} - \boldsymbol{J}_m^T(\boldsymbol{q})\boldsymbol{K}_v\dot{\boldsymbol{r}} + \boldsymbol{g}(\boldsymbol{q}), \tag{2.16}$$

where $\Delta\boldsymbol{r} = \boldsymbol{r} - \boldsymbol{r}_d \in \Re^n$ and $\boldsymbol{r}_d \in \Re^n$ is a desired position in Cartesian space, $\dot{\boldsymbol{r}} \in \Re^n$ is the task-space velocity of the end effector, and $\boldsymbol{K}_p \in \Re^{n\times n}$ and $\boldsymbol{K}_v \in \Re^{n\times n}$ are diagonal and positive definite matrices. In the above Cartesian-space controller, the Cartesian-space feedback quantities $\boldsymbol{r}$ and $\dot{\boldsymbol{r}}$ are usually computed from the forward kinematics equations. The matrix $\boldsymbol{J}_m(\boldsymbol{q}) \in \Re^{n\times n}$ denotes the Jacobian matrix from joint space to Cartesian space. Using the Jacobian matrix, the feedback information in Cartesian space is directly transformed into the torque input of the robot in joint space. A block diagram of the controller in Eq. (2.16) is illustrated in Fig. 2.4.

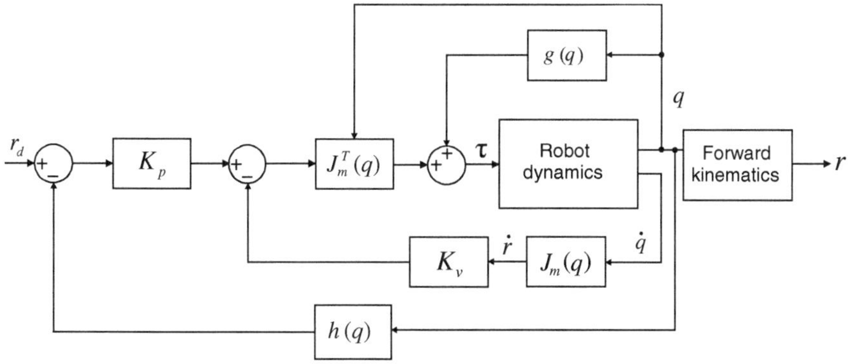

Fig. 2.4 Block diagram of a Cartesian-space setpoint control

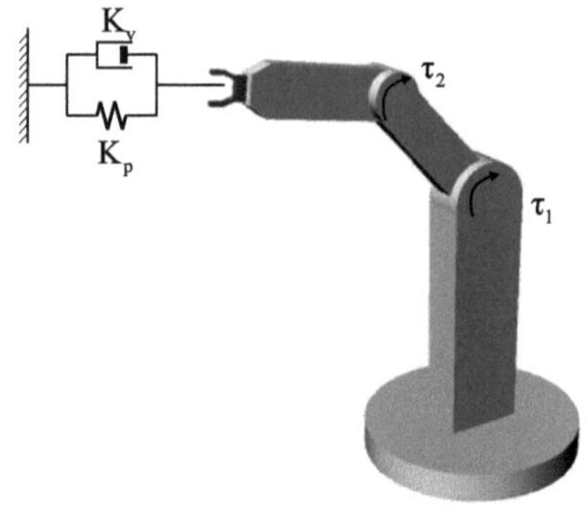

Fig. 2.5 Illustration of virtual forces exerted on the end effector

In the setpoint controller, a virtual restoring spring force $\boldsymbol{F}_s = -\boldsymbol{K}_p\Delta\boldsymbol{r}$ and a damping force $\boldsymbol{F}_d = -\boldsymbol{K}_v\dot{\boldsymbol{r}}$ are generated in Cartesian space, as illustrated in Fig. 2.5.

The closed-loop equation is obtained by substituting the controller in Eq. (2.16) into Eq. (2.9) to yield

$$\boldsymbol{M}(\boldsymbol{q})\ddot{\boldsymbol{q}}+\left(\tfrac{1}{2}\dot{\boldsymbol{M}}(\boldsymbol{q})+\boldsymbol{S}(\boldsymbol{q},\dot{\boldsymbol{q}})\right)\dot{\boldsymbol{q}}+\boldsymbol{J}_m^T(\boldsymbol{q})\boldsymbol{K}_p\Delta\boldsymbol{r}+\boldsymbol{J}_m^T(\boldsymbol{q})\boldsymbol{K}_v\dot{\boldsymbol{r}}=\boldsymbol{0}. \tag{2.17}$$

with the compensation of the gravitational term, $\boldsymbol{J}_m^T(\boldsymbol{q})\boldsymbol{K}_p\Delta\boldsymbol{r}$ is zero when the manipulator stops.

Next, the inner product of Eq. (2.17) and the output $\boldsymbol{y} = \dot{\boldsymbol{q}}$ yields

$$\dot{\boldsymbol{q}}^T\boldsymbol{M}(\boldsymbol{q})\ddot{\boldsymbol{q}}+\dot{\boldsymbol{q}}^T\left(\tfrac{1}{2}\dot{\boldsymbol{M}}(\boldsymbol{q})+\boldsymbol{S}(\boldsymbol{q},\dot{\boldsymbol{q}})\right)\dot{\boldsymbol{q}}+\dot{\boldsymbol{q}}^T\boldsymbol{J}_m^T(\boldsymbol{q})\boldsymbol{K}_p\Delta\boldsymbol{r}+\dot{\boldsymbol{q}}^T\boldsymbol{J}_m^T(\boldsymbol{q})\boldsymbol{K}_v\dot{\boldsymbol{r}}=0. \tag{2.18}$$

That is,

$$\dot{\boldsymbol{q}}^T\boldsymbol{M}(\boldsymbol{q})\ddot{\boldsymbol{q}}+\dot{\boldsymbol{q}}^T\left(\tfrac{1}{2}\dot{\boldsymbol{M}}(\boldsymbol{q})+\boldsymbol{S}(\boldsymbol{q},\dot{\boldsymbol{q}})\right)\dot{\boldsymbol{q}}+\dot{\boldsymbol{r}}^T\boldsymbol{K}_p\Delta\boldsymbol{r}+\dot{\boldsymbol{r}}^T\boldsymbol{K}_v\dot{\boldsymbol{r}}=0, \tag{2.19}$$

which can be written as

$$\tfrac{\mathrm{d}}{\mathrm{d}t}V(\Delta\boldsymbol{r},\dot{\boldsymbol{q}}) = -W(\dot{\boldsymbol{r}}), \tag{2.20}$$

where

$$V(\Delta\boldsymbol{r},\dot{\boldsymbol{q}}) = \tfrac{1}{2}\dot{\boldsymbol{q}}^T\boldsymbol{M}(\boldsymbol{q})\dot{\boldsymbol{q}}+\tfrac{1}{2}\Delta\boldsymbol{r}^T\boldsymbol{K}_p\Delta\boldsymbol{r} \tag{2.21}$$

and

$$W(\dot{\boldsymbol{r}}) = \dot{\boldsymbol{q}}^T\boldsymbol{S}(\boldsymbol{q},\dot{\boldsymbol{q}})\dot{\boldsymbol{q}}+\dot{\boldsymbol{r}}^T\boldsymbol{K}_v\dot{\boldsymbol{r}} \leq 0. \tag{2.22}$$

From *Property* 1.1 in Chap. 1, $\boldsymbol{M}(\boldsymbol{q})$ is positive definite, and hence the function $V(\Delta\boldsymbol{r},\dot{\boldsymbol{q}})$ is positive definite in $\dot{\boldsymbol{q}}$ and $\Delta\boldsymbol{r}$. From *Property* 1.2, the term $\dot{\boldsymbol{q}}^T\boldsymbol{S}(\boldsymbol{q},\dot{\boldsymbol{q}})\dot{\boldsymbol{q}}$ in Eq. (2.22) reduces to zero, and hence $W(\dot{\boldsymbol{r}})$ is positive definite in $\dot{\boldsymbol{q}}$. Therefore, the variables $\dot{\boldsymbol{q}}$ and $\Delta\boldsymbol{r}$ are bounded. From LaSalle's Invariance Theorem, $W(\dot{\boldsymbol{r}}) = 0$ implies that $\dot{\boldsymbol{r}} = \boldsymbol{0}$, which also indicates that $\dot{\boldsymbol{q}} = \boldsymbol{0}$, since it is assumed that $\boldsymbol{J}_m(\boldsymbol{q})$ is nonsingular in a finite workspace such that $\boldsymbol{J}_m^{-1}(\boldsymbol{q})$ exists. Therefore, Eq. (2.17) satisfies $\boldsymbol{J}_m^T(\boldsymbol{q})\boldsymbol{K}_p\Delta\boldsymbol{r} = 0$, that is, the position of the end effector in task space converges to the desired steady-state position in task space, in a finite task space where singularities are avoided.

In general, the positive definiteness of $V(\Delta\boldsymbol{r}, \dot{\boldsymbol{q}})$ in $\Delta\boldsymbol{q}$ and $\dot{\boldsymbol{q}}$ cannot be concluded even if $\boldsymbol{q}_d$ satisfies $\boldsymbol{r}_d = \boldsymbol{h}(\boldsymbol{q}_d)$, because $\boldsymbol{h}(\cdot)$ is periodic in $\boldsymbol{q}$. However, if the manipulator is working in a finite workspace such that the kinematics mapping between Cartesian space and joint space is one-to-one and the Jacobian matrix $\boldsymbol{J}_m(\boldsymbol{q})$ is nonsingular, then asymptotic stability can be concluded.

2.3 Sensory-Space Setpoint Control

In joint-space control methods, exact knowledge of the robot kinematics is required to solve an inverse kinematics problem to generate a desired position in joint space. In the presence of kinematic uncertainty, it is impossible to derive the desired joint angles from the desired end effector's position. Therefore, the assumption of an exact kinematic model means that the robot is unable to react to changes and uncertainties in its kinematics. For example, when the robot picks up objects of different dimensions or with unknown orientations or gripping points, as illustrated in Fig. 2.6, the overall kinematics from the robot base to the tip of the object changes and is therefore difficult to derive exactly. When the control problem is formulated directly in Cartesian space, the inverse kinematics problem is replaced by the transposed Jacobian matrix from joint space to task space in the control law. However, exact parameter values of the kinematics are still required to construct the Jacobian matrix.

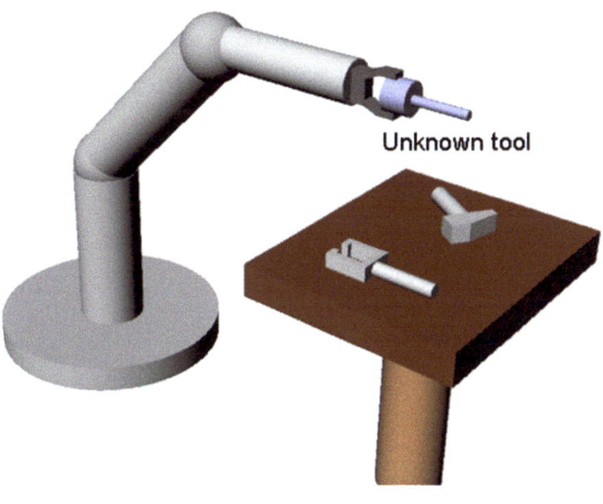

Fig. 2.6 The kinematics model is uncertain when the robot grasps and manipulates an unknown tool

In many modern control applications of robot manipulators, task-space sensory feedback information such as visual feedback is used to monitor the position of the end effector. Using sensory feedback of the position error, the control systems are robust to kinematics uncertainty. Though the position error of the end effector can be measured by task-space sensors, the uncertainty of the Jacobian matrix due to kinematic uncertainty and camera/sensor calibration error poses a challenging dynamic control problem.

2.3.1 Sensory Feedback Control of a Mass-Damper System

To illustrate the concept of sensory feedback control, we first consider a mass–damper system as illustrated in Fig. 2.7. The dynamics equation of the system is obtained from Newton's law as

$$m\ddot{\boldsymbol{r}} + \boldsymbol{B}\dot{\boldsymbol{r}} = \boldsymbol{u}, \tag{2.23}$$

where $\boldsymbol{r} = [r_1, r_2]^T \in \Re^2$ denotes the position of the block in two-dimensional Cartesian space, m denotes the mass of the block, $\boldsymbol{B} = \text{diag}\{b_1, b_2\} \in \Re^{2\times 2}$ denotes the damping matrix, and $\boldsymbol{u} = [u_1, u_2]^T \in \Re^2$ is a vector that represents the control input forces exerted on the system in Cartesian space.

When a sensor is employed to monitor the position or position error of the block for feedback control, the task coordinates are defined in sensory space. When there are rotation and scaling transformations between the sensory space and Cartesian space, the relationship between the Cartesian space and the sensory space is described as

$$\begin{bmatrix} x_1 \\ x_2 \end{bmatrix} = \begin{bmatrix} \cos(\delta) & \sin(\delta) \\ -\sin(\delta) & \cos(\delta) \end{bmatrix} \begin{bmatrix} s_1 & 0 \\ 0 & s_2 \end{bmatrix} \begin{bmatrix} r_1 \\ r_2 \end{bmatrix} - \begin{bmatrix} d_1 \\ d_2 \end{bmatrix}, \tag{2.24}$$

where $\boldsymbol{x} = [x_1, x_2]^T \in \Re^2$ is the position in sensory space, δ is the rotation angle between the Cartesian-space coordinates and the sensory-space coordinates, s_1 and s_2 represent the scaling factors in the two coordinates, and $\boldsymbol{d} = [d_1, d_2]^T \in \Re^2$ denotes the offset between the Cartesian coordinates and the sensory coordinates. An example of a sensor that can be used to measure the position is a camera, as

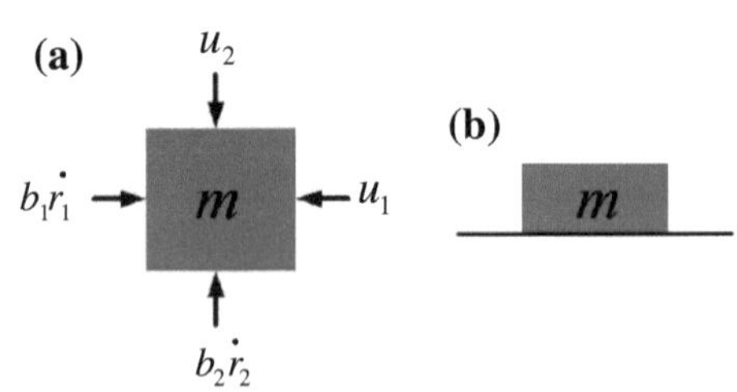

Fig. 2.7 **a** *Top view* of the mass-dmaper system. **b** *Side view* of the mass-damper system

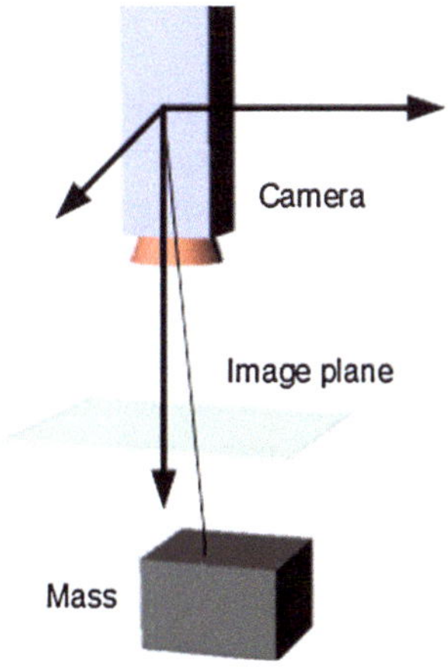

Fig. 2.8 Mass-damper system with an uncalibrated camera

illustrated in Fig. 2.8. The position of the block in Cartesian space is projected onto an image plane, and therefore the sensory space is the image space in pixels. When the camera is not mounted properly, with a rotation error, the relationship between the position in image space and the position in Cartesian space is described by Eq. (2.24).

Differentiating Eq. (2.24) with respect to time, we have

$$\begin{bmatrix} \dot{x}_1 \\ \dot{x}_2 \end{bmatrix} = \begin{bmatrix} \cos(\delta) & \sin(\delta) \\ -\sin(\delta) & \cos(\delta) \end{bmatrix} \begin{bmatrix} s_1 & 0 \\ 0 & s_2 \end{bmatrix} \begin{bmatrix} \dot{r}_1 \\ \dot{r}_2 \end{bmatrix} = \boldsymbol{R} \cdot \boldsymbol{S} \begin{bmatrix} \dot{r}_1 \\ \dot{r}_2 \end{bmatrix}, \tag{2.25}$$

where $\boldsymbol{R} \in \Re^{2\times2}$ and $\boldsymbol{S} \in \Re^{2\times2}$ are the rotation matrix and the scaling matrix. Let

$$\boldsymbol{J}_s = \boldsymbol{R} \cdot \boldsymbol{S}. \tag{2.26}$$

Then $\boldsymbol{J}_s \in \Re^{2\times2}$ denotes the sensory Jacobian matrix from Cartesian space to sensory space. The relationship between Cartesian space and sensory space is shown in Fig. 2.9.

In the presence of estimation or modeling errors in the rotation and scaling transformation, the sensory Jacobian matrix $\boldsymbol{J}_s$ is uncertain and is estimated by an approximate Jacobian matrix as $\hat{\boldsymbol{J}}_s$. Using the approximate Jacobian matrix, the sensory-space feedback controller is designed as

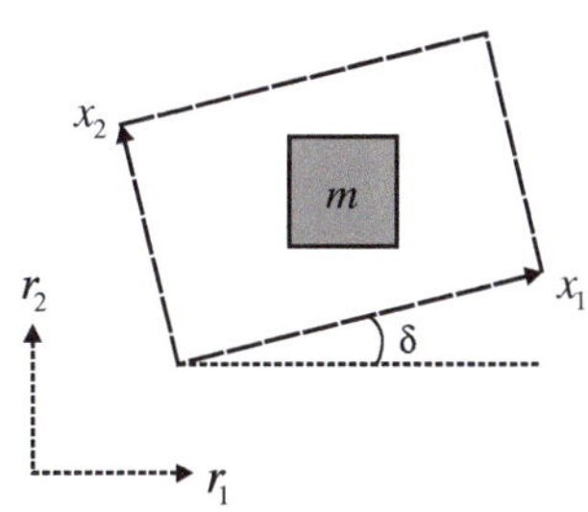

Fig. 2.9 Relationship between Cartesian-space coordinates and sensory-space coordinates

$$u = -\hat{J}_s^T K_v \dot{x} - \hat{J}_s^T K_p \Delta x, \tag{2.27}$$

where $\dot{x} \in \Re^2$ represents the sensory-space velocity, $\Delta x = x - x_d \in \Re^2$ is the position error, $x_d \in \Re^2$ is the desired position in sensory space, and $K_p \in \Re^{2\times2}$ and $K_v \in \Re^{2\times2}$ are diagonal and positive definite matrices. Note that Δx represents the relative distance between the actual position and the desired position in sensory space, which can be directly measured by a sensor such as a camera. A virtual restoring spring force $-K_p \Delta x$ and a damping force $-K_v \dot{x}$ are generated in sensory space using the sensory feedback information. The feedback information in sensor space is directly transformed into the input forces in Cartesian space using the transpose of the approximate sensory Jacobian matrix. Since the relative position error is used, the sensory feedback controller in Eq. (2.27) is robust to the modeling and calibration errors. However, the stability of the system may not be ensured in the presence of uncertainty in the sensory Jacobian matrix. To illustrate this, the following example is considered.

Example 2.1 Consider the mass–damper system described by Eq. (2.23) with a rotation angle θ of $\pi/3$ rad. First, a simulation study was performed on the sensory feedback control system with the exact knowledge of the Jacobian matrix. The dynamic parameters of the mass–damper system were set as $m = 1$ kg and $B = \mathrm{diag}\{0.1, 0.1\}$ kg/s, and the parameters of the Jacobian matrix in Eq. (2.24) were set as $s_1 = 0.7$, $s_2 = 0.8$, $d_1 = 0.1$, and $d_2 = 0.2$. The control parameters in Eq. (2.27) were set as $K_p = \mathrm{diag}\{1, 1\}$, and $K_v = \mathrm{diag}\{1.3, 1.3\}$. The initial position was set as $x(0) = [2.1926, 5.0858]^T$, and the desired position was set as $x_d = [0, 0]^T$. Assuming that the exact Jacobian matrix is known, the path of the block and the position error are shown in Fig. 2.10, where the position error converges to zero at steady state.

Next, the rotation angle was estimated as zero, which results in an approximate or uncertain Jacobian matrix. All the other parameters remained unchanged, and the

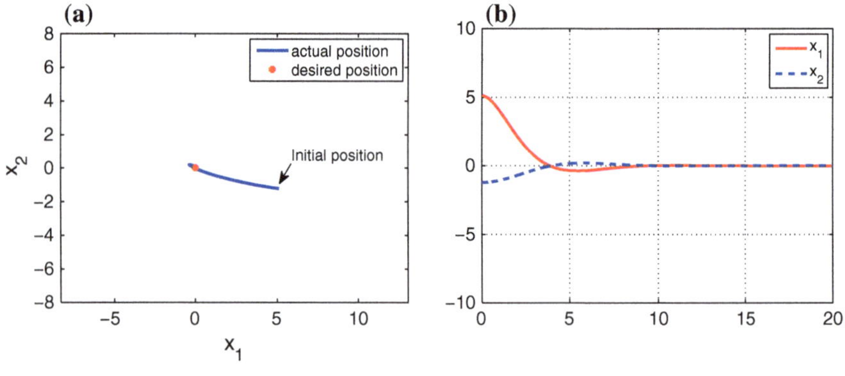

Fig. 2.10 A stable response when the Jacobian matrix is known exactly. **a** Path of block. **b** Position error

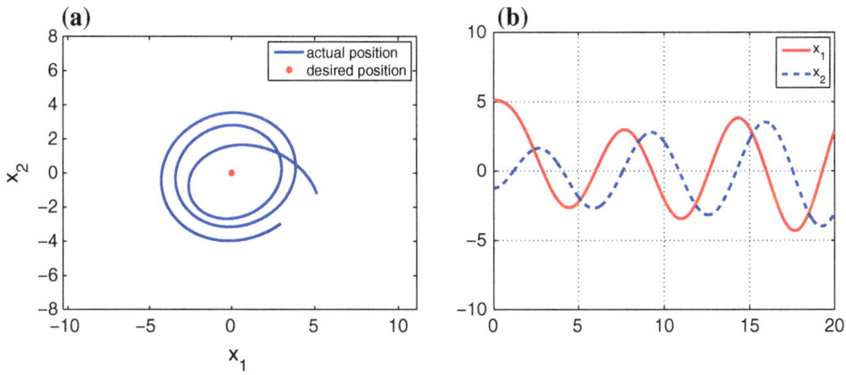

Fig. 2.11 An unstable response when the Jacobian matrix is uncertain. **a** Path of block. **b** Position error

simulation results in Fig. 2.11 show the path of the block and the position error. The closed-loop system is now unstable. □

To analyze the stability of a mass–spring–damper system with uncertain Jacobian or transformation from sensory space to Cartesian space, we substitute Eq. (2.27) into Eq. (2.23) to obtain the following closed-loop equation:

$$\boldsymbol{M}\ddot{\boldsymbol{r}} + \boldsymbol{B}\dot{\boldsymbol{r}} + \hat{\boldsymbol{J}}_s^T \boldsymbol{K}_v \dot{\boldsymbol{x}} + \hat{\boldsymbol{J}}_s^T \boldsymbol{K}_p \Delta \boldsymbol{x} = \boldsymbol{0}. \tag{2.28}$$

The above closed-loop equation describes the motion of the system in Cartesian space with feedback control information obtained in sensory space.

Next, a new output vector is defined as

$$\boldsymbol{y} = \dot{\boldsymbol{r}} + \alpha \hat{\boldsymbol{J}}_s^{-1} \Delta \boldsymbol{x}, \tag{2.29}$$

where α is a positive constant and $\hat{\boldsymbol{J}}_s^{-1}$ is the inverse of the approximate Jacobian matrix. The above output consists of a combination of the velocity in Cartesian space and the position error in sensory space that is being transformed to Cartesian space using the inverse approximate Jacobian matrix. The inner product od this output and the closed-loop equation will eventually lead to a Lyapunov-like function that is positive definite in the task-space error and velocity, and to a derivative of the Lyapunov-like function that is negative definite in the task-space error and velocity. Since the output now contains sensory feedback position error with approximate Jacobian, it can also be used in designing PID setpoint controllers and adaptive setpoint controllers with approximate Jacobian matrix.

The inner product of Eq. (2.28) with the output defined in Eq. (2.29) now yields

$$(\dot{\boldsymbol{r}} + \alpha \hat{\boldsymbol{J}}_s^{-1} \Delta \boldsymbol{x})^T (\boldsymbol{M}\ddot{\boldsymbol{r}} + \boldsymbol{B}\dot{\boldsymbol{r}} + \hat{\boldsymbol{J}}_s^T \boldsymbol{K}_v \dot{\boldsymbol{x}} + \hat{\boldsymbol{J}}_s^T \boldsymbol{K}_p \Delta \boldsymbol{x}) = 0. \tag{2.30}$$

That is,

$$\dot{\boldsymbol{r}}^T \boldsymbol{M}\ddot{\boldsymbol{r}} + \dot{\boldsymbol{r}}^T \boldsymbol{B}\dot{\boldsymbol{r}} + \dot{\boldsymbol{r}}^T \hat{\boldsymbol{J}}_s^T \boldsymbol{K}_v \dot{\boldsymbol{x}} + \dot{\boldsymbol{r}}^T \hat{\boldsymbol{J}}_s^T \boldsymbol{K}_p \Delta\boldsymbol{x} + \alpha\Delta\boldsymbol{x}^T \hat{\boldsymbol{J}}_s^{-T} \boldsymbol{M}\ddot{\boldsymbol{r}} \\ + \alpha\Delta\boldsymbol{x}^T \hat{\boldsymbol{J}}_s^{-T} \boldsymbol{B}\dot{\boldsymbol{r}} + \alpha\Delta\boldsymbol{x}^T \boldsymbol{K}_v \dot{\boldsymbol{x}} + \alpha\Delta\boldsymbol{x}^T \boldsymbol{K}_p \Delta\boldsymbol{x} = 0, \tag{2.31}$$

where $\hat{\boldsymbol{J}}_s^{-T}$ represents the transpose of $\hat{\boldsymbol{J}}_s^{-1}$. Equation (2.31) can be written as

$$\frac{\mathrm{d}}{\mathrm{d}t}\left\{\frac{1}{2}\dot{\boldsymbol{r}}^T \boldsymbol{M}\dot{\boldsymbol{r}} + \alpha\Delta\boldsymbol{x}^T \hat{\boldsymbol{J}}_s^{-T} \boldsymbol{M}\dot{\boldsymbol{r}} + \frac{1}{2}\Delta\boldsymbol{x}^T (\boldsymbol{K}_p + \alpha\boldsymbol{K}_v)\Delta\boldsymbol{x}\right\} \\ + \dot{\boldsymbol{r}}^T \boldsymbol{B}\dot{\boldsymbol{r}} + \dot{\boldsymbol{r}}^T \hat{\boldsymbol{J}}_s^T \boldsymbol{K}_v \dot{\boldsymbol{x}} + \dot{\boldsymbol{r}}^T \hat{\boldsymbol{J}}_s^T \boldsymbol{K}_p \Delta\boldsymbol{x} - \alpha\dot{\boldsymbol{x}}^T \hat{\boldsymbol{J}}_s^{-T} \boldsymbol{M}\dot{\boldsymbol{r}} \\ - \dot{\boldsymbol{r}}^T \boldsymbol{J}_s^T \boldsymbol{K}_p \Delta\boldsymbol{x} + \alpha\Delta\boldsymbol{x}^T \hat{\boldsymbol{J}}_s^{-T} \boldsymbol{B}\dot{\boldsymbol{r}} + \alpha\Delta\boldsymbol{x}^T \boldsymbol{K}_p \Delta\boldsymbol{x} = 0. \tag{2.32}$$

Therefore, we have

$$\frac{\mathrm{d}}{\mathrm{d}t} V(\Delta\boldsymbol{x}, \dot{\boldsymbol{r}}) = -W(\Delta\boldsymbol{x}, \dot{\boldsymbol{r}}), \tag{2.33}$$

where

$$V(\Delta\boldsymbol{x}, \dot{\boldsymbol{r}}) = \frac{1}{2}\dot{\boldsymbol{r}}^T \boldsymbol{M}\dot{\boldsymbol{r}} + \alpha\Delta\boldsymbol{x}^T \hat{\boldsymbol{J}}_s^{-T} \boldsymbol{M}\dot{\boldsymbol{r}} + \frac{1}{2}\Delta\boldsymbol{x}^T (\boldsymbol{K}_p + \alpha\boldsymbol{K}_v)\Delta\boldsymbol{x}, \tag{2.34}$$

and

$$W(\Delta\boldsymbol{x}, \dot{\boldsymbol{r}}) = \dot{\boldsymbol{r}}^T \boldsymbol{B}\dot{\boldsymbol{r}} + \dot{\boldsymbol{r}}^T \hat{\boldsymbol{J}}_s^T \boldsymbol{K}_v \dot{\boldsymbol{x}} + \dot{\boldsymbol{r}}^T \hat{\boldsymbol{J}}_s^T \boldsymbol{K}_p \Delta\boldsymbol{x} - \alpha\dot{\boldsymbol{x}}^T \hat{\boldsymbol{J}}_s^{-T} \boldsymbol{M}\dot{\boldsymbol{r}} \\ - \dot{\boldsymbol{r}}^T \boldsymbol{J}_s^T \boldsymbol{K}_p \Delta\boldsymbol{x} + \alpha\Delta\boldsymbol{x}^T \hat{\boldsymbol{J}}_s^{-T} \boldsymbol{B}\dot{\boldsymbol{r}} + \alpha\Delta\boldsymbol{x}^T \boldsymbol{K}_p \Delta\boldsymbol{x}. \tag{2.35}$$

With output that consists of a combination of the velocity and the position error, the scalar function $V(\Delta\boldsymbol{x}, \dot{\boldsymbol{r}})$ in Eq. (2.34) now contains a coupling term between the velocity and position error, in addition to the kinematic energy in Cartesian space and virtual potential energy in sensory space. To show that $V(\Delta\boldsymbol{x}, \dot{\boldsymbol{r}})$ is positive definite in $\dot{\boldsymbol{r}}$ and $\Delta\boldsymbol{x}$, we express Eq. (2.34) as

$$V(\Delta\boldsymbol{x}, \dot{\boldsymbol{x}}) = \frac{1}{2}\dot{\boldsymbol{x}}^T \boldsymbol{J}_s^{-T} \boldsymbol{M}\boldsymbol{J}_s^{-1}\dot{\boldsymbol{x}} + \alpha\Delta\boldsymbol{x}^T \hat{\boldsymbol{J}}_s^{-T} \boldsymbol{M}\boldsymbol{J}_s^{-1}\dot{\boldsymbol{x}} + \frac{1}{2}\Delta\boldsymbol{x}^T (\boldsymbol{K}_p + \alpha\boldsymbol{K}_v)\Delta\boldsymbol{x} \\ = \frac{1}{4}\dot{\boldsymbol{x}}^T \boldsymbol{M}_x \dot{\boldsymbol{x}} + \frac{1}{4}(\dot{\boldsymbol{x}} + 2\alpha\boldsymbol{J}_s\hat{\boldsymbol{J}}_s^{-1}\Delta\boldsymbol{x})^T \boldsymbol{M}_x (\dot{\boldsymbol{x}} + 2\alpha\boldsymbol{J}_s\hat{\boldsymbol{J}}_s^{-1}\Delta\boldsymbol{x}) \\ + \frac{1}{2}\Delta\boldsymbol{x}^T (\boldsymbol{K}_p + \alpha\boldsymbol{K}_v - 2\alpha^2 \hat{\boldsymbol{J}}_s^{-T} \boldsymbol{M}\hat{\boldsymbol{J}}_s^{-1})\Delta\boldsymbol{x}, \tag{2.36}$$

where $\boldsymbol{M}_x \triangleq \boldsymbol{J}_s^{-T} \boldsymbol{M}\boldsymbol{J}_s^{-1}$ and $\dot{\boldsymbol{r}} = \boldsymbol{J}_s^{-1}\dot{\boldsymbol{x}}$, as seen from Eq. (2.25). Let $\lambda_m \triangleq \lambda_{\max}[\hat{\boldsymbol{J}}_s^{-T} \boldsymbol{M}\hat{\boldsymbol{J}}_s^{-1}]$, and let $\lambda_{\max}[\cdot]$ denote the maximum eigenvalue of the matrix. Then $V(\Delta\boldsymbol{x}, \dot{\boldsymbol{x}})$ is positive definite in $\dot{\boldsymbol{x}}$ and $\Delta\boldsymbol{x}$ when

$$k_{vi} - 2\alpha\lambda_m > 0, \tag{2.37}$$

where k_{vi} denotes the ith diagonal element of $\boldsymbol{K}_v$.

The output also leads to a derivative of $V(\Delta \boldsymbol{x}, \dot{\boldsymbol{x}})$ that contains various terms that are dependent on $\dot{\boldsymbol{x}}$ and $\Delta \boldsymbol{x}$, as seen from Eq. (2.35). Next, we proceed to show that the derivative of $V(\Delta \boldsymbol{x}, \dot{\boldsymbol{x}})$ is negative definite in $\dot{\boldsymbol{x}}$ and $\Delta \boldsymbol{x}$ with proper choices of control gains. Clearly, this is equivalent to showing that $W(\Delta \boldsymbol{x}, \dot{\boldsymbol{x}})$ is positive definite in $\dot{\boldsymbol{x}}$ and $\Delta \boldsymbol{x}$.

Equation (2.35) is first expressed as

$$\begin{aligned} W(\Delta \boldsymbol{x}, \dot{\boldsymbol{x}}) = {} & \dot{\boldsymbol{x}}^T \boldsymbol{J}_s^{-T} \boldsymbol{B} \boldsymbol{J}_s^{-1} \dot{\boldsymbol{x}} + \dot{\boldsymbol{x}}^T \boldsymbol{J}_s^{-T} \hat{\boldsymbol{J}}_s^T \boldsymbol{K}_v \dot{\boldsymbol{x}} + \dot{\boldsymbol{x}}^T \boldsymbol{J}_s^{-T} \hat{\boldsymbol{J}}_s^T \boldsymbol{K}_p \Delta \boldsymbol{x} \\ & - \alpha \dot{\boldsymbol{x}}^T \hat{\boldsymbol{J}}_s^{-T} \boldsymbol{M} \boldsymbol{J}_s^{-1} \dot{\boldsymbol{x}} - \dot{\boldsymbol{x}}^T \boldsymbol{K}_p \Delta \boldsymbol{x} + \alpha \Delta \boldsymbol{x}^T \hat{\boldsymbol{J}}_s^{-T} \boldsymbol{B} \boldsymbol{J}_s^{-1} \dot{\boldsymbol{x}} \\ & + \alpha \Delta \boldsymbol{x}^T \boldsymbol{K}_p \Delta \boldsymbol{x}. \end{aligned} \quad (2.38)$$

To analyze the effects of an uncertain Jacobian on the stability of the system, we define a matrix Δ_J to represent the deviation or estimation error between the actual Jacobian matrix and estimated Jacobian matrix as follows:

$$\Delta_J = \boldsymbol{I}_2 - \boldsymbol{J}_s^{-T} \hat{\boldsymbol{J}}_s^T, \quad (2.39)$$

where $\boldsymbol{I}_2 \in \Re^{2\times 2}$ is the 2×2 identity matrix. Using Δ_J and letting $\boldsymbol{B}_x \triangleq \boldsymbol{J}_s^{-T} \boldsymbol{B} \boldsymbol{J}_s^{-1}$, Eq. (2.38) can be written as

$$\begin{aligned} W(\Delta \boldsymbol{x}, \dot{\boldsymbol{x}}) = {} & \dot{\boldsymbol{x}}^T \boldsymbol{B}_x \dot{\boldsymbol{x}} + \dot{\boldsymbol{x}}^T (\boldsymbol{I}_2 - \Delta_J) \boldsymbol{K}_v \dot{\boldsymbol{x}} - \dot{\boldsymbol{x}}^T \Delta_J \boldsymbol{K}_p \Delta \boldsymbol{x} \\ & + \alpha \Delta \boldsymbol{x}^T \boldsymbol{K}_p \Delta \boldsymbol{x} - \alpha h(\Delta \boldsymbol{x}, \dot{\boldsymbol{x}}), \end{aligned} \quad (2.40)$$

where $h(\Delta \boldsymbol{x}, \dot{\boldsymbol{x}}) = \dot{\boldsymbol{x}}^T \hat{\boldsymbol{J}}_s^{-T} \boldsymbol{M} \boldsymbol{J}_s^{-1} \dot{\boldsymbol{x}} - \Delta \boldsymbol{x}^T \hat{\boldsymbol{J}}_s^{-T} \boldsymbol{B} \boldsymbol{J}_s^{-1} \dot{\boldsymbol{x}}$ contains the terms that are independent of the proportional control gain $\boldsymbol{K}_p$ and the derivative control gain $\boldsymbol{K}_v$. Since $h(\Delta \boldsymbol{x}, \dot{\boldsymbol{x}})$ is dependent on $\dot{\boldsymbol{x}}$ and $\Delta \boldsymbol{x}$, there exist positive constants c_0, c_1 such that

$$-\alpha h(\Delta \boldsymbol{x}, \dot{\boldsymbol{x}}) \geq -\alpha c_0 ||\dot{\boldsymbol{x}}||^2 - \alpha c_1 ||\Delta \boldsymbol{x}||^2. \quad (2.41)$$

Substituting the above inequality into Eq. (2.40) yields

$$\begin{aligned} W(\Delta \boldsymbol{x}, \dot{\boldsymbol{x}}) \geq {} & \dot{\boldsymbol{x}}^T (\boldsymbol{B}_x + (\boldsymbol{I}_2 - \Delta_J) \boldsymbol{K}_v - \alpha c_0 \boldsymbol{I}_2) \dot{\boldsymbol{x}} \\ & - \dot{\boldsymbol{x}}^T \Delta_J \boldsymbol{K}_p \Delta \boldsymbol{x} + \alpha \Delta \boldsymbol{x}^T (\boldsymbol{K}_p - c_1 \boldsymbol{I}_2) \Delta \boldsymbol{x}. \end{aligned} \quad (2.42)$$

Let $\hat{\boldsymbol{J}}_s$ be chosen such that $||\Delta_J|| = ||\boldsymbol{I}_2 - \boldsymbol{J}_s^{-T} \hat{\boldsymbol{J}}_s^T|| \leq p$, where p is a positive constant. We proceed to derive a sufficient condition to guarantee that $W(\Delta \boldsymbol{x}, \dot{\boldsymbol{x}})$ is positive definite in $\dot{\boldsymbol{x}}$ and $\Delta \boldsymbol{x}$ or that $-W(\Delta \boldsymbol{x}, \dot{\boldsymbol{x}})$ is negative definite in $\dot{\boldsymbol{x}}$ and $\Delta \boldsymbol{x}$. From Eq. (2.42), we obtain

$$\begin{aligned} W(\Delta \boldsymbol{x}, \dot{\boldsymbol{x}}) \geq {} & (\lambda_{\min}[\boldsymbol{K}_v] - p\lambda_{\max}[\boldsymbol{K}_v] - \alpha c_0) \|\dot{\boldsymbol{x}}\|^2 - p\lambda_{\max}[\boldsymbol{K}_p] ||\Delta \boldsymbol{x}|| \cdot ||\dot{\boldsymbol{x}}|| \\ & + \alpha(\lambda_{\min}[\boldsymbol{K}_p] - c_1) ||\Delta \boldsymbol{x}||^2. \end{aligned} \quad (2.43)$$

Note that $-p\lambda_{\max}[\boldsymbol{K}_p]||\Delta\boldsymbol{x}|| \cdot ||\dot{\boldsymbol{x}}|| \geq -\frac{1}{2}p\lambda_{\max}[\boldsymbol{K}_p](||\Delta\boldsymbol{x}||^2 + ||\dot{\boldsymbol{x}}||^2)$. Hence,

$$W(\Delta\boldsymbol{x}, \dot{\boldsymbol{x}}) \geq \lambda_{\max}[\boldsymbol{K}_v]l_1 \|\dot{\boldsymbol{x}}\|^2 + a\lambda_{\max}[\boldsymbol{K}_v]l_2||\Delta\boldsymbol{x}||^2, \tag{2.44}$$

where $l_1 = \frac{\lambda_{\min}[\boldsymbol{K}_v]}{\lambda_{\max}[\boldsymbol{K}_v]} - \frac{\alpha c_0}{\lambda_{\max}[\boldsymbol{K}_v]} - p(1 + \frac{a}{2})$, $l_2 = \alpha\frac{\lambda_{\min}[\boldsymbol{K}_p]}{\lambda_{\max}[\boldsymbol{K}_p]} - \frac{\alpha c_1}{\lambda_{\max}[\boldsymbol{K}_p]} - \frac{p}{2}$, and $a = \frac{\lambda_{\max}[\boldsymbol{K}_p]}{\lambda_{\max}[\boldsymbol{K}_v]}$.

When the control parameters are chosen such that

$$\min\left\{\frac{2}{2+a}\left(\frac{\lambda_{\min}[\boldsymbol{K}_v]}{\lambda_{\max}[\boldsymbol{K}_v]} - \frac{\alpha c_0}{\lambda_{\max}[\boldsymbol{K}_v]}\right), 2\alpha\left(\frac{\lambda_{\min}[\boldsymbol{K}_p]}{\lambda_{\max}[\boldsymbol{K}_p]} - \frac{c_1}{\lambda_{\max}[\boldsymbol{K}_p]}\right)\right\} > p, \tag{2.45}$$

and $\boldsymbol{K}_v$ and $\boldsymbol{K}_p$ are also chosen large enough that

$$\begin{aligned} \frac{\lambda_{\min}[\boldsymbol{K}_v]}{\lambda_{\max}[\boldsymbol{K}_v]} - \frac{\alpha c_0}{\lambda_{\max}[\boldsymbol{K}_v]} > 0, \\ \frac{\lambda_{\min}[\boldsymbol{K}_p]}{\lambda_{\max}[\boldsymbol{K}_p]} - \frac{c_1}{\lambda_{\max}[\boldsymbol{K}_p]} > 0, \end{aligned} \tag{2.46}$$

then $W(\Delta\boldsymbol{x}, \dot{\boldsymbol{x}}) > 0$. We can now state the following Theorem:

Theorem 2.1 *The task-space sensory feedback controller* (2.27) *for the mass–damper system* (2.23) *gives rise to the convergence of the task-space error and velocity such that* $\Delta\boldsymbol{x} \to \boldsymbol{0}$ *and* $\dot{\boldsymbol{x}} \to \boldsymbol{0}$ *as* $t \to \infty$ *if the feedback gains* $\boldsymbol{K}_p$ *and* $\boldsymbol{K}_v$ *are chosen to satisfy conditions* (2.37), (2.45), *and* (2.46), *and* $\hat{\boldsymbol{J}}_s$ *is chosen such that*

$$\|\boldsymbol{I}_2 - \boldsymbol{J}_s^{-T}\hat{\boldsymbol{J}}_s^T\| \leq p. \tag{2.47}$$

Proof When conditions (2.37), (2.45), and (2.46) are satisfied and $\|\boldsymbol{I}_2 - \boldsymbol{J}_s^{-T}\hat{\boldsymbol{J}}_s^T\| \leq p$, we have that $V(\Delta\boldsymbol{x}, \dot{\boldsymbol{x}})$ and $W(\Delta\boldsymbol{x}, \dot{\boldsymbol{x}})$ are both positive definite in $\dot{\boldsymbol{x}}$ and $\Delta\boldsymbol{x}$. Therefore, $\dot{\boldsymbol{x}}$ and $\Delta\boldsymbol{x}$ are bounded. The boundedness of $\dot{\boldsymbol{x}}$ ensures the boundedness of $\dot{\boldsymbol{r}}$, since it is assumed that $\boldsymbol{J}_S^{-1}$ exists. From Eq. (2.28), it is seen that $\ddot{\boldsymbol{r}}$ is bounded, which also implies that $\ddot{\boldsymbol{x}}$ is bounded, since $\ddot{\boldsymbol{x}} = \boldsymbol{J}_s\ddot{\boldsymbol{r}}$. Since $\ddot{\boldsymbol{x}}$ and $\dot{\boldsymbol{x}}$ are bounded, both $\dot{\boldsymbol{x}}$ and $\Delta\boldsymbol{x}$ are uniformly continuous. From Eqs. (2.33) and (2.44), it is seen that $\dot{\boldsymbol{x}}, \Delta\boldsymbol{x} \in L_2(0, \infty)$. Therefore, $\dot{\boldsymbol{x}} \to \boldsymbol{0}$ and $\Delta\boldsymbol{x} \to \boldsymbol{0}$ as $t \to \infty$. △△△

From Eq. (2.24), note that

$$\Delta\boldsymbol{x} = \boldsymbol{J}_s\Delta\boldsymbol{r}, \tag{2.48}$$

where $\Delta\boldsymbol{r} = \boldsymbol{r} - \boldsymbol{r}_d$, and $\boldsymbol{r}_d$ is the desired position in Cartesian space that corresponds to the desired position x_d in sensory space. The desired position $\boldsymbol{r}_d$ is unknown, since the sensory Jacobian matrix is uncertain. Similarly, the relationship between $\dot{\boldsymbol{x}}$ and $\dot{\boldsymbol{r}}$ is described in Eq. (2.25) as

$$\dot{\boldsymbol{x}} = \boldsymbol{J}_s\dot{\boldsymbol{r}}. \tag{2.49}$$

Since $\boldsymbol{J}_s$ as defined in Eq. (2.25) is always of full rank, $V(\Delta\boldsymbol{x}, \dot{\boldsymbol{r}})$ and $W(\Delta\boldsymbol{x}, \dot{\boldsymbol{r}})$ are also positive definite in $\dot{\boldsymbol{r}}$ and $\Delta\boldsymbol{r}$, and thus asymptotic stability can also be concluded using the Lyapunov method directly.

To gain further insight into the condition (2.45), we let $\boldsymbol{K}_v = k_v \boldsymbol{I}_2$, $\boldsymbol{K}_p = ak_v \boldsymbol{I}_2$. Then condition (2.45) becomes

$$p < \min\left\{\frac{2}{a+2}\left(1 - \frac{\alpha c_0}{k_v}\right), 2\alpha\left(1 - \frac{c_1}{ak_v}\right)\right\}. \tag{2.50}$$

When k_v is so large that $\frac{\alpha c_0}{k_v}$ and $\frac{c_1}{ak_v}$ are negligible, condition (2.50) reduces to

$$p < \min\left\{\frac{2}{a+2}, 2\alpha\right\}. \tag{2.51}$$

From condition (2.51), when $\alpha \geq \frac{1}{2}$, we have

$$p < \frac{2}{a+2} < 1. \tag{2.52}$$

This implies that the sufficient allowable bound for the Jacobian uncertainty is less than 1 and that a can be chosen small enough to guarantee the stability of the system if p is large. If p is small, then from condition (2.51), α can be small, because

$$p < 2\alpha, \qquad 0 < a \leq \bar{a}, \tag{2.53}$$
$$p < \frac{2}{a+2}, \qquad \bar{a} < a < \infty, \tag{2.54}$$

where $\alpha < \frac{1}{2}$ and $\bar{a} = \frac{1}{\alpha} - 2$. Hence, a larger range of a can be chosen for a smaller p, as illustrated in Fig. 2.12.

Therefore, condition (2.51) implies that if a is increased, then the allowable bound p of the Jacobian uncertainty $\Delta_J = \boldsymbol{I}_2 - \boldsymbol{J}_s^{-T}\hat{\boldsymbol{J}}_s^T$ is reduced. Hence, a should be kept small so that the allowable bound of the Jacobian uncertainty is larger. This can be done easily by increasing $\boldsymbol{K}_v$.

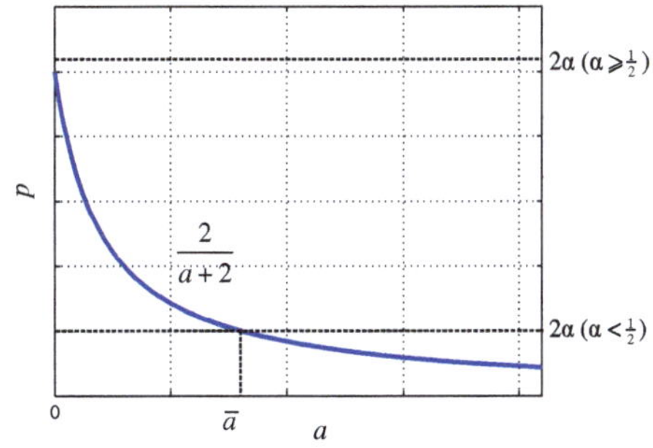

Fig. 2.12 Variation of p with a (k_v is very large)

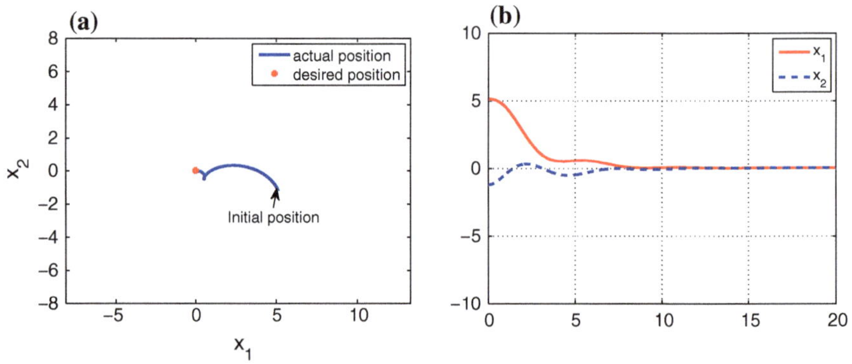

Fig. 2.13 The closed-loop system is stable after the control gains are adjusted. **a** Path of block. **b** Position error

Using the results illustrated in Fig. 2.12, the control parameter was adjusted to reduce a by increasing from $\boldsymbol{K}_v = \text{diag}\{1.3, 1.3\}$ to $\boldsymbol{K}_v = \text{diag}\{2.4, 2.4\}$. The system was stabilized as shown in Fig. 2.13.

2.3.2 Sensory Feedback Control of a Robot Manipulator

Consider now a robot manipulator with n degrees of freedom for which the position or position error of the end effector is measured by a sensor. Let $\boldsymbol{r} \in \Re^p$ represent the position of the end effector in Cartesian space, and let $\boldsymbol{x} \in \Re^m$ represent the position obtained by the sensor in sensory space. The velocity vector $\dot{\boldsymbol{x}}$ in sensory space is related to $\dot{\boldsymbol{r}}$ in Cartesian space by

$$\dot{\boldsymbol{x}} = \boldsymbol{J}_s(\boldsymbol{r})\dot{\boldsymbol{r}}, \tag{2.55}$$

where $\boldsymbol{J}_s(\boldsymbol{r}) \in \Re^{m \times p}$ is the sensory Jacobian matrix, as illustrated in Fig. 2.14. From Eq. (1.24) in Chap. 1, the velocity vector $\dot{\boldsymbol{r}}$ in Cartesian space is related to $\dot{\boldsymbol{q}}$ in joint space by

$$\dot{\boldsymbol{r}} = \boldsymbol{J}_m(\boldsymbol{q})\dot{\boldsymbol{q}}, \tag{2.56}$$

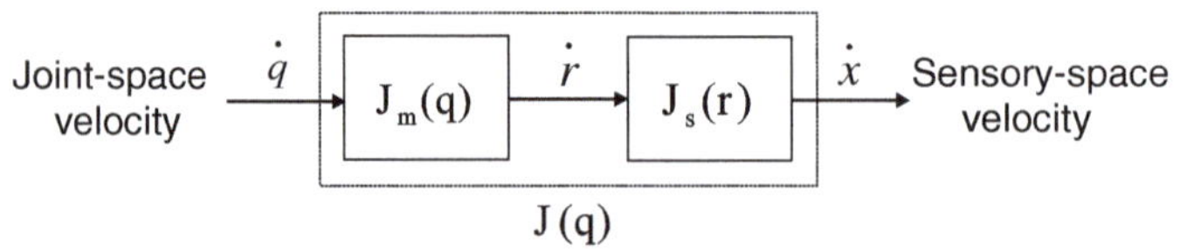

Fig. 2.14 The sensory Jacobian matrix is denoted by $\boldsymbol{J}_s(\boldsymbol{r})$, and the manipulator Jacobian matrix is denoted by $\boldsymbol{J}_m(\boldsymbol{q})$

where $\boldsymbol{J}_m(\boldsymbol{q}) \in \Re^{p \times n}$ is the manipulator Jacobian matrix of the mapping from joint space to Cartesian space.

The overall relationship between the velocity vector $\dot{\boldsymbol{x}}$ in sensory space and $\dot{\boldsymbol{q}}$ in joint space is therefore given by

$$\dot{\boldsymbol{x}} = \boldsymbol{J}_s(\boldsymbol{r})\dot{\boldsymbol{r}} = \boldsymbol{J}_s(\boldsymbol{r})\boldsymbol{J}_m(\boldsymbol{q})\dot{\boldsymbol{q}}, \tag{2.57}$$

and the overall Jacobian of the mapping from joint space to sensory space is defined as

$$\boldsymbol{J}(\boldsymbol{q}) = \boldsymbol{J}_s(\boldsymbol{r})\boldsymbol{J}_m(\boldsymbol{q}). \tag{2.58}$$

In the presence of uncertain kinematics or calibration errors, either the manipulator Jacobian matrix or the sensory Jacobian matrix is uncertain, and the overall Jacobian matrix is estimated as

$$\hat{\boldsymbol{J}}(\boldsymbol{q}) = \hat{\boldsymbol{J}}_s(\boldsymbol{r})\hat{\boldsymbol{J}}_m(\boldsymbol{q}), \tag{2.59}$$

where $\hat{\boldsymbol{J}}(\boldsymbol{q}) \in \Re^{m \times n}$ is an approximate matrix for $\boldsymbol{J}(\boldsymbol{q}) \in \Re^{m \times n}$.

When cameras are used to monitor the position of the end effector, the task coordinates are defined as image coordinates. The pinhole camera model can be used to represent the mapping from Cartesian space to image space. From Eq. (1.42) in Chap. 1, the relationship between image space and Cartesian space is represented by

$$\dot{\boldsymbol{x}} = \boldsymbol{Z}^{-1}(\boldsymbol{q})\boldsymbol{L}(\boldsymbol{x})\dot{\boldsymbol{r}}, \tag{2.60}$$

where

$$\boldsymbol{Z}^{-1}(\boldsymbol{q}) = \begin{bmatrix} \frac{1}{z_1(\boldsymbol{q})} & \boldsymbol{0} & \cdots & 0 \\ \boldsymbol{0} & \frac{1}{z_2(\boldsymbol{q})} & \cdots & \boldsymbol{0} \\ \vdots & \vdots & \ddots & \vdots \\ \boldsymbol{0} & \boldsymbol{0} & \cdots & \frac{1}{z_m(\boldsymbol{q})} \end{bmatrix} \tag{2.61}$$

is a diagonal matrix that contains the depth information of the feature points with respect to the camera image frame, and $\boldsymbol{L}(\boldsymbol{x}) \in \Re^{m \times p}$ is a Jacobian matrix. The overall matrix $\boldsymbol{Z}^{-1}(\boldsymbol{q})\boldsymbol{L}(\boldsymbol{x})$ is the image Jacobian matrix or sensory Jacobian matrix.

The velocity of the image feature is then related to the joint-space velocity as

$$\dot{\boldsymbol{x}} = \underbrace{\boldsymbol{Z}^{-1}(\boldsymbol{q})\boldsymbol{L}(\boldsymbol{x})}_{\boldsymbol{J}_s}\boldsymbol{J}_m(\boldsymbol{q})\dot{\boldsymbol{q}}. \tag{2.62}$$

In the presence of uncertain kinematics and uncalibrated camera, the estimated image-space velocity of the end effector $\dot{\hat{x}}$ is obtained as

$$\dot{\hat{x}} = \hat{Z}^{-1}(q)\hat{L}(x)\hat{J}_m(q)\dot{q}, \tag{2.63}$$

where $\hat{Z}^{-1}(q)$ is the approximate depth matrix, and $\hat{L}(x)$ and $\hat{J}_m(q)$ represent the approximate matrices of $L(x)$ and $J_m(q)$ respectively.

When a position sensor such as a PSD camera is used to measure the end effector's position directly in Cartesian space, the task coordinates are defined as the Cartesian coordinates. In this case,

$$J(q) = J_m(q). \tag{2.64}$$

If the kinematics is uncertain, the Jacobian matrix is also uncertain, which is specified as

$$\hat{J}(q) = \hat{J}_m(q), \tag{2.65}$$

where $\hat{J}_m(q)$ is an estimated matrix of $J_m(q)$.

Therefore, the task-space coordinates include the image coordinates and the Cartesian coordinates. In this section, we consider a setpoint control for a nonredundant robot with $m = n$ and a sensory feedback controller with approximate Jacobian matrix designed as

$$\tau = -\hat{J}^T(q)(K_p s(\Delta x) + K_v \dot{x}) + g(q), \tag{2.66}$$

where $\dot{x} \in \Re^n$ is the velocity of the end effector in sensory space, $\hat{J}(q) \in \Re^{n \times n}$ is the uncertain Jacobian matrix from joint-space to sensory space, $s(\cdot) \in \Re^n$ is a saturated function to be defined, and $K_p \in \Re^{n \times n}$ and $K_v \in \Re^{n \times n}$ are diagonal and positive definite matrices. A block diagram for the controller in Eq. (2.66) is shown in Fig. 2.15.

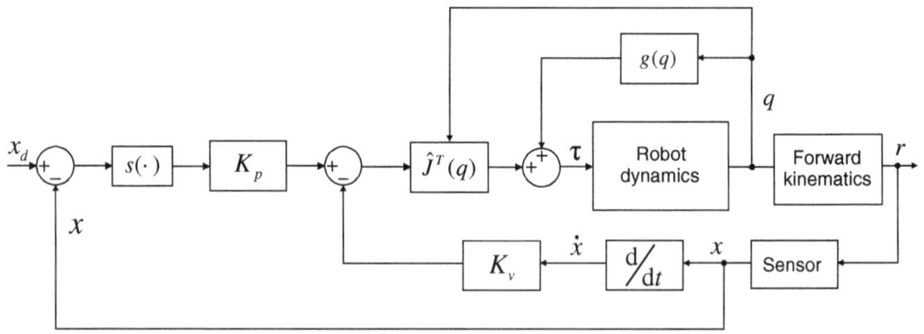

Fig. 2.15 Block diagram of the task-space sensory feedback control

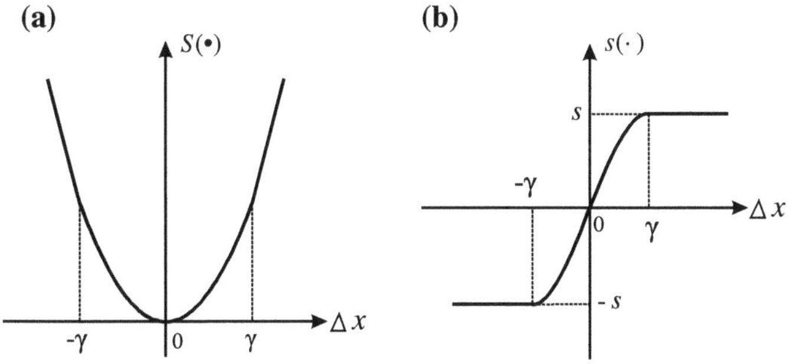

Fig. 2.16 **a** Quasinatural potential: $S(\Delta \boldsymbol{x})$. **b** Derivative of $S(\Delta \boldsymbol{x})$: $s(\Delta \boldsymbol{x})$

The approximate Jacobian matrix $\hat{\boldsymbol{J}}^T(\boldsymbol{q})$ is also chosen so that

$$\|\boldsymbol{I}_n - \boldsymbol{J}^{-T}(\boldsymbol{q})\hat{\boldsymbol{J}}^T(\boldsymbol{q})\| \leq \bar{p}, \tag{2.67}$$

where $\boldsymbol{I}_n \in \Re^{n\times n}$ is the identity matrix, and $\bar{p}$ is a positive constant to be defined later.

First, let us define a scalar potential function $S_i(\Delta x_i)$ and its derivative $s_i(\Delta x_i)$, where Δx_i denotes the ith diagonal element of $\Delta \boldsymbol{x}$, as shown in Fig. 2.16, with the following properties:

1. $S_i(\Delta x_i) > 0$ for $\Delta x_i \neq 0$ and $S_i(0) = 0$.
2. $S_i(\Delta x_i)$ is twice continuously differentiable, and the derivative $s_i(\Delta x_i) = \frac{\mathrm{d}S_i(\Delta x_i)}{\mathrm{d}\Delta x_i}$ is strictly increasing in Δx_i for $|\Delta x_i| < \gamma_i$ with some γ_i and is saturated for $|\Delta x_i| \geq \gamma_i$, i.e., $s_i(\Delta x_i) = \pm s_i$ for $\Delta x_i \geq \gamma_i$ and $\Delta x_i \leq -\gamma_i$ respectively.
3. There are constants $\bar{c}_i > 0$, $d_i > 0$, $\bar{d}_i(> d_i) > 0$ such that

$$\bar{d}_i s_i^2(\Delta x_i) \geq \Delta x_i s_i(\Delta x_i) \geq d_i s_i^2(\Delta x_i) > 0, \quad S_i(\Delta x_i) \geq \bar{c}_i s_i^2(\Delta x_i), \tag{2.68}$$

for $\Delta x_i \neq 0$.

The saturation function is used to bound the position error $\Delta \boldsymbol{x}$ in the sensory feedback controller described by Eq. (2.66). Its role is quite similar to visually guided reaching tasks of human beings in the sense that when a person reaches for an object, only the directional information is required at the beginning, when the position error is large. That is, the magnitude is not important at the beginning stage and thus can be saturated, and only the directional information is focused on. The magnitude of the position error is monitored only when the target is near.

Next, the output is defined as

$$\boldsymbol{y} = \dot{\boldsymbol{q}} + \alpha \hat{\boldsymbol{J}}^{-1}(\boldsymbol{q}) \boldsymbol{s}(\Delta \boldsymbol{x}). \tag{2.69}$$

where joint-space velocity and sensory task-space position error are used, and an inverse of the approximate Jacobian matrix is used to transform the sensory feedback error to joint space.

Substituting Eq. (2.66) into Eq. (2.9), we have,

$$\boldsymbol{M}(\boldsymbol{q})\ddot{\boldsymbol{q}}+\left(\tfrac{1}{2}\dot{\boldsymbol{M}}(\boldsymbol{q})+\boldsymbol{S}(\boldsymbol{q},\dot{\boldsymbol{q}})\right)\dot{\boldsymbol{q}}+\hat{\boldsymbol{J}}^T(\boldsymbol{q})(\boldsymbol{K}_p\boldsymbol{s}(\Delta\boldsymbol{x})+\boldsymbol{K}_v\dot{\boldsymbol{x}})=\boldsymbol{0}. \tag{2.70}$$

Similarly, the inner product of the output $\boldsymbol{y}$ described by Eqs. (2.69) and (2.70) yields

$$\tfrac{\mathrm{d}}{\mathrm{d}t}V(\boldsymbol{s}(\Delta\boldsymbol{x}),\dot{\boldsymbol{q}})+W(\boldsymbol{s}(\Delta\boldsymbol{x}),\dot{\boldsymbol{q}})=0, \tag{2.71}$$

where

$$\begin{aligned}V(\boldsymbol{s}(\Delta\boldsymbol{x}),\dot{\boldsymbol{q}})=\;&\tfrac{1}{2}\dot{\boldsymbol{q}}^T\boldsymbol{M}(\boldsymbol{q})\dot{\boldsymbol{q}}+\alpha\boldsymbol{s}(\Delta\boldsymbol{x})^T\hat{\boldsymbol{J}}^{-T}(\boldsymbol{q})\boldsymbol{M}(\boldsymbol{q})\dot{\boldsymbol{q}}\\&+\sum_{i=1}^{n}(k_{pi}+\alpha k_{vi})S_i(\Delta x_i)\end{aligned} \tag{2.72}$$

and

$$\begin{aligned}W(\boldsymbol{s}(\Delta\boldsymbol{x}),\dot{\boldsymbol{q}})=\;&\dot{\boldsymbol{q}}^T\hat{\boldsymbol{J}}^T(\boldsymbol{q})\boldsymbol{K}_v\dot{\boldsymbol{x}}+\dot{\boldsymbol{q}}^T\hat{\boldsymbol{J}}^T(\boldsymbol{q})\boldsymbol{K}_p\boldsymbol{s}(\Delta\boldsymbol{x})-\dot{\boldsymbol{x}}^T\boldsymbol{K}_p\boldsymbol{s}(\Delta\boldsymbol{x})\\&+\alpha\{\boldsymbol{s}(\Delta\boldsymbol{x})^T\hat{\boldsymbol{J}}^{-T}(\boldsymbol{q})(\boldsymbol{S}(\boldsymbol{q},\dot{\boldsymbol{q}})-\tfrac{1}{2}\dot{\boldsymbol{M}}(\boldsymbol{q}))\dot{\boldsymbol{q}}-\dot{\boldsymbol{s}}(\Delta\boldsymbol{x})^T\hat{\boldsymbol{J}}^{-T}(\boldsymbol{q})\boldsymbol{M}(\boldsymbol{q})\dot{\boldsymbol{q}}\\&-\boldsymbol{s}(\Delta\boldsymbol{x})^T\dot{\hat{\boldsymbol{J}}}^{-T}(\boldsymbol{q})\boldsymbol{M}(\boldsymbol{q})\dot{\boldsymbol{q}}\}+\alpha\boldsymbol{s}(\Delta\boldsymbol{x})^T\boldsymbol{K}_p\boldsymbol{s}(\Delta\boldsymbol{x}),\end{aligned} \tag{2.73}$$

where $\dot{\boldsymbol{q}}^T\boldsymbol{S}(\boldsymbol{q},\dot{\boldsymbol{q}})\dot{\boldsymbol{q}}$ equals zero by *Property* 1.2 in Chap. 1, and k_{pi}, k_{vi} denote the ith diagonal elements of $\boldsymbol{K}_p$ and $\boldsymbol{K}_v$ respectively. Since

$$\begin{aligned}&\tfrac{1}{4}\dot{\boldsymbol{q}}^T\boldsymbol{M}(\boldsymbol{q})\dot{\boldsymbol{q}}+\alpha\boldsymbol{s}(\Delta\boldsymbol{x})^T\hat{\boldsymbol{J}}^{-T}(\boldsymbol{q})\boldsymbol{M}(\boldsymbol{q})\dot{\boldsymbol{q}}+\sum_{i=1}^{n}(k_{pi}+\alpha k_{vi})S_i(\Delta x_i)\\&=\tfrac{1}{4}(\dot{\boldsymbol{x}}+2\alpha\boldsymbol{J}(\boldsymbol{q})\hat{\boldsymbol{J}}^{-1}(\boldsymbol{q})\boldsymbol{s}(\Delta\boldsymbol{x}))^T\boldsymbol{M}_x(\boldsymbol{q})(\dot{\boldsymbol{x}}+2\alpha\boldsymbol{J}(\boldsymbol{q})\hat{\boldsymbol{J}}^{-1}(\boldsymbol{q})\boldsymbol{s}(\Delta\boldsymbol{x}))\\&-\alpha^2\boldsymbol{s}(\Delta\boldsymbol{x})^T(\boldsymbol{J}(\boldsymbol{q})\hat{\boldsymbol{J}}^{-1}(\boldsymbol{q}))^T\boldsymbol{M}_x(\boldsymbol{q})\boldsymbol{J}(\boldsymbol{q})\hat{\boldsymbol{J}}^{-1}(\boldsymbol{q})\boldsymbol{s}(\Delta\boldsymbol{x})+\sum_{i=1}^{n}(k_{pi}+\alpha k_{vi})S_i(\Delta x_i)\\&\geq\sum_{i=1}^{n}(k_{pi}+\alpha k_{vi})\bar{c}_i s_i^2(\Delta x_i)-\alpha^2\boldsymbol{s}(\Delta\boldsymbol{x})^T(\hat{\boldsymbol{J}}^{-T}(\boldsymbol{q})\boldsymbol{M}(\boldsymbol{q})\hat{\boldsymbol{J}}^{-1}(\boldsymbol{q}))\boldsymbol{s}(\Delta\boldsymbol{x})\\&\geq\sum_{i=1}^{n}\{k_{pi}+\alpha(k_{vi}\bar{c}_i-\alpha\lambda_m)\}s_i^2(\Delta x_i),\end{aligned} \tag{2.74}$$

where $\boldsymbol{M}_x(\boldsymbol{q})=\boldsymbol{J}^{-T}(\boldsymbol{q})\boldsymbol{M}(\boldsymbol{q})\boldsymbol{J}^{-1}(\boldsymbol{q})$, $\lambda_m\triangleq\lambda_{\max}[\hat{\boldsymbol{J}}^{-T}(\boldsymbol{q})\boldsymbol{M}(\boldsymbol{q})\hat{\boldsymbol{J}}^{-1}(\boldsymbol{q})]$, and $\lambda_{\max}[\cdot]$ denotes the maximum eigenvalue of the matrix. Note from *Property* 1.1 in Chap. 1, $\boldsymbol{M}(\boldsymbol{q})$ is positive definite and hence $\boldsymbol{M}_x(\boldsymbol{q})$ is also positive definite. Substituting this into Eq. (2.72), we have

$$V(\boldsymbol{s}(\Delta\boldsymbol{x}),\dot{\boldsymbol{x}})\geq\tfrac{1}{4}\dot{\boldsymbol{x}}^T\boldsymbol{M}_x(\boldsymbol{q})\dot{\boldsymbol{x}}+\sum_{i=1}^{n}\{k_{pi}+\alpha(k_{vi}\bar{c}_i-\alpha\lambda_m)\}s_i^2(\Delta x_i)\geq 0, \tag{2.75}$$

when

$$k_{vi}\bar{c}_i - \alpha\lambda_m > 0. \tag{2.76}$$

By letting $\Delta_J = \boldsymbol{I}_n - \boldsymbol{J}^{-T}(\boldsymbol{q})\hat{\boldsymbol{J}}^T(\boldsymbol{q})$, Eq. (2.73) becomes

$$\begin{aligned} W(\boldsymbol{s}(\Delta\boldsymbol{x}), \dot{\boldsymbol{x}}) &= \dot{\boldsymbol{x}}^T((\boldsymbol{I}_n - \Delta_J)\boldsymbol{K}_v)\dot{\boldsymbol{x}} + \alpha\boldsymbol{s}(\Delta\boldsymbol{x})^T\boldsymbol{K}_p\boldsymbol{s}(\Delta\boldsymbol{x}) \\ &\quad - \dot{\boldsymbol{x}}^T(\Delta_J\boldsymbol{K}_p)\boldsymbol{s}(\Delta\boldsymbol{x}) - \alpha h(\boldsymbol{x}, \dot{\boldsymbol{q}}), \end{aligned} \tag{2.77}$$

where

$$\begin{aligned} h(\boldsymbol{x}, \dot{\boldsymbol{q}}) = \boldsymbol{s}(\Delta\boldsymbol{x})^T\hat{\boldsymbol{J}}^{-T}(\boldsymbol{q})\left(\tfrac{1}{2}\dot{\boldsymbol{M}}(\boldsymbol{q}) - \boldsymbol{S}(\boldsymbol{q}, \dot{\boldsymbol{q}})\right)\dot{\boldsymbol{q}} + \dot{\boldsymbol{s}}(\Delta\boldsymbol{x})^T\hat{\boldsymbol{J}}^{-T}(\boldsymbol{q})\boldsymbol{M}(\boldsymbol{q})\dot{\boldsymbol{q}} \\ + \boldsymbol{s}(\Delta\boldsymbol{x})^T\dot{\hat{\boldsymbol{J}}}^{-T}(\boldsymbol{q})\boldsymbol{M}(\boldsymbol{q})\dot{\boldsymbol{q}}. \end{aligned} \tag{2.78}$$

Since the position error is bounded by the saturation function, i.e., $\boldsymbol{s}(\Delta\boldsymbol{x})$ is bounded, and the Jacobian matrix is nonsingular, there exists a constant c_0 such that

$$\alpha h(\boldsymbol{x}, \dot{q}) \geq -\alpha c_0\|\dot{\boldsymbol{x}}\|^2. \tag{2.79}$$

Substitution of this into Eq. (2.73) yields

$$\begin{aligned} W(\boldsymbol{s}(\Delta\boldsymbol{x}), \dot{\boldsymbol{x}}) \geq \dot{\boldsymbol{x}}^T((\boldsymbol{I}_n - \Delta_J)\boldsymbol{K}_v - \alpha c_0\boldsymbol{I}_n)\dot{\boldsymbol{x}} - \dot{\boldsymbol{x}}^T(\Delta_J\boldsymbol{K}_p)\boldsymbol{s}(\Delta\boldsymbol{x}) \\ + \alpha\boldsymbol{s}(\Delta\boldsymbol{x})^T\boldsymbol{K}_p\boldsymbol{s}(\Delta\boldsymbol{x}). \end{aligned} \tag{2.80}$$

The existence of Δ_J such that $W(\boldsymbol{s}(\Delta\boldsymbol{x}), \dot{\boldsymbol{x}}) \geq 0$ can be clearly seen from Eq. (2.80). In the following development, we derived a sufficient condition to guarantee $W(\boldsymbol{s}(\Delta\boldsymbol{x}), \dot{\boldsymbol{x}}) \geq 0$. From Eq. (2.80), we have

$$\begin{aligned} W(\boldsymbol{s}(\Delta\boldsymbol{x}), \dot{\boldsymbol{x}}) \geq (\lambda_{\min}[\boldsymbol{K}_v] - \bar{p}\lambda_{\max}[\boldsymbol{K}_v] - \alpha c_0)\|\dot{\boldsymbol{x}}\|^2 - \bar{p}\lambda_{\max}[\boldsymbol{K}_p]\|\boldsymbol{s}(\Delta\boldsymbol{x})\|\|\dot{\boldsymbol{x}}\| \\ + \alpha\lambda_{\min}[\boldsymbol{K}_p]\|\boldsymbol{s}(\Delta\boldsymbol{x})\|^2. \end{aligned} \tag{2.81}$$

Note that,

$$-\bar{p}\lambda_{\max}[\boldsymbol{K}_p]\|\boldsymbol{s}(\Delta\boldsymbol{x})\| \cdot \|\dot{\boldsymbol{x}}\| \geq -\tfrac{1}{2}\bar{p}\lambda_{\max}[\boldsymbol{K}_p](\|\boldsymbol{s}(\Delta\boldsymbol{x})\|^2 + \|\dot{\boldsymbol{x}}\|^2). \tag{2.82}$$

Hence,

$$W(\boldsymbol{s}(\Delta\boldsymbol{x}), \dot{\boldsymbol{x}}) \geq \lambda_{\max}[\boldsymbol{K}_v]l_1\|\dot{\boldsymbol{x}}\|^2 + a\lambda_{\max}[\boldsymbol{K}_v]l_2\|\boldsymbol{s}(\Delta\boldsymbol{x})\|^2, \tag{2.83}$$

where

$$l_1 = \frac{\lambda_{\min}[\boldsymbol{K}_v]}{\lambda_{\max}[\boldsymbol{K}_v]} - \frac{\alpha c_0}{\lambda_{\max}[\boldsymbol{K}_v]} - \bar{p}(1 + \frac{a}{2}),$$

$$l_2 = \alpha \frac{\lambda_{\min}[\boldsymbol{K}_p]}{\lambda_{\max}[\boldsymbol{K}_p]} - \frac{\bar{p}}{2} \tag{2.84}$$

and $a = \frac{\lambda_{\max}[\boldsymbol{K}_p]}{\lambda_{\max}[\boldsymbol{K}_v]}$. Hence if

$$\min\left\{\tfrac{2}{2+a}\left(\tfrac{\lambda_{\min}[\boldsymbol{K}_v]}{\lambda_{\max}[\boldsymbol{K}_v]} - \tfrac{\alpha c_0}{\lambda_{\max}[\boldsymbol{K}_v]}\right), 2\alpha \tfrac{\lambda_{\min}[\boldsymbol{K}_p]}{\lambda_{\max}[\boldsymbol{K}_p]}\right\} > \bar{p}, \tag{2.85}$$

and $\frac{\lambda_{\min}[\boldsymbol{K}_v]}{\lambda_{\max}[\boldsymbol{K}_v]} - \frac{\alpha c_0}{\lambda_{\max}[\boldsymbol{K}_v]} > 0$, then $l_1 > 0$ and $l_2 > 0$, and hence $W(\boldsymbol{s}(\Delta \boldsymbol{x}), \dot{\boldsymbol{x}})$ is positive definite in $\dot{\boldsymbol{x}}$ and $\boldsymbol{s}(\Delta \boldsymbol{x})$.

Similarly, we let $\boldsymbol{K}_v = k_v \boldsymbol{I}_n$, $\boldsymbol{K}_p = a k_v \boldsymbol{I}_n$. Then condition (2.85) becomes

$$\bar{p} < \min\left\{\tfrac{2}{a+2}\left(1 - \tfrac{\alpha c_0}{k_v}\right), 2\alpha\right\}. \tag{2.86}$$

The relationship between $\bar{p}$ and a described by inequality (2.86) is shown in Fig. 2.17.

From Fig. 2.17, it can be seen that *both the feedback-gains ratio* $a = k_p/k_v$ *and the absolute value of* k_v *play an important role in stabilizing the system in the presence of uncertain kinematics*. As seen from the figure, increasing k_v will shift the curve $2(1 - \frac{\alpha c_0}{k_v})/(a+2)$ upward, resulting in higher $\bar{p}$ for the same feedback ratio a. That is, larger k_v would allow a higher margin of uncertainty of the approximate Jacobian matrix for the same a. When k_v is so large that $\alpha c_0/k_v$ is negligible, condition (2.86) reduces to

$$\min\left\{\tfrac{2}{a+2}, 2\alpha\right\} > \bar{p}. \tag{2.87}$$

Conversely, if $\bar{p}$ is small, a smaller controller gain k_v is required. The relationship between $\bar{p}$ and a described by inequality (2.87) is shown in Fig. 2.18.

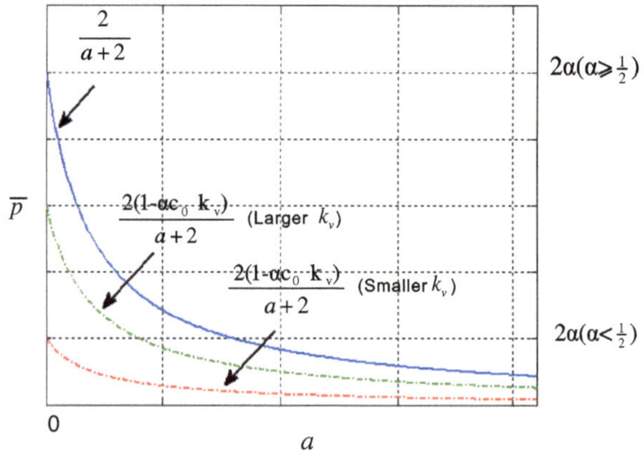

Fig. 2.17 Variation of $\bar{p}$ with a

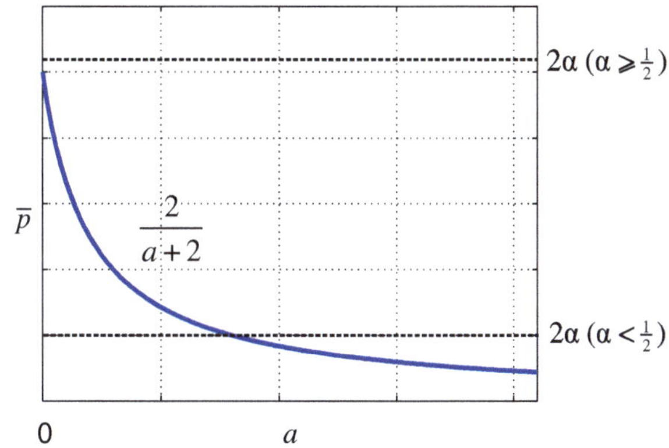

Fig. 2.18 Variation of $\bar{p}$ with a (k_v is very large)

Next, it is interesting to note the effect of a on the stability of the system. As seen from Fig. 2.17, a larger range of a can be chosen for a smaller $\bar{p}$. In fact, in the extreme case $\bar{p} = 0$, a could be chosen as any value. However, when $\bar{p}$ is larger, a narrower range of a is allowed. Hence, while for the case of perfect kinematics, increasing $\boldsymbol{K}_p$ improves steady-state performance by reducing error, this is no longer true for the case of uncertain kinematics. As seen from the figure, for any chosen $\boldsymbol{K}_v$, increasing $\boldsymbol{K}_p$ increases a and hence results in a smaller bound on the Jacobian uncertainty $\bar{p}$. This means that the robot system with uncertain Jacobian matrix could become unstable if a is too large. Therefore, a should be kept smaller so that the allowable bound of the Jacobian uncertainty is larger. This can be achieved by either increasing $\boldsymbol{K}_v$ or reducing $\boldsymbol{K}_p$. Though the condition is a sufficient condition, it is natural and reasonable because it simply means that

1. if you are not sure, then do it at a slower speed (i.e., if $\bar{p}$ is large, then more damping is required);
2. if you are sure, then you can do it at any speed (i.e., if $\bar{p}$ is small, then a wide range a is allowed).

Another important and practical conclusion of the result is that when the system is unstable, redesign of $\hat{\boldsymbol{J}}(\boldsymbol{q})$ or calibrations may not be necessary, since the instability may be due to the feedback gain k_v or the feedback gain ratio a not being tuned properly. In practice, we should therefore try to stabilize the system or increase the margin of stability first by increasing feedback gains or reducing a.

The following theorem states the results:

Theorem 2.2 *The sensory feedback controller with uncertain Jacobian matrix* (2.66) *for the robot system* (2.9) *gives rise to the convergence of the sensory-space position error and velocity such that* $\Delta\boldsymbol{x} \to \boldsymbol{0}$ *and* $\dot{\boldsymbol{x}} \to \boldsymbol{0}$ *as* $t \to \infty$ *if the feedback gains* $\boldsymbol{K}_p$ *and* $\boldsymbol{K}_v$ *are chosen to satisfy conditions* (2.76), (2.85), *and* $\hat{\boldsymbol{J}}(\boldsymbol{q})$ *is chosen so that*

$$\|\boldsymbol{I}_n - \boldsymbol{J}^{-T}(\boldsymbol{q})\hat{\boldsymbol{J}}^T(\boldsymbol{q})\| \le \bar{p}. \tag{2.88}$$

Proof Since $V(s(\Delta x), \dot{x})$ and $W(s(\Delta x), \dot{x})$ are positive definite in $s(\Delta x)$ and $\dot{x}$, both $s(\Delta x)$ and $\dot{x}$ are bounded. The boundedness of $\dot{x}$ ensures the boundedness of $\dot{q}$, since it is assumed that $J^{-1}(q)$ exists. From Eq. (2.70), it is seen that $\ddot{q}$ is bounded. The boundedness of $\ddot{q}$ also ensures the boundedness of $\ddot{x}$, since $\ddot{x} = J(q)\ddot{q} + \dot{J}(q)\dot{q}$. Since both $\ddot{x}$ and $\dot{x}$ are bounded, $\dot{x}$ and $s(\Delta x)$ are uniformly continuous. From Eqs. (2.71) and (2.83), we have $\dot{x}, s(\Delta x) \in L_2(0, +\infty)$. Therefore, from Barbalat's emma, $s(\Delta x) \to \mathbf{0}$ and $\dot{x} \to \mathbf{0}$ as $t \to \infty$, and hence $\Delta x \to \mathbf{0}$. △△△

Given the desired end effector's position x_d, the manipulator has its own configuration q_d at joint space. The positive definiteness of V in the state variables $\Delta q = q - q_d$ and $\dot{q}$ cannot be concluded in general. However, for a finite task space such that the kinematics mapping between task space and joint space is one-to-one and the Jacobian matrix $J(q)$ is nonsingular, then the Lyapunov method can be used directly to conclude the asymptotic stability of the equilibrium state.

When cameras are used to specify the position of the robot end effector, the sensory space is defined as the image coordinates. In the presence of an uncalibrated camera, the image Jacobian matrix is uncertain, as indicated by $\hat{Z}^{-1}(q)\hat{L}(x)\hat{J}_m(q)$ in Eq. (2.63). A vision-based controller using the approximate Jacobian matrix can be designed as follows:

$$\tau = -\hat{J}_m^T(q)\hat{L}^T(x)\hat{Z}^{-1}(q)(K_p s(\Delta x) + K_v \dot{x}) + g(q). \tag{2.89}$$

If the approximate image Jacobian matrix is chosen such that

$$\| J_m^T(q) L^T(x) Z^{-1}(q) - \hat{J}_m^T(q)\hat{L}^T(x)\hat{Z}^{-1}(q) \| \le \bar{p}, \tag{2.90}$$

then it can be proved similarly that the vision-based controller in Eq. (2.89) ensures the convergence of the image-space error.

2.3.3 Sensory Feedback Control with Uncertain Gravitational Force

The control input (2.66) requires exact knowledge of the gravitational force $g(q)$ or else could be implemented on a gravity-free robot. In principle, the parameters of the gravitational force can be identified offline by commanding the robot to move with arbitrary joint motions without solving the inverse kinematics problem. However, when the robot is required to handle objects of various lengths and masses, the gravitational force could vary according to different tasks. Therefore, it is important to compensate the gravitational force online without exact knowledge of the kinematics and Jacobian matrix.

In this section, we consider the sensory feedback control problem with uncertain gravitational force compensation and with uncertain Jacobian using the concept of gravity regressor. Note that the gravity term can be completely characterized by a set of parameters $\boldsymbol{\theta}_g = (\theta_{g1}, \cdots, \theta_{gn_g})^T \in \Re^{n_g}$ (see *Property* 1.4) as

$$g(q) = Y_g(q)\theta_g, \tag{2.91}$$

where $Y_g(q) \in \Re^{n\times n_g}$ is the gravity regressor. Consider the following control input:

$$\tau = -\hat{J}^T(q)(K_p s(\Delta x) + K_v \dot{x}) + Y_g(q)\hat{\theta}_g, \tag{2.92}$$

$$\dot{\hat{\theta}}_g = L_g Y_g^T(q) y, \tag{2.93}$$

where $y = \dot{q} + \alpha \hat{J}^{-1}(q)s(\Delta x)$ is the output, and $L_g \in \Re^{n_g\times n_g}$ is a positive definite matrix. Note that the output vector defined in Eq. (2.69) is now used in the design of the update law in Eq. (2.93). A block diagram of the controller in Eq. (2.92) is shown in Fig. 2.19.

Substituting Eqs. (2.92) and (2.91) into Eq. (2.9) yields

$$M(q)\ddot{q}+\left(\tfrac{1}{2}\dot{M}(q)+S(q,\dot{q})\right)\dot{q}+\hat{J}^T(q)(K_p s(\Delta x)+K_v\dot{x})+Y_g(q)\Delta\theta_g=0, \tag{2.94}$$

where $\Delta\theta_g = \theta_g - \hat{\theta}_g \in \Re^{n_g\times n_g}$.

The convergence of the task-space position error and velocity is specified by the following theorem:

Theorem 2.3 *The sensory feedback controller with uncertain gravitational force* (2.93) *for the robot system* (2.9) *gives rise to the convergence of the sensory-space position error and velocity such that* $\Delta x \to 0$ *and* $\dot{x} \to 0$ *as* $t \to \infty$ *if the feedback gains* K_p *and* K_v *are chosen to satisfy conditions* (2.76), (2.85), *and* $\hat{J}(q)$ *is chosen to satisfy condition* (2.67).

Proof Multiplying both sides of Eq. (2.94) by $\hat{J}^{-T}(q)$, taking the inner product with the output y, and using Eq. (2.93) leads to

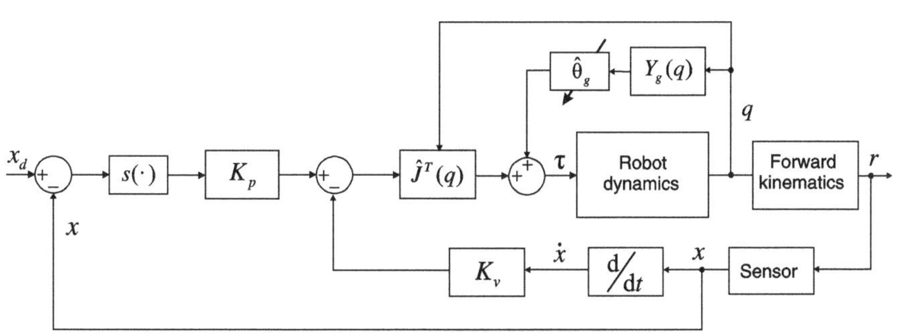

Fig. 2.19 A block diagram of a sensory feedback control with uncertain gravitational force

$$\frac{\mathrm{d}}{\mathrm{d}t}\Bigg\{\tfrac{1}{2}\dot{\boldsymbol{q}}^T\boldsymbol{M}(\boldsymbol{q})\dot{\boldsymbol{q}} + \alpha\boldsymbol{s}(\Delta\boldsymbol{x})^T\hat{\boldsymbol{J}}^{-T}(\boldsymbol{q})\boldsymbol{M}(\boldsymbol{q})\dot{\boldsymbol{q}} + \sum_{i=1}^{n}(k_{pi}+\alpha k_{vi})S_i(\Delta x_i)$$
$$+\tfrac{1}{2}\Delta\boldsymbol{\theta}_g^T\boldsymbol{L}_g^{-1}\Delta\boldsymbol{\theta}_g\Bigg\} + \dot{\boldsymbol{x}}^T\boldsymbol{J}^{-T}(\boldsymbol{x})\hat{\boldsymbol{J}}^T(\boldsymbol{q})\boldsymbol{K}_v\dot{\boldsymbol{x}} + \alpha\boldsymbol{s}(\Delta\boldsymbol{x})^T\boldsymbol{K}_p\boldsymbol{s}(\Delta\boldsymbol{x})$$
$$-\dot{\boldsymbol{x}}^T(\boldsymbol{I}_n - \boldsymbol{J}^{-T}(\boldsymbol{q})\hat{\boldsymbol{J}}^T(\boldsymbol{q}))\boldsymbol{K}_p\boldsymbol{s}(\Delta\boldsymbol{x}) + \alpha\Bigg\{\boldsymbol{s}(\Delta\boldsymbol{x})^T\hat{\boldsymbol{J}}^{-T}(\boldsymbol{q})\left(\tfrac{1}{2}\dot{\boldsymbol{M}}(\boldsymbol{q})\right.$$
$$\left.-\boldsymbol{S}(\boldsymbol{q},\dot{\boldsymbol{q}})\right)\dot{\boldsymbol{q}} - \dot{\boldsymbol{s}}(\Delta\boldsymbol{x})^T\hat{\boldsymbol{J}}^{-T}(\boldsymbol{q})\boldsymbol{M}(\boldsymbol{q})\dot{\boldsymbol{q}} - \boldsymbol{s}(\Delta\boldsymbol{x})^T\dot{\hat{\boldsymbol{J}}}^{-T}(\boldsymbol{q})\boldsymbol{M}(\boldsymbol{q})\dot{\boldsymbol{q}}\Bigg\} = 0, \quad (2.95)$$

which is equal to

$$\frac{\mathrm{d}}{\mathrm{d}t}(V(\boldsymbol{s}(\Delta\boldsymbol{x}),\dot{\boldsymbol{x}}) + \tfrac{1}{2}\Delta\boldsymbol{\theta}_g^T\boldsymbol{L}_g^{-1}\Delta\boldsymbol{\theta}_g) = -W(\boldsymbol{s}(\Delta\boldsymbol{x}),\dot{\boldsymbol{x}}). \quad (2.96)$$

Both $V(\boldsymbol{s}(\Delta\boldsymbol{x}),\dot{\boldsymbol{x}})$ and $W(\boldsymbol{s}(\Delta\boldsymbol{x}),\dot{\boldsymbol{x}})$ are positive definite in $\boldsymbol{s}(\Delta\boldsymbol{x})$ and $\dot{\boldsymbol{x}}$. From Eq. (2.96), $\boldsymbol{s}(\Delta\boldsymbol{x})$, $\dot{\boldsymbol{x}}$, and $\Delta\boldsymbol{\theta}_g$ are bounded. The boundedness of $\dot{\boldsymbol{x}}$ ensures the boundedness of $\dot{\boldsymbol{q}}$, since it is assumed that $\boldsymbol{J}^{-1}(\boldsymbol{q})$ exists. The boundedness of $\dot{\boldsymbol{q}}$ and $\boldsymbol{s}(\Delta\boldsymbol{x})$ ensures the boundedness of the output $\boldsymbol{y}$. From Eq. (2.94), it is seen that $\ddot{\boldsymbol{q}}$ is bounded. The boundedness of $\ddot{\boldsymbol{q}}$ also ensures the boundedness of $\ddot{\boldsymbol{x}}$, since $\ddot{\boldsymbol{x}} = \boldsymbol{J}(\boldsymbol{q})\ddot{\boldsymbol{q}} + \dot{\boldsymbol{J}}(\boldsymbol{q})\dot{\boldsymbol{q}}$. Since both $\ddot{\boldsymbol{x}}$ and $\dot{\boldsymbol{x}}$ are bounded, $\dot{\boldsymbol{x}}$ and $\boldsymbol{s}(\Delta\boldsymbol{x})$ are uniformly continuous. From Eqs. (2.95) and (2.96), we have $\dot{\boldsymbol{x}}$, $\boldsymbol{s}(\Delta\boldsymbol{x})\in L_2(0,+\infty)$. Therefore, from Barbalat's lemma, $\boldsymbol{s}(\Delta\boldsymbol{x}) \to \boldsymbol{0}$ and $\dot{\boldsymbol{x}}\to\boldsymbol{0}$ as $t\to\infty$, and hence $\Delta\boldsymbol{x}\to\boldsymbol{0}$. $\Delta\Delta\Delta$

2.4 Sensory-Space Setpoint Control with Joint-Space Damping

Note that the controllers described by Eqs. (2.66) and (2.92) require the task-space velocity $\dot{\boldsymbol{x}}$. The task-space velocity is obtained from numerical differentiation of the task-space position which is usually noisy. In addition, in the case of redundant robot, the convergence of the task-space velocity to zero does not necessary ensure the convergence of the joint-space velocity to zero, even if the Jacobian matrix is nonsingular.

2.4.1 Sensory Feedback Control with Exact Gravity Compensation

Due to the possible existence of self motions in the case of redundant robot, the joint-space damping is used directly to damp out the joint motion. First, we consider a simple approximate Jacobian feedback controller by using joint-space damping as follows:

$$\boldsymbol{\tau} = -\hat{\boldsymbol{J}}^T(\boldsymbol{q})\boldsymbol{K}_p\boldsymbol{s}(\Delta\boldsymbol{x}) - \boldsymbol{K}_v\dot{\boldsymbol{q}} + \boldsymbol{g}(\boldsymbol{q}), \quad (2.97)$$

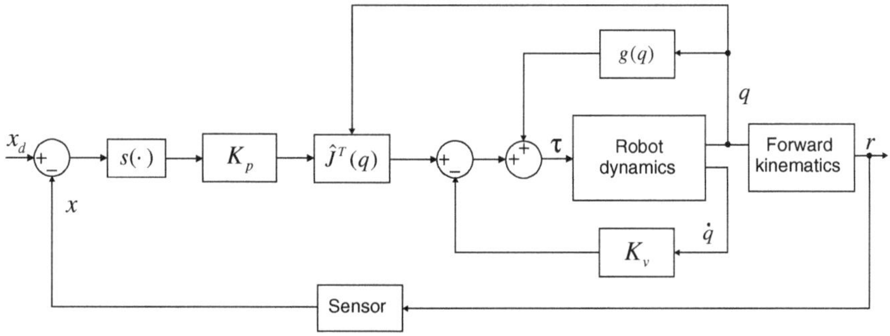

Fig. 2.20 A block diagram of the approximate Jacobian feedback control with joint-space damping and gravitational force compensation

where $\boldsymbol{K}_p = k_p \boldsymbol{I}_n$, $\boldsymbol{K}_v = k_v \boldsymbol{I}_n$, and $\hat{\boldsymbol{J}}^T(\boldsymbol{q})$ is chosen so that

$$\|\boldsymbol{J}^T(\boldsymbol{q}) - \hat{\boldsymbol{J}}^T(\boldsymbol{q})\| \le p. \tag{2.98}$$

A block diagram of the controller in Eq. (2.97) is shown in Fig. 2.20.

The closed-loop equation of the system is obtained by substituting Eq. (2.97) into Eq. (2.9),

$$\boldsymbol{M}(\boldsymbol{q})\ddot{\boldsymbol{q}} + \left(\tfrac{1}{2}\dot{\boldsymbol{M}}(\boldsymbol{q}) + \boldsymbol{S}(\boldsymbol{q}, \dot{\boldsymbol{q}})\right)\dot{\boldsymbol{q}} + \hat{\boldsymbol{J}}^T(\boldsymbol{q})\boldsymbol{K}_p \boldsymbol{s}(\Delta \boldsymbol{x}) + \boldsymbol{K}_v \dot{\boldsymbol{q}} = \boldsymbol{0}. \tag{2.99}$$

Let us define an output vector $\boldsymbol{y}$ of the form

$$\boldsymbol{y} = \dot{\boldsymbol{q}} + \alpha \hat{\boldsymbol{J}}^T(\boldsymbol{q})\boldsymbol{s}(\Delta \boldsymbol{x}). \tag{2.100}$$

where joint-space velocity and sensory task-space position error are used and a transpose of the approximate Jacobian is now used to transform the task-space position error to joint space. The above output does not require the use of the inverse or pseudo-inverse of the approximate Jacobian matrix and hence is simpler and more useful.

To carry out the stability analysis for the closed-loop system with the approximate Jacobian controller, we take the inner product between the output vector $\boldsymbol{y}$ described by Eq. (2.100) and the closed-loop Eq. (2.99) and make use of *Properties* 1.1 and 1.2 of the robot dynamics to give,

$$\tfrac{\mathrm{d}}{\mathrm{d}t} V(\boldsymbol{s}(\Delta \boldsymbol{x}), \dot{\boldsymbol{q}}) + W(\boldsymbol{s}(\Delta \boldsymbol{x}), \dot{\boldsymbol{q}}) = 0, \tag{2.101}$$

where

$$\begin{aligned} V(\boldsymbol{s}(\Delta \boldsymbol{x}), \dot{\boldsymbol{q}}) = {} & \tfrac{1}{2}\dot{\boldsymbol{q}}^T \boldsymbol{M}(\boldsymbol{q})\dot{\boldsymbol{q}} + \alpha \dot{\boldsymbol{q}}^T \boldsymbol{M}(\boldsymbol{q})\hat{\boldsymbol{J}}^T(\boldsymbol{q})\boldsymbol{s}(\Delta \boldsymbol{x}) \\ & + \sum_{i=1}^{m} (k_p + \alpha k_v) S_i(\Delta x_i), \end{aligned} \tag{2.102}$$

and

$$
\begin{aligned}
W(\boldsymbol{s}(\Delta\boldsymbol{x}), \dot{\boldsymbol{q}}) = k_v \dot{\boldsymbol{q}}^T \dot{\boldsymbol{q}} + \alpha k_p \boldsymbol{s}^T(\Delta\boldsymbol{x}) \hat{\boldsymbol{J}}(\boldsymbol{q}) \hat{\boldsymbol{J}}^T(\boldsymbol{q}) \boldsymbol{s}(\Delta\boldsymbol{x}) \\
-(k_p + \alpha k_v) \dot{\boldsymbol{q}}^T (\boldsymbol{J}^T(\boldsymbol{q}) - \hat{\boldsymbol{J}}^T(\boldsymbol{q})) \boldsymbol{s}(\Delta\boldsymbol{x}) \\
-\alpha \Big\{ \boldsymbol{s}^T(\Delta\boldsymbol{x}) \hat{\boldsymbol{J}}(\boldsymbol{q}) \left(\tfrac{1}{2}\dot{\boldsymbol{M}}(\boldsymbol{q}) - \boldsymbol{S}(\boldsymbol{q}, \dot{\boldsymbol{q}})\right) \dot{\boldsymbol{q}} + \dot{\boldsymbol{s}}^T(\Delta\boldsymbol{x}) \hat{\boldsymbol{J}}(\boldsymbol{q}) \boldsymbol{M}(\boldsymbol{q}) \dot{\boldsymbol{q}} \\
+ \boldsymbol{s}^T(\Delta\boldsymbol{x}) \dot{\hat{\boldsymbol{J}}}(\boldsymbol{q}) \boldsymbol{M}(\boldsymbol{q}) \dot{\boldsymbol{q}}. \Big\}
\end{aligned}
\tag{2.103}
$$

To show that $V(\boldsymbol{s}(\Delta\boldsymbol{x}), \dot{\boldsymbol{q}})$ in Eq. (2.102) is positive definite in $\dot{\boldsymbol{q}}$ and $\boldsymbol{s}(\Delta\boldsymbol{x})$, we note that

$$
\begin{aligned}
&\tfrac{1}{4}\dot{\boldsymbol{q}}^T \boldsymbol{M}(\boldsymbol{q}) \dot{\boldsymbol{q}} + \alpha \dot{\boldsymbol{q}}^T \boldsymbol{M}(\boldsymbol{q}) \hat{\boldsymbol{J}}^T(\boldsymbol{q}) \boldsymbol{s}(\Delta\boldsymbol{x}) + \sum_{i=1}^{m} (k_p + \alpha k_v) S_i(e_i) \\
&= \tfrac{1}{4}(\dot{\boldsymbol{q}} + 2\alpha \hat{\boldsymbol{J}}^T(\boldsymbol{q}) \boldsymbol{s}(\Delta\boldsymbol{x}))^T \boldsymbol{M}(\boldsymbol{q}) (\dot{\boldsymbol{q}} + 2\alpha \hat{\boldsymbol{J}}^T(\boldsymbol{q}) \boldsymbol{s}(\Delta\boldsymbol{x})) \\
&\quad -\alpha^2 \boldsymbol{s}^T(\Delta\boldsymbol{x}) \hat{\boldsymbol{J}}(\boldsymbol{q}) \boldsymbol{M}(\boldsymbol{q}) \hat{\boldsymbol{J}}^T(\boldsymbol{q}) \boldsymbol{s}(\Delta\boldsymbol{x}) + \sum_{i=1}^{m} (k_p + \alpha k_v) S_i(e_i) \\
&\geq \sum_{i=1}^{m} \{k_p \bar{c}_i + \alpha(k_v \bar{c}_i - \alpha\lambda_m)\} s_i^2(\Delta x_i)
\end{aligned}
\tag{2.104}
$$

where $\lambda_m \triangleq \lambda_{\max}[\hat{\boldsymbol{J}}(\boldsymbol{q}) \boldsymbol{M}(\boldsymbol{q}) \hat{\boldsymbol{J}}^T(\boldsymbol{q})]$. Substituting the above equation into Eq. (2.102) yields:

$$
V(\boldsymbol{s}(\Delta\boldsymbol{x}), \dot{\boldsymbol{q}}) \geq \tfrac{1}{4}\dot{\boldsymbol{q}}^T \boldsymbol{M}(\boldsymbol{q}) \dot{\boldsymbol{q}} + \sum_{i=1}^{m} \{k_p \bar{c}_i + \alpha(k_v \bar{c}_i - \alpha\lambda_m)\} s_i^2(\Delta x_i) \geq 0, \tag{2.105}
$$

where k_v and α can be chosen so that

$$
k_v \bar{c}_i - \alpha\lambda_m > 0. \tag{2.106}
$$

Therefore, the function $V(\boldsymbol{s}(\Delta\boldsymbol{x}), \dot{\boldsymbol{q}})$ is positive definite in $\boldsymbol{s}(\Delta\boldsymbol{x})$ and $\dot{\boldsymbol{q}}$ when k_v is chosen sufficiently large and α is set sufficiently small.

Next, we proceed to show that the time derivative of the function $V(\boldsymbol{s}(\Delta\boldsymbol{x}), \dot{\boldsymbol{q}})$ is negative definite in $\boldsymbol{s}(\Delta\boldsymbol{x})$ and $\dot{\boldsymbol{q}}$. As seen from Eq. (2.101), this is to show that $W(\boldsymbol{s}(\Delta\boldsymbol{x}), \dot{\boldsymbol{q}})$ is positive definite in $\boldsymbol{s}(\Delta\boldsymbol{x})$ and $\dot{\boldsymbol{q}}$. From the last term on the right-hand side of Eq. (2.103), since $\boldsymbol{s}(\Delta\boldsymbol{x})$ is bounded, there exist a constant $c_0 > 0$ such that:

$$
\begin{aligned}
\alpha \left| \boldsymbol{s}^T(\Delta\boldsymbol{x}) \hat{\boldsymbol{J}}(\boldsymbol{q}) \left(\tfrac{1}{2}\dot{\boldsymbol{M}}(\boldsymbol{q}) - \boldsymbol{S}(\boldsymbol{q}, \dot{\boldsymbol{q}})\right) \dot{\boldsymbol{q}} + \dot{\boldsymbol{s}}^T(\Delta\boldsymbol{x}) \hat{\boldsymbol{J}}(\boldsymbol{q}) \boldsymbol{M}(\boldsymbol{q}) \dot{\boldsymbol{q}} + \boldsymbol{s}^T(\Delta\boldsymbol{x}) \dot{\hat{\boldsymbol{J}}}(\boldsymbol{q}) \boldsymbol{M}(\boldsymbol{q}) \dot{\boldsymbol{q}} \right| \\
\leq \alpha c_0 \|\dot{\boldsymbol{q}}\|^2.
\end{aligned}
\tag{2.107}
$$

Substituting inequality (2.107) into Eq. (2.103) yields

$$W(\boldsymbol{s}(\Delta\boldsymbol{x}), \dot{\boldsymbol{q}}) \geq \dot{\boldsymbol{q}}^T(k_v\boldsymbol{I}_n - \alpha c_0\boldsymbol{I}_n)\dot{\boldsymbol{q}} + \alpha k_p \boldsymbol{s}^T(\Delta\boldsymbol{x})\hat{\boldsymbol{J}}(\boldsymbol{q})\hat{\boldsymbol{J}}^T(\boldsymbol{q})\boldsymbol{s}(\Delta\boldsymbol{x})$$
$$-(k_p + \alpha k_v)\dot{\boldsymbol{q}}^T(\boldsymbol{J}^T(\boldsymbol{q}) - \hat{\boldsymbol{J}}^T(\boldsymbol{q}))\boldsymbol{s}(\Delta\boldsymbol{x}). \quad (2.108)$$

Now, letting $\bar{\Delta}_J \overset{\Delta}{=} \boldsymbol{J}^T(\boldsymbol{q}) - \hat{\boldsymbol{J}}^T(\boldsymbol{q})$, we have

$$W(\boldsymbol{s}(\Delta\boldsymbol{x}), \dot{\boldsymbol{q}}) \geq \dot{\boldsymbol{q}}^T(k_v\boldsymbol{I}_n - \alpha c_0\boldsymbol{I}_n)\dot{\boldsymbol{q}} + \alpha k_p \boldsymbol{s}^T(\Delta\boldsymbol{x})\hat{\boldsymbol{J}}(\boldsymbol{q})\hat{\boldsymbol{J}}^T(\boldsymbol{q})\boldsymbol{s}(\Delta\boldsymbol{x})$$
$$-(k_p + \alpha k_v)\dot{\boldsymbol{q}}^T\bar{\Delta}_J\boldsymbol{s}(\Delta\boldsymbol{x}). \quad (2.109)$$

The existence of a $\bar{\Delta}_J$ ensures that $W(\boldsymbol{s}(\Delta\boldsymbol{x}), \dot{\boldsymbol{q}})$ is positive definite in $\boldsymbol{s}(\Delta\boldsymbol{x})$ and $\dot{\boldsymbol{q}}$, as seen from Eq. (2.109). In the following development, a sufficient condition is derived to guarantee the positive definiteness of $W(\boldsymbol{s}(\Delta\boldsymbol{x}), \dot{\boldsymbol{q}})$ in $\boldsymbol{s}(\Delta\boldsymbol{x})$ and $\dot{\boldsymbol{q}}$. Note that

$$W(\boldsymbol{s}(\Delta\boldsymbol{x}), \dot{\boldsymbol{q}}) \geq (k_v - \alpha c_0)\|\dot{\boldsymbol{q}}\|^2 - p(k_p + \alpha k_v)\|\boldsymbol{s}(\Delta\boldsymbol{x})\| \cdot \|\dot{\boldsymbol{q}}\|$$
$$+\alpha k_p \lambda_{\hat{J}}\|\boldsymbol{s}(\Delta\boldsymbol{x})\|^2, \quad (2.110)$$

where $\lambda_{\hat{J}} = \lambda_{\min}[\hat{\boldsymbol{J}}(\boldsymbol{q})\hat{\boldsymbol{J}}^T(\boldsymbol{q})]$. Next, note that

$$-\|\boldsymbol{s}(\Delta\boldsymbol{x})\| \cdot \|\dot{\boldsymbol{q}}\| \geq -\tfrac{1}{2}(\|\boldsymbol{s}(\Delta\boldsymbol{x})\|^2 + \|\dot{\boldsymbol{q}}\|^2). \quad (2.111)$$

Substituting inequality (2.111) into Eq. (2.110) gives,

$$W(\boldsymbol{s}(\Delta\boldsymbol{x}), \dot{\boldsymbol{q}}) \geq k_v l_1\|\dot{\boldsymbol{q}}\|^2 + k_v l_2\|\boldsymbol{s}(\Delta\boldsymbol{x})\|^2, \quad (2.112)$$

where

$$l_1 = 1 - \tfrac{\alpha c_0}{k_v} - \tfrac{p}{2}(\bar{a} + \alpha), \quad l_2 = \alpha\bar{a}\lambda_{\hat{J}} - \tfrac{p}{2}(\bar{a} + \alpha), \quad (2.113)$$

and $\bar{a} = \frac{k_p}{k_v}$. Hence, if

$$\min\left\{\frac{2(1-\frac{\alpha c_0}{k_v})}{\bar{a}+\alpha}, \frac{2\bar{a}\alpha\lambda_{\hat{J}}}{\bar{a}+\alpha}\right\} > p, \quad (2.114)$$

and $k_v > \alpha c_o$, then $l_1 > 0$ and $l_2 > 0$, and hence $W(\boldsymbol{s}(\Delta\boldsymbol{x}), \dot{\boldsymbol{q}})$ is positive definite in $\dot{\boldsymbol{q}}$ and $\boldsymbol{s}(\Delta\boldsymbol{x})$. The relationship between p and $\bar{a}$ can be similarly illustrated in the following (Fig. 2.21).

We are now in a position to prove the following theorem:

Theorem 2.4 *The sensory feedback controller in Eq.* (2.97) *for the robot system described by Eq.* (2.9) *ensures the convergence of sensory-space position error and*

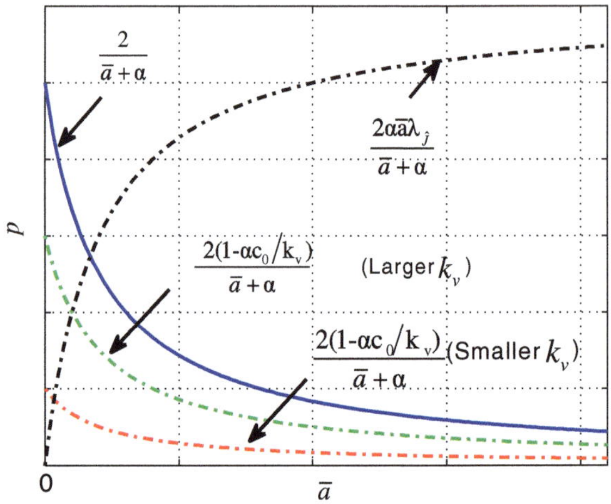

Fig. 2.21 Variation of p with $\bar{a}$

the joint-space velocity is guaranteed, if the feedback gains $\boldsymbol{K}_p$ and $\boldsymbol{K}_v$ are chosen to satisfy conditions (2.106), (2.114)*, and $\hat{\boldsymbol{J}}(\boldsymbol{q})$ is chosen to satisfy condition* (2.98).

Proof Since both $V(\boldsymbol{s}(\Delta\boldsymbol{x}), \dot{\boldsymbol{q}})$ and $W(\boldsymbol{s}(\Delta\boldsymbol{x}), \dot{\boldsymbol{q}})$ are positive definite in $\dot{\boldsymbol{q}}$ and $\boldsymbol{s}(\Delta\boldsymbol{x})$, from Eq. (2.101), we have

$$\frac{\mathrm{d}}{\mathrm{d}t}V(\boldsymbol{s}(\Delta\boldsymbol{x}), \dot{\boldsymbol{q}}) = -W(\boldsymbol{s}(\Delta\boldsymbol{x}), \dot{\boldsymbol{q}}) < 0. \tag{2.115}$$

Hence, $V(\boldsymbol{s}(\Delta\boldsymbol{x}), \dot{\boldsymbol{q}})$ is a Lyapunov-like function whose time derivative is negative in $\boldsymbol{s}(\Delta\boldsymbol{x})$ and $\dot{\boldsymbol{q}}$. Therefore, both $\boldsymbol{s}(\Delta\boldsymbol{x})$ and $\dot{\boldsymbol{q}}$ are bounded. From Eq. (2.99), it is seen that $\ddot{\boldsymbol{q}}$ is bounded. The boundedness of $\dot{\boldsymbol{q}}$ also ensures the boundedness of $\dot{\boldsymbol{x}}$ since $\boldsymbol{J}(\boldsymbol{q})$ are trigonometric functions of joint angles. Since both $\ddot{\boldsymbol{q}}$ and $\dot{\boldsymbol{x}}$ are bounded, $\dot{\boldsymbol{q}}$ and $\boldsymbol{s}(\Delta\boldsymbol{x})$ are uniformly continuous. From Eqs. (2.101) and (2.112), we have $\dot{\boldsymbol{q}}, \boldsymbol{s}(\Delta\boldsymbol{x}) \in L_2(0, +\infty)$. Therefore, from Barbalat's lemma, $\boldsymbol{s}(\Delta\boldsymbol{x}) \to \boldsymbol{0}$ and $\dot{\boldsymbol{q}} \to \boldsymbol{0}$ as $t \to \infty$, and hence $\Delta\boldsymbol{x} \to \boldsymbol{0}$. △△△

If the robot is non-redundant so that the dimension of task space is equal to the dimension of joint space and the kinematics mapping between task space and joint space is one-to-one, then Lyapunov method can be used directly to conclude the asymptotical stability if the Jacobian matrix $\boldsymbol{J}(\boldsymbol{q})$ is nonsingular.

When cameras are used to specified the position of the robot end effector, the sensory space is defined as image coordinates. In the presence of uncalibrated camera, the image Jaocbian matrix is uncertain as denoted as $\hat{\boldsymbol{Z}}^{-1}(\boldsymbol{q})\hat{\boldsymbol{L}}(\boldsymbol{x})\hat{\boldsymbol{J}}_m(\boldsymbol{q})$ in Eq. (2.63). A vision-based controller by using the joint-space damping can be designed as follows:

$$\boldsymbol{\tau} = -\hat{\boldsymbol{J}}_m^T(\boldsymbol{q})\hat{\boldsymbol{L}}^T(\boldsymbol{x})\hat{\boldsymbol{Z}}^{-1}(\boldsymbol{q})\boldsymbol{K}_p\boldsymbol{s}(\Delta\boldsymbol{x}) - \boldsymbol{K}_v\dot{\boldsymbol{q}} + \boldsymbol{g}(\boldsymbol{q}). \tag{2.116}$$

If the approximate Jacobian matrix is chosen such that:

$$\|\boldsymbol{J}_m^T(\boldsymbol{q})\boldsymbol{L}^T(\boldsymbol{x})\boldsymbol{Z}^{-1}(\boldsymbol{q}) - \hat{\boldsymbol{J}}_m^T(\boldsymbol{q})\hat{\boldsymbol{L}}^T(\boldsymbol{x})\hat{\boldsymbol{Z}}^{-1}(\boldsymbol{q})\| \le p, \tag{2.117}$$

then the vision-based controller in Eq. (2.116) ensures the convergence of image-space error and joint velocity to zero.

2.4.2 Sensory Feedback Control in Presence of Uncertain Gravitational Force

Next, consider the problem of sensory task-space setpoint control with uncertain gravitational force compensation and Jacobian matrix. The control input is designed as

$$\tau = -\hat{J}^T(q)K_p s(\Delta x) - K_v \dot{q} + Y_g(q)\hat{\theta}_g, \tag{2.118}$$

$$\dot{\hat{\theta}}_g = -LY_g^T(q)y, \tag{2.119}$$

where $y = \dot{q} + \alpha \hat{J}^T(q)s(\Delta x)$ is the output defined in Eq. (2.100). Note that the above controller does not require the use of generalized inverse of the approximate Jacobian matrix, which could become very large when the robot approaches any singularity configuration. A block diagram of the controller in Eq. (2.118) is shown in Fig. 2.22.

Substituting Eq. (2.118) into Eq. (2.9), we have the closed-loop equation,

$$M(q)\ddot{q}+\left(\tfrac{1}{2}\dot{M}(q)+S(q,\dot{q})\right)\dot{q}+\hat{J}^T(q)K_p s(\Delta x)+K_v\dot{q}+Y_g(q)\Delta\theta_g=\mathbf{0}, \tag{2.120}$$

The following theorem states the result of sensory-space setpoint control with uncertain gravitational force compensation:

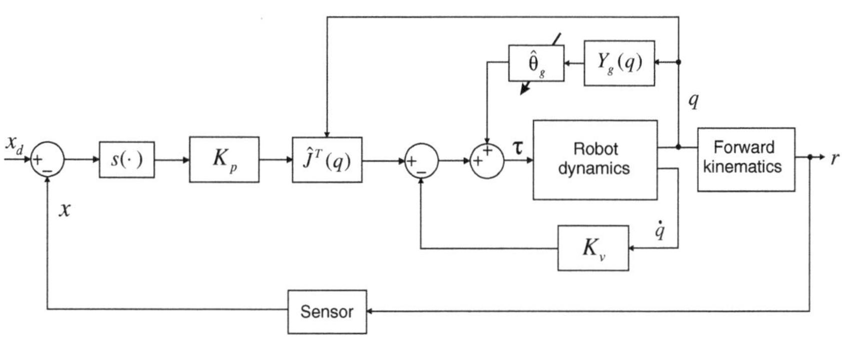

Fig. 2.22 A block diagram of the approximate Jacobian feedback control with joint-space damping and update of uncertain parameters in the estimated gravitational force

Theorem 2.5 *The sensory feedback controller with uncertain gravitational force* (2.118) *and the update law* (2.119) *for the robot system* (2.9) *ensures the convergence of the sensory-space position error and the joint-space velocity, if the feedback gains* $\boldsymbol{K}_p$, $\boldsymbol{K}_v$ *and* $\hat{\boldsymbol{J}}(\boldsymbol{q})$ *are chosen as in Theorem* 2.4.

Proof Taking the inner product of Eq. (2.120) with $\boldsymbol{y} = \dot{\boldsymbol{q}} + \alpha \hat{\boldsymbol{J}}^T(\boldsymbol{q})\boldsymbol{s}(\Delta \boldsymbol{x})$ and using Eq. (2.119), we have

$$\frac{\mathrm{d}}{\mathrm{d}t}(V(\boldsymbol{s}(\Delta \boldsymbol{x}), \dot{\boldsymbol{q}}) + \frac{1}{2}\Delta\boldsymbol{\theta}_g^T \boldsymbol{L}_g^{-1}\Delta\boldsymbol{\theta}_g) = -W(\boldsymbol{s}(\Delta \boldsymbol{x}), \dot{\boldsymbol{q}}) \leq 0. \qquad (2.121)$$

Hence, $V(\boldsymbol{s}(\Delta \boldsymbol{x}), \dot{\boldsymbol{q}})$ is a Lyapunov-like function whose time derivative is negative in $\boldsymbol{s}(\Delta \boldsymbol{x})$ and $\dot{\boldsymbol{q}}$. Therefore, $\boldsymbol{s}(\Delta \boldsymbol{x})$, $\dot{\boldsymbol{q}}$, and $\Delta\boldsymbol{\theta}_g$ are bounded. The boundedness of $\dot{\boldsymbol{q}}$ and $\boldsymbol{s}(\Delta \boldsymbol{x})$ ensures the boundedness of the output $\boldsymbol{y}$. From Eq. (2.120), it is seen that $\ddot{\boldsymbol{q}}$ is bounded. Since both $\ddot{\boldsymbol{q}}$ and $\dot{\boldsymbol{x}}$ are bounded, $\dot{\boldsymbol{q}}$ and $\boldsymbol{s}(\Delta \boldsymbol{x})$ are uniformly continuous. Similarly, it can be shown that $\dot{\boldsymbol{q}}, \boldsymbol{s}(\Delta \boldsymbol{x}) \in L_2(0, +\infty)$. Therefore, $\boldsymbol{s}(\Delta \boldsymbol{x}) \rightarrow \boldsymbol{0}$ and $\dot{\boldsymbol{q}} \rightarrow \boldsymbol{0}$ as $t \rightarrow \infty$, and hence $\Delta \boldsymbol{x} \rightarrow \boldsymbol{0}$. $\triangle\triangle\triangle$

2.5 Unified Analysis of Sensory-Space Setpoint Control

Task-space regulation of robots is classified into two basic approaches, namely transpose Jacobian regulation and inverse Jacobian regulation. In the previous sections, various transpose Jacobian setpoint controllers using task-space sensory feedback are considered. The basic idea is to form a control force of the robot end-effector using the sensory task-space error, and then transform it into joint torques by using the transpose of the approximate Jacobian matrix. In the inverse Jacobian regulation method, the task-space error is transformed into the joint error directly by means of the inverse Jacobian matrix. The basic idea is to relate an infinitesimal displacement in joint space to an infinitesimal displacement in task space directly, by using the inverse Jacobian matrix. An illustration of the concept is shown in Fig. 2.23. In this section, a unified approach for the analysis and design of both the transpose Jacobian and inverse Jacobian setpoint controllers using task-space sensory feedback is introduced. Based on the unified analysis, a fundamental property in the task-space regulation problem, namely the duality property, is presented.

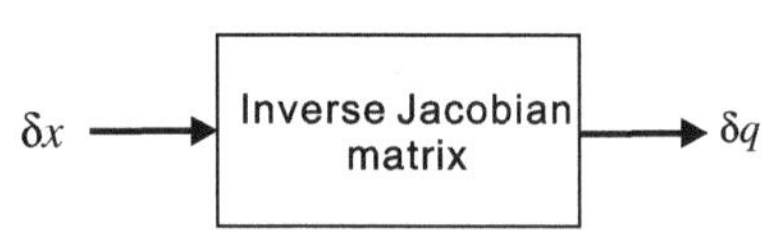

Fig. 2.23 The infinitesimal displacement in joint space is related to an infinitesimal displacement in task space by using the inverse Jacobian matrix

2.5.1 Generalized Approximate Jacobian Setpoint Control

In this section, we consider a generalised approximate Jacobian controller with gravitational force compensation as:

$$\tau = -k_p \hat{D}_1(q)\Delta x - k_v \hat{D}_2(q)\dot{\hat{x}} + g(q) \tag{2.122}$$

where $\hat{D}_1(q)$ and $\hat{D}_2(q)$ are approximate transformations in the presence of uncertain kinematics, $\dot{\hat{x}} = \hat{J}(q)\dot{q}$ is an estimated task-space velocity, $\hat{J}(q) \in \Re^{n\times n}$ is an approximate Jacobian matrix for non-redundant robot where $m = n$.

The matrices $\hat{D}_1(q)$ and $\hat{D}_2(q)$ can be either the approximate transpose Jacobian matrix $\hat{J}^T(q)$ or the approximate inverse Jacobian matrix $\hat{J}^{-1}(q)$. That is,

$$\begin{aligned} \hat{D}_1(q) &: \hat{J}^T(q) \Longleftrightarrow \hat{J}^{-1}(q) \\ \hat{D}_2(q) &: \hat{J}^T(q) \Longleftrightarrow \hat{J}^{-1}(q) \end{aligned} \tag{2.123}$$

There exists four combinations of approximate Jacobian controllers as shown in Table 2.1. Therefore, there is a duality property in task-space regulation in the sense that the transpose Jacobian matrix $\hat{J}^T(q)$ can be replaced by the inverse Jacobian matrix $\hat{J}^{-1}(q)$ and vice versa. The two transformations, the transpose Jacobian and the inverse Jacobian, are said to be dual.

The closed-loop equation of the system is obtained by substituting Eq. (2.122) into Eq. (2.9),

$$M(q)\ddot{q} + \left(\tfrac{1}{2}\dot{M}(q) + S(q,\dot{q})\right)\dot{q} + k_p \hat{D}_1(q)\Delta x + k_v \hat{D}_2(q)\dot{\hat{x}} = \mathbf{0}. \tag{2.124}$$

To prove the stability, a Lyapunov-like candidate is defined as:

$$\begin{aligned} V(\Delta x, \dot{q}) = {} & \tfrac{1}{2}\dot{q}^T M(q)\dot{q} + \alpha \Delta x^T \hat{D}_2^{-1}(q) M(q)\dot{q} \\ & + \tfrac{1}{2}k_p \Delta x^T \hat{D}_1^T(q_d)\hat{J}^{-1}(q_d)\Delta x + \tfrac{1}{2}\alpha k_v \Delta x^T \Delta x, \end{aligned} \tag{2.125}$$

where $\hat{D}_2^{-1}(q)$ is the inverse matrix of $\hat{D}_2(q)$, and q_d is the desired joint configuration that corresponds to x_d. Note that q_d is not required in the controller (2.122) and is defined for analysis only. Similarly, Eq. (2.125) can be expressed as:

Table 2.1 Various combinations of approximate Jacobian controllers with exact gravitational force

	$\hat{D}_1(q)$	$\hat{D}_2(q)$	Controller
$(B1)$	$\hat{J}^T(q)$	$\hat{J}^T(q)$	$\tau = -k_p \hat{J}^T(q)\Delta x - k_v \hat{J}^T(q)\dot{\hat{x}} + g(q)$
$(B2)$	$\hat{J}^T(q)$	$\hat{J}^{-1}(q)$	$\tau = -k_p \hat{J}^T(q)\Delta x - k_v \dot{q} + g(q)$
$(B3)$	$\hat{J}^{-1}(q)$	$\hat{J}^T(q)$	$\tau = -k_p \hat{J}^{-1}(q)\Delta x - k_v \hat{J}^T(q)\dot{\hat{x}} + g(q)$
$(B4)$	$\hat{J}^{-1}(q)$	$\hat{J}^{-1}(q)$	$\tau = -k_p \hat{J}^{-1}(q)\Delta x - k_v \dot{q} + g(q)$

$$
\begin{aligned}
V(\Delta x, \dot{q}) = &\tfrac{1}{4}\dot{q}^T M(q)\dot{q} + \tfrac{1}{4}(\dot{q} + 2\alpha(\hat{D}_2^{-1}(q))^T \Delta x)^T M(q)(\dot{q} + 2\alpha(\hat{D}_2^{-1}(q))^T \Delta x) \\
&- \alpha^2 \Delta x^T \hat{D}_2^{-1}(q) M(q) (\hat{D}_2^{-1}(q))^T \Delta x + \tfrac{1}{2} k_p \Delta x^T \hat{D}_1^T(q_d) \hat{J}^{-1}(q_d) \Delta x + \tfrac{1}{2}\alpha k_v \Delta x^T \Delta x \\
&\geq \tfrac{1}{4}\dot{q}^T M(q)\dot{q} + \tfrac{1}{2} k_p \Delta x^T \hat{D}_1^T(q_d) \hat{J}^{-1}(q_d) \Delta x + \alpha\left(\tfrac{k_v}{2} - \alpha\lambda_m\right) \|\Delta x\|^2
\end{aligned} \tag{2.126}
$$

where $\lambda_m \triangleq \lambda_{\max}[\hat{D}_2^{-1}(q) M(q) (\hat{D}_2^{-1}(q))^T]$. Since $\hat{D}_1(q)$ is either $\hat{J}^T(q)$ or $\hat{J}^{-1}(q)$, the matrix $\hat{D}_1^T(q_d)\hat{J}^{-1}(q_d)$ in (2.125) is either an identity matrix I_n or $(\hat{J}^{-1}(q_d))^T \hat{J}^{-1}(q_d)$. Hence it is always symmetric and positive definite if the approximate Jacobian matrix is non singular at the desired position. Therefore k_v must be chosen so that

$$
k_v - 2\alpha\lambda_m > 0. \tag{2.127}
$$

Differentiating Eq. (2.125) with respect to time and substituting $M(q)\ddot{q}$ from Eq. (2.124) into it yields,

$$
\tfrac{\mathrm{d}}{\mathrm{d}t} V(\Delta x, \dot{q}) = -W(\Delta x, \dot{q}), \tag{2.128}
$$

where

$$
\begin{aligned}
W(\Delta x, \dot{q}) = &\, k_v \dot{q}^T \hat{D}_2(q) \hat{J}(q) \dot{q} + \alpha k_p \Delta x^T \hat{D}_2^{-1}(q) \hat{D}_1(q) \Delta x \\
&+ k_p \Delta x^T (\hat{D}_1^T(q) - \hat{D}_1^T(q_d) \hat{J}^{-1}(q_d) J(q)) \dot{q} + \alpha k_v \Delta x^T (\hat{J}(q) - J(q)) \dot{q} \\
&- \alpha \Big\{ \Delta x^T \hat{D}_2^{-1}(q) \left(\tfrac{1}{2}\dot{M}(q) - S(q, \dot{q})\right) \dot{q} + \dot{q}^T J^T(q) \hat{D}_2^{-1}(q) M(q) \dot{q} \\
&+ \Delta x^T \dot{\hat{D}}_2^{-1}(q) M(q) \dot{q} \Big\}.
\end{aligned} \tag{2.129}
$$

Here, it is assumed that x is bounded due to the finite workspace of robot manipulator. From the last term on the right-hand side of Eq. (2.129), there exist a constant $c_1 > 0$ such that:

$$
\begin{aligned}
\alpha\left\{ \Delta x^T \hat{D}_2^{-1}(q) \left(\tfrac{1}{2}\dot{M}(q) - S(q, \dot{q})\right) \dot{q} + \dot{q}^T J^T(q) \hat{D}_2^{-1}(q) M(q) \dot{q} + \Delta x^T \dot{\hat{D}}_2^{-1}(q) M(q) \dot{q} \right\} \\
\leq c_1 \|\dot{q}\|^2.
\end{aligned} \tag{2.130}
$$

Substituting inequality (2.130) into Eq. (2.129) yields:

$$
\begin{aligned}
W(\Delta x, \dot{q}) \geq &\, \dot{q}^T (k_v \hat{D}_2(q) \hat{J}(q) - \alpha c_1 I_n) \dot{q} + \alpha k_p \Delta x^T \hat{D}_2^{-1}(q) \hat{D}_1(q) \Delta x \\
&+ k_p \Delta x^T (\hat{D}_1^T(q) - \hat{D}_1^T(q_d) \hat{J}^{-1}(q_d) J(q)) \dot{q} \\
&+ \alpha k_v \Delta x^T (\hat{J}(q) - J(q)) \dot{q}.
\end{aligned} \tag{2.131}
$$

As the approximate transformation $\hat{D}_2(q)$ is either $\hat{J}^T(q)$ or $\hat{J}^{-1}(q)$, the matrix $\hat{D}_2(q)\hat{J}(q)$ in (2.131) is either $\hat{J}^T(q)\hat{J}(q)$ or I_n and is always symmetric and positive definite if the approximate Jacobian matrix is non singular. Similarly, $\hat{D}_2^{-1}(q)\hat{D}_1(q)$ is either I_n, $\hat{J}(q)\hat{J}^T(q)$ or $(\hat{J}^{-1}(q))^T\hat{J}^{-1}(q)$ which is also symmetric and positive definite. From Eq. (2.131), we have

$$W(\Delta x, \dot{q}) \geq (k_v\lambda_3 - \alpha c_1)\|\dot{q}\|^2 - (k_p\rho_1 + \alpha k_v\rho_2)\|\Delta x\| \cdot \|\dot{q}\| + \alpha k_p\lambda_4\|\Delta x\|^2, \quad (2.132)$$

where

$$\|\hat{D}_1^T(q) - \hat{D}_1^T(q_d)\hat{J}^{-1}(q_d)J(q)\| \leq \rho_1 \quad (2.133)$$

$$\|\hat{J}(q) - J(q)\| \leq \rho_2 \quad (2.134)$$

and $\lambda_3 \triangleq \lambda_{\min}[\hat{D}_2(q)\hat{J}(q)]$ and $\lambda_4 \triangleq \lambda_{\min}[\hat{D}_2^{-1}(q)\hat{D}_1(q)]$. Note that $-\|\Delta x\| \cdot \|\dot{q}\| \geq \frac{1}{2}(\|\Delta x\|^2 + \|\dot{q}\|^2)$. Therefore, we have:

$$W(\Delta x, \dot{q}) \geq \left(k_v\lambda_3 - \tfrac{k_p}{2}\rho_1 - \alpha\tfrac{k_v}{2}\rho_2 - \alpha c_1\right)\|\dot{q}\|^2 + \left(\alpha k_p\lambda_4 - \tfrac{k_p}{2}\rho_1 - \alpha\tfrac{k_v}{2}\rho_2\right)\|\Delta x\|^2, \quad (2.135)$$

Let $\rho_a = \max(\rho_1, \rho_2)$, if

$$\rho_a < \frac{2\left(\lambda_3 - \frac{\alpha c_1}{k_v}\right)}{a + \alpha}$$

$$\rho_a < \frac{2\alpha\lambda_4 a}{a + \alpha} \quad (2.136)$$

then $W(\Delta x, \dot{q})$ is positive definite in Δx and $\dot{q}$.

The stability of the closed-loop control system is specified by the following theorem:

Theorem 2.6 *Consider the generalised task-space PD controller* (2.122), *the closed-loop system gives rise to the convergence of the sensory-space position error and the joint-space velocity, if the feedback gains* k_p *and* k_v *are chosen to satisfy conditions* (2.127) *and* (2.136) *and the approximate Jacobian matrix is chosen to satisfy conditions* (2.133) *and* (2.134).

Proof If the feedback gains k_p and k_v are chosen to satisfy conditions (2.127), (2.136), then both $V(\Delta x, \dot{q})$ and $W(\Delta x, \dot{q})$ are positive definite in $\dot{q}$ and Δx. From Eq. (2.128), we have

$$\tfrac{\mathrm{d}}{\mathrm{d}t}V(\Delta x, \dot{q}) = -W(\Delta x, \dot{q}) < 0. \quad (2.137)$$

Hence, $V(\Delta \boldsymbol{x}, \dot{\boldsymbol{q}})$ is a Lyapunov-like candidate whose time derivative is negative definite in $\Delta \boldsymbol{x}$ and $\dot{\boldsymbol{q}}$. Therefore, both $\Delta \boldsymbol{x}$ and $\dot{\boldsymbol{q}}$ are bounded. The boundedness of $\dot{\boldsymbol{q}}$ ensures the boundedness of $\dot{\hat{\boldsymbol{x}}}$. From Eq. (2.124), it is seen that $\ddot{\boldsymbol{q}}$ is bounded. The boundedness of $\dot{\boldsymbol{q}}$ also ensures the boundedness of $\dot{\boldsymbol{x}}$ since $\boldsymbol{J}(\boldsymbol{q})$ are trigonometric functions of joint angles. Since both $\ddot{\boldsymbol{q}}$ and $\dot{\boldsymbol{x}}$ are bounded, $\dot{\boldsymbol{q}}$ and $\Delta \boldsymbol{x}$ are uniformly continuous. From Eqs. (2.128) and (2.135), we have $\dot{\boldsymbol{q}}, \Delta \boldsymbol{x} \in L_2(0, +\infty)$. Therefore, from Barbalat's lemma, $\Delta \boldsymbol{x} \to \boldsymbol{0}$ and $\dot{\boldsymbol{q}} \to \boldsymbol{0}$ as $t \to \infty$. △△△

Similarly, the approximate transformations, transpose Jacobian, and inverse Jacobian matrix are said to be dual, as specified by the following corollary:

Corollary 2.1 *For the approximate Jacobian controller* (2.122)*, a duality property exists in the sense that the approximate Jacobian transpose* $\hat{\boldsymbol{J}}^T(\boldsymbol{q})$ *can be replaced by the approximate inverse Jacobian* $\hat{\boldsymbol{J}}^{-1}(\boldsymbol{q})$ *and conversely.*

Proof Let $\bar{\lambda}_m = \max\{\lambda_{\max}[\hat{\boldsymbol{J}}(\boldsymbol{q})\boldsymbol{M}(\boldsymbol{q})\hat{\boldsymbol{J}}^T(\boldsymbol{q})], \lambda_{\max}[\hat{\boldsymbol{J}}^{-T}(\boldsymbol{q})\boldsymbol{M}(\boldsymbol{q})\hat{\boldsymbol{J}}^{-1}(\boldsymbol{q})]\}$. Then from Eq. (2.126), it is possible to choose k_v and α such that

$$k_v - 2\alpha\bar{\lambda}_m > 0 \tag{2.138}$$

for both approximate inverse and transpose Jacobian matrices. Next, we let $\underline{\lambda}_3 = \min\{\lambda_{\min}[\hat{\boldsymbol{J}}^T(\boldsymbol{q})\hat{\boldsymbol{J}}(\boldsymbol{q})], 1\}$ and $\underline{\lambda}_4 = \min\{\lambda_{\min}[\hat{\boldsymbol{J}}(\boldsymbol{q})\hat{\boldsymbol{J}}^T(\boldsymbol{q})], \lambda_{\min}[(\hat{\boldsymbol{J}}^{-1}(\boldsymbol{q}))^T\hat{\boldsymbol{J}}^{-1}(\boldsymbol{q})], 1\}$. Then Eq. (2.135) becomes

$$\begin{aligned} W(\Delta \boldsymbol{x}, \dot{\boldsymbol{q}}) \geq & \left(k_v\underline{\lambda}_3 - \tfrac{k_p}{2}\bar{\rho}_a - \alpha\tfrac{k_v}{2}\bar{\rho}_a - \alpha c_1\right)\|\dot{\boldsymbol{q}}\|^2 \\ & + \left(\alpha k_p\underline{\lambda}_4 - \tfrac{k_p}{2}\bar{\rho}_a - \alpha\tfrac{k_v}{2}\bar{\rho}_a\right)\|\Delta \boldsymbol{x}\|^2, \end{aligned} \tag{2.139}$$

where $\bar{\rho}_a = \max(\rho_1, \rho_2)$ and

$$\|(\hat{\boldsymbol{J}}^{-1}(\boldsymbol{q}))^T - (\hat{\boldsymbol{J}}^{-1}(\boldsymbol{q}_d))^T\hat{\boldsymbol{J}}^{-1}(\boldsymbol{q}_d)\boldsymbol{J}(\boldsymbol{q})\| \leq \rho_1 \tag{2.140}$$

$$\|\hat{\boldsymbol{J}}(\boldsymbol{q}) - \boldsymbol{J}(\boldsymbol{q})\| \leq \rho_2. \tag{2.141}$$

Hence if

$$\begin{aligned} \bar{\rho}_a &< \frac{2\left(\underline{\lambda}_3 - \frac{\alpha c_1}{k_v}\right)}{a + \alpha} \\ \bar{\rho}_a &< \frac{2\alpha\underline{\lambda}_4 a}{a + \alpha}, \end{aligned} \tag{2.142}$$

then $W(\Delta \boldsymbol{x}, \dot{\boldsymbol{q}})$ is positive definite in $\dot{\boldsymbol{q}}$ and $\Delta \boldsymbol{x}$ for both approximate inverse Jacobian and approximate transpose Jacobian. The controller gains can therefore be chosen such that $\hat{\boldsymbol{J}}^T(\boldsymbol{q})$ can be replaced by $\hat{\boldsymbol{J}}^{-1}(\boldsymbol{q})$ and conversely. △△△

2.5.2 Generalized Approximate Jacobian Adaptive Setpoint Control

Next, a generalized approximate Jacobian adaptive setpoint controller with uncertain gravitational force is considered as follows:

$$\tau = -k_p\hat{D}_1(q)\Delta x - k_v\hat{D}_2(q)\dot{x} + Y_g(q)\hat{\theta}_g, \tag{2.143}$$

$$\dot{\hat{\theta}}_g = -\bar{L}_g Y_g^T(q)(\dot{q} + \alpha\hat{D}_3(q)\Delta x), \tag{2.144}$$

where $\bar{L}_g \in \Re^{n_g\times n_g}$ is a positive definite matrix, $Y_g(q) \in \Re^{n\times n_g}$ is the gravity regressor (see *Property* 1.4), $\hat{\theta}_g$ is an estimated parameter updated by (2.144), and $\hat{D}_3(q)$ is an approximate transformation that is defined similarly to $\hat{D}_1(q)$ and $\hat{D}_2(q)$. As shown in Table 2.2, there are eight combinations of the approximate Jacobian adaptive controllers.

The closed-loop equation of the system is obtained by substituting Eq. (2.143) into Eq. (2.9),

$$M(q)\ddot{q}+\left(\frac{1}{2}\dot{M}(q)+S(q,\dot{q})\right)\dot{q}+k_p\hat{D}_1(q)\Delta x+k_v\hat{D}_2(q)\dot{x} + Y_g(q)\Delta\theta_g=\mathbf{0}, \tag{2.145}$$

Table 2.2 Various combinations of approximate Jacobian adaptive control

	$\hat{D}_1(q)$	$\hat{D}_2(q)$	$\hat{D}_3(q)$	Controller
$(C1)$	$\hat{J}^T(q)$	$\hat{J}^T(q)$	$\hat{J}^T(q)$	$\tau = -k_p\hat{J}^T(q)\Delta x - k_v\hat{J}^T(q)\dot{x} + Y_g(q)\hat{\theta}_g$ $\dot{\hat{\theta}}_g = -\bar{L}_g Y_g^T(q)(\dot{q} + \alpha\hat{J}^T(q)\Delta x)$
$(C2)$	$\hat{J}^T(q)$	$\hat{J}^T(q)$	$\hat{J}^{-1}(q)$	$\tau = -k_p\hat{J}^T(q)\Delta x - k_v\hat{J}^T(q)\dot{x} + Y_g(q)\hat{\theta}_g$ $\dot{\hat{\theta}}_g = -\bar{L}_g Y_g^T(q)(\dot{q} + \alpha\hat{J}^{-1}(q)\Delta x)$
$(C3)$	$\hat{J}^T(q)$	$\hat{J}^{-1}(q)$	$\hat{J}^T(q)$	$\tau = -k_p\hat{J}^T(q)\Delta x - k_v\dot{q} + Y_g(q)\hat{\theta}_g$ $\dot{\hat{\theta}}_g = -\bar{L}_g Y_g^T(q)(\dot{q} + \alpha\hat{J}^T(q)\Delta x)$
$(C4)$	$\hat{J}^T(q)$	$\hat{J}^{-1}(q)$	$\hat{J}^{-1}(q)$	$\tau = -k_p\hat{J}^T(q)\Delta x - k_v\dot{q} + Y_g(q)\hat{\theta}_g$ $\dot{\hat{\theta}}_g = -\bar{L}_g Y_g^T(q)(\dot{q} + \alpha\hat{J}^{-1}(q)\Delta x)$
$(C5)$	$\hat{J}^{-1}(q)$	$\hat{J}^T(q)$	$\hat{J}^T(q)$	$\tau = -k_p\hat{J}^{-1}(q)\Delta x - k_v\hat{J}^T(q)\dot{x} + Y_g(q)\hat{\theta}_g$ $\dot{\hat{\theta}}_g = -\bar{L}_g Y_g^T(q)(\dot{q} + \alpha\hat{J}^T(q)\Delta x)$
$(C6)$	$\hat{J}^{-1}(q)$	$\hat{J}^T(q)$	$\hat{J}^{-1}(q)$	$\tau = -k_p\hat{J}^{-1}(q)\Delta x - k_v\hat{J}^T(q)\dot{x} + Y_g(q)\hat{\theta}_g$ $\dot{\hat{\theta}}_g = -\bar{L}_g Y_g^T(q)(\dot{q} + \alpha\hat{J}^{-1}(q)\Delta x)$
$(C7)$	$\hat{J}^{-1}(q)$	$\hat{J}^{-1}(q)$	$\hat{J}^T(q)$	$\tau = -k_p\hat{J}^{-1}(q)\Delta x - k_v\dot{q} + Y_g(q)\hat{\theta}_g$ $\dot{\hat{\theta}}_g = -\bar{L}_g Y_g^T(q)(\dot{q} + \alpha\hat{J}^T(q)\Delta x)$
$(C8)$	$\hat{J}^{-1}(q)$	$\hat{J}^{-1}(q)$	$\hat{J}^{-1}(q)$	$\tau = -k_p\hat{J}^{-1}(q)\Delta x - k_v\dot{q} + Y_g(q)\hat{\theta}_g$ $\dot{\hat{\theta}}_g = -\bar{L}_g Y_g^T(q)(\dot{q} + \alpha\hat{J}^{-1}(q)\Delta x)$

where $\Delta\boldsymbol{\theta}_g = \boldsymbol{\theta}_g - \hat{\boldsymbol{\theta}}_g$. We define a generalized Lyapunov function candidate as

$$V(\Delta\boldsymbol{x}, \dot{\boldsymbol{q}}, \Delta\boldsymbol{\theta}_g) = \tfrac{1}{2}\dot{\boldsymbol{q}}^T\boldsymbol{M}(\boldsymbol{q})\dot{\boldsymbol{q}} + \alpha\Delta\boldsymbol{x}^T\hat{\boldsymbol{D}}_3^T(\boldsymbol{q})\boldsymbol{M}(\boldsymbol{q})\dot{\boldsymbol{q}}$$
$$+\tfrac{1}{2}k_p\Delta\boldsymbol{x}^T\hat{\boldsymbol{D}}_1^T(\boldsymbol{q}_d)\hat{\boldsymbol{J}}^{-1}(\boldsymbol{q}_d)\Delta\boldsymbol{x}+\tfrac{1}{2}\alpha k_v\Delta\boldsymbol{x}^T\hat{\boldsymbol{D}}_3^T(\boldsymbol{q}_d)\hat{\boldsymbol{D}}_2(\boldsymbol{q}_d)\Delta\boldsymbol{x}+\tfrac{1}{2}\Delta\boldsymbol{\theta}_g^T\bar{\boldsymbol{L}}_g^{-1}\Delta\boldsymbol{\theta}_g. \quad (2.146)$$

Again, since $\hat{\boldsymbol{D}}_2(\boldsymbol{q})$ and $\hat{\boldsymbol{D}}_3(\boldsymbol{q})$ are either $\hat{\boldsymbol{J}}^T(\boldsymbol{q})$ or $\hat{\boldsymbol{J}}^{-1}(\boldsymbol{q})$, it can be easily verified that $\hat{\boldsymbol{D}}_3^T(\boldsymbol{q}_d)\hat{\boldsymbol{D}}_2(\boldsymbol{q}_d)$ is always symmetric and positive definite if the approximate Jacobian matrix is nonsingular at the desired position. Following a similar argument to one in previous subsections, we have

$$V(\Delta\boldsymbol{x}, \dot{\boldsymbol{q}}, \Delta\boldsymbol{\theta}_g) \geq \tfrac{1}{4}\dot{\boldsymbol{q}}^T\boldsymbol{M}(\boldsymbol{q})\dot{\boldsymbol{q}} + \tfrac{1}{2}k_p\Delta\boldsymbol{x}^T\hat{\boldsymbol{D}}_1^T(\boldsymbol{q}_d)\hat{\boldsymbol{J}}^{-1}(\boldsymbol{q}_d)\Delta\boldsymbol{x}$$
$$+\alpha\left(\tfrac{1}{2}k_v\lambda_5 - \alpha\lambda_{m2}\right)\|\Delta\boldsymbol{x}\|^2 + \tfrac{1}{2}\Delta\boldsymbol{\theta}_g^T\bar{\boldsymbol{L}}_g^{-1}\Delta\boldsymbol{\theta}_g, \quad (2.147)$$

where $\lambda_5 \triangleq \lambda_{\min}[\hat{\boldsymbol{D}}_3^T(\boldsymbol{q}_d)\hat{\boldsymbol{D}}_2(\boldsymbol{q}_d)]$ and $\lambda_m \triangleq \lambda_{\max}[\hat{\boldsymbol{D}}_3^T(\boldsymbol{q})\boldsymbol{M}(\boldsymbol{q})\hat{\boldsymbol{D}}_3(\boldsymbol{q})]$. Hence k_v and α must be chosen such that

$$k_v\lambda_5 - 2\alpha\lambda_m > 0. \quad (2.148)$$

Differentiating Eq. (2.146) with respect to time and substituting Eqs. (2.145) and (2.144) into it yields

$$\tfrac{\mathrm{d}}{\mathrm{d}t}V(\Delta\boldsymbol{x}, \dot{\boldsymbol{q}}, \Delta\boldsymbol{\theta}_g) = -W(\Delta\boldsymbol{x}, \dot{\boldsymbol{q}}), \quad (2.149)$$

where

$$W(\Delta\boldsymbol{x}, \dot{\boldsymbol{q}}) = k_v\dot{\boldsymbol{q}}^T\hat{\boldsymbol{D}}_2(\boldsymbol{q})\hat{\boldsymbol{J}}(\boldsymbol{q})\dot{\boldsymbol{q}}+\alpha k_p\Delta\boldsymbol{x}^T\hat{\boldsymbol{D}}_3^T(\boldsymbol{q})\hat{\boldsymbol{D}}_1(\boldsymbol{q})\Delta\boldsymbol{x}+k_p\Delta\boldsymbol{x}^T(\hat{\boldsymbol{D}}_1^T(\boldsymbol{q})$$
$$-\hat{\boldsymbol{D}}_1^T(\boldsymbol{q}_d)\hat{\boldsymbol{J}}^{-1}(\boldsymbol{q}_d)\boldsymbol{J}(\boldsymbol{q}))\dot{\boldsymbol{q}}+\alpha k_v\Delta\boldsymbol{x}^T(\hat{\boldsymbol{D}}_3^T(\boldsymbol{q})\hat{\boldsymbol{D}}_2(\boldsymbol{q})\hat{\boldsymbol{J}}(\boldsymbol{q})-\hat{\boldsymbol{D}}_3^T(\boldsymbol{q}_d)\hat{\boldsymbol{D}}_2(\boldsymbol{q}_d)\boldsymbol{J}(\boldsymbol{q}))\dot{\boldsymbol{q}}$$
$$-\alpha\Big\{\Delta\boldsymbol{x}^T\hat{\boldsymbol{D}}_3^T(\boldsymbol{q})\left(\tfrac{1}{2}\dot{\boldsymbol{M}}(\boldsymbol{q}) - \boldsymbol{S}(\boldsymbol{q}, \dot{\boldsymbol{q}})\right)\dot{\boldsymbol{q}}$$
$$+\dot{\boldsymbol{q}}^T\boldsymbol{J}^T(\boldsymbol{q})\hat{\boldsymbol{D}}_3^T(\boldsymbol{q})\boldsymbol{M}(\boldsymbol{q})\dot{\boldsymbol{q}} + \Delta\boldsymbol{x}^T\dot{\hat{\boldsymbol{D}}}_3(\boldsymbol{q})\boldsymbol{M}(\boldsymbol{q})\dot{\boldsymbol{q}}\Big\}. \quad (2.150)$$

From the last term on the right-hand side of Eq. (2.150), there exists a constant $c_2 > 0$ such that

$$\alpha\Big\{\Delta\boldsymbol{x}^T\hat{\boldsymbol{D}}_3^T(\boldsymbol{q})\left(\tfrac{1}{2}\dot{\boldsymbol{M}}(\boldsymbol{q}) - \boldsymbol{S}(\boldsymbol{q}, \dot{\boldsymbol{q}})\right)\dot{\boldsymbol{q}} + \dot{\boldsymbol{q}}^T\boldsymbol{J}^T(\boldsymbol{q})\hat{\boldsymbol{D}}_3^T(\boldsymbol{q})\boldsymbol{M}(\boldsymbol{q})\dot{\boldsymbol{q}}$$
$$+\Delta\boldsymbol{x}^T\dot{\hat{\boldsymbol{D}}}_3(\boldsymbol{q})\boldsymbol{M}(\boldsymbol{q})\dot{\boldsymbol{q}}\Big\} \leq \alpha c_2\|\dot{\boldsymbol{q}}\|^2. \quad (2.151)$$

Hence

$$W(\Delta \boldsymbol{x}, \dot{\boldsymbol{q}}) \geq (k_v\lambda_3 - \alpha c_2)\|\dot{\boldsymbol{q}}\|^2 - (k_p\rho_1 + \alpha k_v \rho_3)\|\Delta \boldsymbol{x}\| \cdot \|\dot{\boldsymbol{q}}\| + \alpha k_p \lambda_6 \|\Delta \boldsymbol{x}\|^2, \quad (2.152)$$

where

$$\|\hat{\boldsymbol{D}}_1^T(\boldsymbol{q}) - \hat{\boldsymbol{D}}_1^T(\boldsymbol{q}_d)\hat{\boldsymbol{J}}^{-1}(\boldsymbol{q}_d)\boldsymbol{J}(\boldsymbol{q})\| \leq \rho_1 \quad (2.153)$$

$$\|\hat{\boldsymbol{D}}_3^T(\boldsymbol{q})\hat{\boldsymbol{D}}_2(\boldsymbol{q})\hat{\boldsymbol{J}}(\boldsymbol{q}) - \hat{\boldsymbol{D}}_3^T(\boldsymbol{q}_d)\hat{\boldsymbol{D}}_2(\boldsymbol{q}_d)\boldsymbol{J}(\boldsymbol{q})\| \leq \rho_3 \quad (2.154)$$

and $\lambda_6 \stackrel{\triangle}{=} \lambda_{\min}[\hat{\boldsymbol{D}}_3^T(\boldsymbol{q})\hat{\boldsymbol{D}}_1(\boldsymbol{q})]$. Next, note that $-\|\Delta \boldsymbol{x}\| \cdot \|\dot{\boldsymbol{q}}\| \geq \frac{1}{2}(\|\Delta \boldsymbol{x}\|^2 + \|\dot{\boldsymbol{q}}\|^2)$, and therefore,

$$\begin{aligned} W(\Delta \boldsymbol{x}, \dot{\boldsymbol{q}}) \geq & \left(k_v\lambda_3 - \frac{k_p}{2}\rho_1 - \alpha\frac{k_v}{2}\rho_3 - \alpha c_2\right)\|\dot{\boldsymbol{q}}\|^2 \\ & + \left(\alpha k_p \lambda_6 - \frac{k_p}{2}\rho_1 - \alpha\frac{k_v}{2}\rho_3\right)\|\Delta \boldsymbol{x}\|^2. \end{aligned} \quad (2.155)$$

Let $\bar{\rho}_2 = \max(\rho_1, \rho_3)$. If

$$\begin{aligned} \bar{\rho}_2 &< \frac{2\left(\lambda_3 - \frac{\alpha c_2}{k_v}\right)}{a + \alpha} \\ \bar{\rho}_2 &< \frac{2\alpha\lambda_6 a}{a + \alpha}, \end{aligned} \quad (2.156)$$

then $W(\Delta \boldsymbol{x}, \dot{\boldsymbol{q}})$ is positive definite in $\dot{\boldsymbol{q}}$ and $\Delta \boldsymbol{x}$.

We are now ready to state the following theorem:

Theorem 2.7 *For the generalized approximate Jacobian adaptive controller describedby Eqs.* (2.143) *and* (2.144)*, the closed-loop system gives rise to the convergence of* $\boldsymbol{x}_d - \boldsymbol{x} \to \boldsymbol{0}$, $\dot{\boldsymbol{q}} \to \boldsymbol{0}$ *as* $t \to \infty$ *if the feedback gains* k_p *and* k_v *are chosen to satisfy conditions* (2.148) *and* (2.156) *and the approximate Jacobian matrix is chosen to satisfy conditions* (2.153) *and* (2.154).

Proof Since $V(\Delta \boldsymbol{x}, \dot{\boldsymbol{q}}, \Delta \boldsymbol{\theta}_g)$ is positive definite in $\Delta \boldsymbol{x}$, $\dot{\boldsymbol{q}}$ and $\Delta \boldsymbol{\theta}_g$, and $W(\Delta \boldsymbol{x}, \dot{\boldsymbol{q}})$ is positive definite in $\Delta \boldsymbol{x}$, $\dot{\boldsymbol{q}}$, from Eq. (2.149), we have

$$\frac{\mathrm{d}}{\mathrm{d}t}V(\Delta \boldsymbol{x}, \dot{\boldsymbol{q}}) = -W(\Delta \boldsymbol{x}, \dot{\boldsymbol{q}}) \leq 0. \quad (2.157)$$

Therefore, $\Delta \boldsymbol{x}$, $\dot{\boldsymbol{q}}$, and $\Delta \boldsymbol{\theta}_g$ are bounded. The boundedness of $\dot{\boldsymbol{q}}$ also ensures the boundedness of $\dot{\hat{\boldsymbol{x}}}$. From Eq. (2.145), it is seen that $\ddot{\boldsymbol{q}}$ is bounded. Since both $\ddot{\boldsymbol{q}}$ and $\dot{\boldsymbol{x}}$ are bounded, $\dot{\boldsymbol{q}}$ and $\Delta \boldsymbol{x}$ are uniformly continuous. From Eqs. (2.149) and (2.155), it is seen that $\dot{\boldsymbol{q}}, \Delta \boldsymbol{x} \in L_2(0, +\infty)$. Therefore, $\Delta \boldsymbol{x} \to \boldsymbol{0}$ and $\dot{\boldsymbol{q}} \to \boldsymbol{0}$ as $t \to \infty$. △△△

Similarly, the approximate transformations, transpose Jacobian $\hat{\boldsymbol{J}}^T(\boldsymbol{q})$, and inverse Jacobian matrix in the adaptive controllers are said to be dual, as specified by the following corollary:

Corollary 2.2 *For the generalized approximate Jacobian adaptive controller described by Eqs.* (2.143) *and* (2.144), *a duality property exists in the sense that the approximate Jacobian transpose* $\hat{\boldsymbol{J}}^T(\boldsymbol{q})$ *can be replaced by the approximate inverse Jacobian* $\hat{\boldsymbol{J}}^{-1}(\boldsymbol{q})$ *and conversely.*

Proof Let $\underline{\lambda}_5 = \min\{1, \lambda_{\min}[\hat{\boldsymbol{J}}^{-T}(\boldsymbol{q}_d)\hat{\boldsymbol{J}}^{-1}(\boldsymbol{q}_d)], \lambda_{\min}[\hat{\boldsymbol{J}}(\boldsymbol{q}_d)\hat{\boldsymbol{J}}^T(\boldsymbol{q}_d)]\}$, and $\bar{\lambda}_{m2} = \max\{\lambda_{\max}[\hat{\boldsymbol{J}}^{-T}(\boldsymbol{q})\boldsymbol{M}(\boldsymbol{q})\hat{\boldsymbol{J}}^{-1}(\boldsymbol{q})], \lambda_{\max}[\hat{\boldsymbol{J}}(\boldsymbol{q})\boldsymbol{M}(\boldsymbol{q})\hat{\boldsymbol{J}}^T(\boldsymbol{q})]\}$. Then from Eq. (2.147), it is possible to chose k_v and α such that

$$k_v\underline{\lambda}_5 - 2\alpha\bar{\lambda}_{m2} > 0. \tag{2.158}$$

Next, let $\underline{\lambda}_3 = \min\{\lambda_{\min}[\hat{\boldsymbol{J}}^T(\boldsymbol{q})\hat{\boldsymbol{J}}(\boldsymbol{q})], 1\}$ and $\underline{\lambda}_6 = \min\{\lambda_{\min}[\hat{\boldsymbol{J}}(\boldsymbol{q})\hat{\boldsymbol{J}}^T(\boldsymbol{q})], \lambda_{\min}[(\hat{\boldsymbol{J}}^{-1}(\boldsymbol{q}))^T\hat{\boldsymbol{J}}^{-1}(\boldsymbol{q})], 1\}$. Then Eq. (2.155) becomes

$$\begin{aligned} W(\Delta\boldsymbol{x}, \dot{\boldsymbol{q}}) \geq & \left(k_v\underline{\lambda}_3 - \frac{k_p}{2}\bar{\rho}_a - \alpha\frac{k_v}{2}\bar{\rho}_a - \alpha c_2\right)\|\dot{\boldsymbol{q}}\|^2 \\ & + \left(\alpha k_p\underline{\lambda}_6 - \frac{k_p}{2}\bar{\rho}_a - \alpha\frac{k_v}{2}\bar{\rho}_a\right)\|\Delta\boldsymbol{x}\|^2, \end{aligned} \tag{2.159}$$

where $\bar{\rho}_a = \max(\rho_1, \rho_3)$, and if

$$\begin{aligned} \bar{\rho}_a &< \frac{2\left(\underline{\lambda}_3 - \frac{\alpha c_2}{k_v}\right)}{a+\alpha} \\ \bar{\rho}_a &< \frac{2\alpha\underline{\lambda}_6 a}{a+\alpha}, \end{aligned} \tag{2.160}$$

then $W(\Delta\boldsymbol{x}, \dot{\boldsymbol{q}})$ is positive definite in $\dot{\boldsymbol{q}}$ and $\Delta\boldsymbol{x}$ for both approximate inverse Jacobian and approximate transpose Jacobian. The controller gains can therefore be chosen such that $\hat{\boldsymbol{J}}^T(\boldsymbol{q})$ can be replaced by $\hat{\boldsymbol{J}}^{-1}(\boldsymbol{q})$ and conversely. △△△

2.6 Summary and Notes

The dynamics of a robot manipulator is highly nonlinear with couplings between joints. It was first shown in Takegaki and Arimoto (1981); Arimoto and Miyazaki (1985); Arimoto (1996) using the Lyapunov method that a simple PD controller with gravity compensation is effective for setpoint control of a robot manipulator. The gravity compensation term can also be computed offline using the desired position (Tomei 1991; Kelly 1993; Arimoto 1996). A main drawback of the PD plus gravity controller is that exact knowledge of the gravitational force of the robot manipulator is required. The existence of modeling uncertainty in the gravitational term results in steady-state position error. One way to alleviate the problem is to increase the P control gain, which may lead to saturation of the control torques. A common practice to remove the steady-state error is to use a PID controller (Arimoto and Miyazaki 1983; Arimoto 1996).

A linear PID robot control system is asymptotically stable in a local sense. To achieve global asymptotic stability, a saturated or bounded nonlinear function of the position error can be used (Arimoto et al. 1994; Arimoto 1996). Other regulators for robot manipulators that do not use the exact knowledge of the gravitational force can be found in Kelly (1998), Ortega et al. (1995). If the structure of the gravitational force is known, an adaptive PD controller with gravity regressor can also be used to eliminate the steady-state position error in the presence of uncertain parameters (Tomei 1991; Kelly 1993; Arimoto 1996). The aforementioned controllers were formulated in joint space. This book focuses on sensory task-space feedback control of robot manipulators, and more detail on joint-space control can be found in Lewis et al. (1993), Arimoto (1996), Kelly et al. (2005). The use of the inner product for the derivation of Lyapunov functions can be found in Arimoto (1996).

Task-space feedback control laws with an imperfect Jacobian matrix were first considered in Miyazaki and Masutani (1990), where a sensor coordinate is defined as the task space and the exact knowledge of the Jacobian matrix of the mapping from Cartesian space to sensor space is not required. However, the exact model of the manipulator Jacobian matrix from joint space to Cartesian space is assumed to be known exactly. A fundamental benefit of task-space feedback control is its ability to deal with kinematic uncertainty. Setpoint controllers for robot manipulators with uncertain kinematics and dynamics were first proposed in Cheah et al. (1998). The approximate Jacobian setpoint controller (2.66) with task-space damping was proposed in Cheah et al. (1999). The controller requires the measurement of a task-space position using a task-space sensor such as a vision system. The task-space velocity is obtained by numerical differentiation of the position. In Cheah et al. (2003), an approximate Jacobian feedback controller using only joint-space damping was developed. The joint-space velocity is obtained by numerical differentiation of the joint position, which is usually less noisy than that in task space. In the presence of uncertainty in the gravitational force, the update law (2.93) or the integration control term Cheah et al. (1999) also requires the use of the inverse Jacobian matrix. An approximate Jacobian controller with adaptive gravity compensation was developed in Cheah et al. (2003). The main advantages of the controller are that the task-space velocity and the inverse of the approximate Jacobian matrix are not required. An update law can also be used to update the kinematic parameters of the estimated Jacobian matrix online (Cheah 2003; Dixon 2004; Dixon 2007) for setpoint control, and simpler stability conditions than those obtained using a fixed approximate Jacobian matrix can be obtained.

The duality property in task-space regulation of nonredundant robots was analyzed in Cheah (2008a), and it was shown that the transpose Jacobian matrix can be replaced by the inverse Jacobian matrix and conversely.

Chapter 3
Sensory Task-Space Adaptive Jacobian Control

A human being is able to adapt to kinematic uncertainty by learning from previous experience. For example, we can manipulate a tool more skillfully after using it several times. This chapter introduces several basic concepts and design techniques in adaptive Jacobian control of robot manipulators using sensory task-space feedback. This approach allows us to design adaptive controllers with concurrent adaptation to both kinematic and dynamic uncertainties. Several sensory task-space adaptive controllers are introduced for tracking the control of a robot in the presence of uncertain transformations from sensory space to joint space. The design concept is also a useful modular tool for studying robot or nonlinear systems containing multiple uncertainties. Update laws for various uncertain parameters such as actuator model and depth information can be added and updated concurrently while preserving the structure and properties of the original adaptive control systems and ensuring the stability of the overall system.

3.1 Joint-Space Adaptive Control

In tracking control applications, it is necessary to specify the motion of the end effector in great detail rather than simply stating the desired final position. In this section, the standard joint-space adaptive control problem for a robot manipulator is first reviewed. The robot kinematics is assumed to be known, so that a desired trajectory $\boldsymbol{q}_d(t)$ can be obtained in joint space by solving the inverse kinematics problem.

To design the adaptive control, a reference vector is defined in joint space as follows:

$$\dot{\boldsymbol{q}}_r = \dot{\boldsymbol{q}}_d - \alpha \Delta \boldsymbol{q}, \tag{3.1}$$

where $\dot{\boldsymbol{q}}_d \in \Re^n$ is the desired velocity in joint space, $\Delta \boldsymbol{q} = \boldsymbol{q} - \boldsymbol{q}_d \in \Re^n$ denotes the tracking error, and α is a positive constant.

C.C. Cheah and X. Li, *Task-Space Sensory Feedback Control of Robot Manipulators*, Intelligent Systems, Control and Automation: Science and Engineering 73,
DOI 10.1007/978-981-287-062-9_3

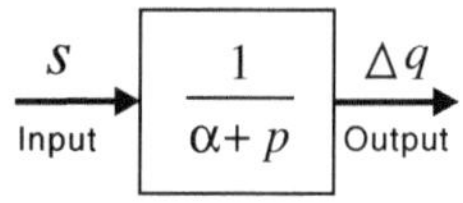

Fig. 3.1 A stable first-order system, where p denotes the Laplace variable

Using the reference vector defined in Eq. (3.1), a sliding vector $\boldsymbol{s} \in \Re^n$ is introduced as

$$\boldsymbol{s} = \dot{\boldsymbol{q}} - \dot{\boldsymbol{q}}_r = \Delta\dot{\boldsymbol{q}} + \alpha\Delta\boldsymbol{q}, \tag{3.2}$$

where $\Delta\dot{\boldsymbol{q}} = \dot{\boldsymbol{q}} - \dot{\boldsymbol{q}}_d$. The sliding vector is a weighted sum of the position error and the velocity error. Using the sliding vector, the problem of trajectory tracking becomes equivalent to that of remaining on the sliding surface. The above equation described by (3.2) can be interpreted as a stable first-order system with input $\boldsymbol{s}$ and output $\Delta\boldsymbol{q}$, as illustrated in Fig. 3.1. Therefore, the problem of tracking the desired trajectory $\boldsymbol{q}_d$ can be converted to that of keeping the sliding vector $\boldsymbol{s}$ at zero.

By differentiating Eq. (3.2) with respect to time, the time derivative of the sliding vector is obtained as

$$\dot{\boldsymbol{s}} = \ddot{\boldsymbol{q}} - \ddot{\boldsymbol{q}}_r = \Delta\ddot{\boldsymbol{q}} + \alpha\Delta\dot{\boldsymbol{q}}. \tag{3.3}$$

From Eq. (1.5) in Chap. 1, the nonlinear robot dynamics is described as

$$\boldsymbol{M}(\boldsymbol{q})\ddot{\boldsymbol{q}} + \left(\tfrac{1}{2}\dot{\boldsymbol{M}}(\boldsymbol{q}) + \boldsymbol{S}(\boldsymbol{q},\dot{\boldsymbol{q}})\right)\dot{\boldsymbol{q}} + \boldsymbol{g}(\boldsymbol{q}) = \boldsymbol{\tau}. \tag{3.4}$$

Using Eqs. (3.2) and (3.3), we can express the joint velocity and acceleration in terms of the sliding vector and the reference vector as

$$\dot{\boldsymbol{q}} = \boldsymbol{s} + \dot{\boldsymbol{q}}_r, \tag{3.5}$$

$$\ddot{\boldsymbol{q}} = \dot{\boldsymbol{s}} + \ddot{\boldsymbol{q}}_r. \tag{3.6}$$

Substituting Eqs. (3.5) and (3.6) into Eq. (3.4), we have

$$\boldsymbol{M}(\boldsymbol{q})\dot{\boldsymbol{s}} + \left(\tfrac{1}{2}\dot{\boldsymbol{M}}(\boldsymbol{q}) + \boldsymbol{S}(\boldsymbol{q},\dot{\boldsymbol{q}})\right)\boldsymbol{s} + \boldsymbol{M}(\boldsymbol{q})\ddot{\boldsymbol{q}}_r + \left(\tfrac{1}{2}\dot{\boldsymbol{M}}(\boldsymbol{q}) + \boldsymbol{S}(\boldsymbol{q},\dot{\boldsymbol{q}})\right)\dot{\boldsymbol{q}}_r = \boldsymbol{\tau}. \tag{3.7}$$

From *Property* 1.3, it is obtained that

$$\boldsymbol{M}(\boldsymbol{q})\ddot{\boldsymbol{q}}_r + \left(\tfrac{1}{2}\dot{\boldsymbol{M}}(\boldsymbol{q}) + \boldsymbol{S}(\boldsymbol{q},\dot{\boldsymbol{q}})\right)\dot{\boldsymbol{q}}_r + \boldsymbol{g}(\boldsymbol{q}) = \boldsymbol{Y}_d(\boldsymbol{q},\dot{\boldsymbol{q}},\dot{\boldsymbol{q}}_r,\ddot{\boldsymbol{q}}_r)\boldsymbol{\theta}_d, \tag{3.8}$$

where $\boldsymbol{Y}_d(\boldsymbol{q},\dot{\boldsymbol{q}},\dot{\boldsymbol{q}}_r,\ddot{\boldsymbol{q}}_r) \in \Re^{n\times n_d}$ is a known regressor matrix, and $\boldsymbol{\theta}_d \in \Re^{n_d}$ is an unknown parameter vector. Note that the derivative of the reference vector does not contain the acceleration vector.

Substituting Eq. (3.8) into Eq. (3.7) yields

$$\boldsymbol{M}(\boldsymbol{q})\dot{\boldsymbol{s}} + \left(\tfrac{1}{2}\dot{\boldsymbol{M}}(\boldsymbol{q}) + \boldsymbol{S}(\boldsymbol{q},\dot{\boldsymbol{q}})\right)\boldsymbol{s} + \boldsymbol{Y}_d(\boldsymbol{q},\dot{\boldsymbol{q}},\dot{\boldsymbol{q}}_r,\ddot{\boldsymbol{q}}_r)\boldsymbol{\theta}_d = \boldsymbol{\tau}. \tag{3.9}$$

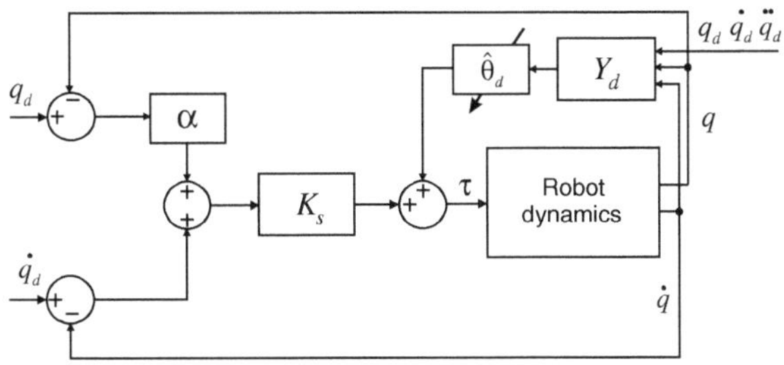

Fig. 3.2 A block diagram of the adaptive control system

Next, the adaptive tracking control scheme is designed in joint space, which consists of a PD feedback using the sliding vector and a full dynamics compensation as

$$\boldsymbol{\tau} = -\boldsymbol{K}_s\boldsymbol{s} + \boldsymbol{Y}_d(\boldsymbol{q}, \dot{\boldsymbol{q}}, \dot{\boldsymbol{q}}_r, \ddot{\boldsymbol{q}}_r)\hat{\boldsymbol{\theta}}_d, \tag{3.10}$$

where $\boldsymbol{K}_s \in \Re^{n\times n}$ is a positive definite matrix, and $\hat{\boldsymbol{\theta}}_d \in \Re^{n_d}$ represent the estimated dynamic parameters, which are updated by the following update law:

$$\dot{\hat{\boldsymbol{\theta}}}_d = -\boldsymbol{L}_d\boldsymbol{Y}_d^T(\boldsymbol{q}, \dot{\boldsymbol{q}}, \dot{\boldsymbol{q}}_r, \ddot{\boldsymbol{q}}_r)\boldsymbol{s}, \tag{3.11}$$

where $\boldsymbol{L}_d \in \Re^{n_d\times n_d}$ is a symmetric positive definite matrix. A block diagram of the controller in Eq. (3.10) is illustrated in Fig. 3.2.

The closed-loop equation for the robotic manipulator is obtained by substituting the adaptive joint-space controller in Eq. (3.10) into the dynamic equation (3.9) as

$$\boldsymbol{M}(\boldsymbol{q})\dot{\boldsymbol{s}} + \left(\tfrac{1}{2}\dot{\boldsymbol{M}}(\boldsymbol{q}) + \boldsymbol{S}(\boldsymbol{q}, \dot{\boldsymbol{q}})\right)\boldsymbol{s} + \boldsymbol{K}_s\boldsymbol{s} + \boldsymbol{Y}_d(\boldsymbol{q}, \dot{\boldsymbol{q}}, \dot{\boldsymbol{q}}_r, \ddot{\boldsymbol{q}}_r)\Delta\boldsymbol{\theta}_d = \boldsymbol{0}, \tag{3.12}$$

where $\Delta\boldsymbol{\theta}_d = \boldsymbol{\theta}_d - \hat{\boldsymbol{\theta}}_d \in \Re^{n_d}$.

To prove stability, consider the following Lyapunov-like candidate:

$$V = \tfrac{1}{2}\boldsymbol{s}^T\boldsymbol{M}(\boldsymbol{q})\boldsymbol{s} + \tfrac{1}{2}\Delta\boldsymbol{\theta}_d^T\boldsymbol{L}_d^{-1}\Delta\boldsymbol{\theta}_d. \tag{3.13}$$

Differentiating Eq. (3.13) with respect to time and substituting Eq. (3.12) into it, we have

$$\begin{aligned}\dot{V} &= \boldsymbol{s}^T\boldsymbol{M}(\boldsymbol{q})\dot{\boldsymbol{s}} + \tfrac{1}{2}\boldsymbol{s}^T\dot{\boldsymbol{M}}(\boldsymbol{q})\boldsymbol{s} - \dot{\hat{\boldsymbol{\theta}}}_d^T\boldsymbol{L}_d^{-1}\Delta\boldsymbol{\theta}_d \\ &= -\boldsymbol{s}^T\boldsymbol{S}(\boldsymbol{q}, \dot{\boldsymbol{q}})\boldsymbol{s} - \boldsymbol{s}^T\boldsymbol{K}_s\boldsymbol{s} - \boldsymbol{s}^T\boldsymbol{Y}_d(\boldsymbol{q}, \dot{\boldsymbol{q}}, \dot{\boldsymbol{q}}_r, \ddot{\boldsymbol{q}}_r)\Delta\boldsymbol{\theta}_d - \dot{\hat{\boldsymbol{\theta}}}_d^T\boldsymbol{L}_d^{-1}\Delta\boldsymbol{\theta}_d. \end{aligned} \tag{3.14}$$

Using *Property* 1.2 and substituting Eq. (3.11) into Eq. (3.14), $\dot{V}$ is simplified as

$$\dot{V} = -\boldsymbol{s}^T\boldsymbol{K}_s\boldsymbol{s}. \tag{3.15}$$

As $\boldsymbol{M}(\boldsymbol{q})$ and $\boldsymbol{L}_d$ are positive definite, V is positive definite in $\boldsymbol{s}$ and $\Delta\boldsymbol{\theta}_d$. Since $\dot{V}$ is negative definite in $\boldsymbol{s}$, $\boldsymbol{s}$ and $\Delta\boldsymbol{\theta}_d$ are bounded. Hence, from closed-loop equation (3.12), the boundedness of $\boldsymbol{s}$ and $\Delta\boldsymbol{\theta}_d$ ensures the boundedness of $\dot{\boldsymbol{s}}$. Therefore, $\dot{V}$ is uniformly continuous. Using Barbalat's Lemma or the related Lyapunov-Like Lemma, it then follows that $\dot{V}$ goes to zero as $t \rightarrow \infty$, so it can be concluded that $\boldsymbol{s}$ converges to zero. The convergence of $\boldsymbol{s}$ yields the convergence of $\Delta\boldsymbol{q}$ and $\Delta\dot{\boldsymbol{q}}$ to zero.

3.2 Adaptive Jacobian Tracking Control

In the presence of kinematic uncertainty, inverse kinematics cannot be used to derive the desired trajectory in joint space. In Chap. 2, we studied the sensory task-space setpoint control problem of a robot manipulator with a fixed approximate Jacobian matrix. The structures of the sensory task-space setpoint controllers are simple, but the controllers are not effective for trajectory-tracking tasks especially at high speed. The kinematic parameters of the approximate Jacobian matrix are also not updated to improve the performance. This chapter introduces the concepts of adaptive Jacobian matrix and adaptive task-space sliding vector for trajectory tracking control of a robot manipulator using task-space sensory feedback information.

To understand the problem, let us first consider a two-link nonredundant robot holding an object. The sensory-space velocity $\dot{\boldsymbol{x}}$ is related to the joint-space velocity $\dot{\boldsymbol{q}}$ by the Jacobian matrix as

$$\dot{\boldsymbol{x}} = \boldsymbol{J}(\boldsymbol{q})\dot{\boldsymbol{q}}, \tag{3.16}$$

where the Jacobian matrix $\boldsymbol{J}(\boldsymbol{q})$ from joint space to sensory space can be derived as

$$\boldsymbol{J}(\boldsymbol{q}) = \begin{bmatrix} -(l_1 s_1 + l_2 s_{12} + l_o s_{12o}) & -(l_2 s_{12} + l_o s_{12o}) \\ l_1 c_1 + l_2 c_{12} + l_o c_{12o} & l_2 c_{12} + l_o c_{12o} \end{bmatrix}, \tag{3.17}$$

where l_1, l_2 are the link lengths, l_0 and q_0 are respectively the length and grasping angle of the object, $s_1 = \sin(q_1)$, $s_{12} = \sin(q_1 + q_2)$, $s_{12o} = \sin(q_1 + q_2 + q_o)$, $c_1 = \cos(q_1)$, $c_{12} = \cos(q_1 + q_2)$, $c_{12o} = \cos(q_1 + q_2 + q_o)$. In the presence of kinematic uncertainty, the parameters l_1, l_2, l_o, q_0 are unknown. The inverse Jacobian matrix $\boldsymbol{J}^{-1}(\boldsymbol{q})$ can be obtained as

$$\boldsymbol{J}^{-1}(\boldsymbol{q}) = \frac{1}{l_1 l_2 s_2 + l_1 l_o s_{2o}} \begin{bmatrix} l_2 c_{12} + l_o c_{12o} & l_2 s_{12} + l_o s_{12o} \\ -(l_1 c_1 + l_2 c_{12} + l_o c_{12o}) & -(l_1 s_1 + l_2 s_{12} + l_o s_{12o}) \end{bmatrix}, \tag{3.18}$$

which is therefore nonlinear in the unknown parameters l_1, l_2, l_o, q_0 (Fig. 3.3).

Differentiating Eq. (3.16) with respect to me, we have

$$\ddot{\boldsymbol{x}} = \boldsymbol{J}(\boldsymbol{q})\ddot{\boldsymbol{q}} + \dot{\boldsymbol{J}}(\boldsymbol{q})\dot{\boldsymbol{q}}, \tag{3.19}$$

where $\ddot{\boldsymbol{x}}$ and $\ddot{\boldsymbol{q}}$ represent the acceleration of robot end effector in sensory space and joint space respectively, and $\dot{\boldsymbol{J}}(\boldsymbol{q})$ denotes the time-derivative of the Jacobian matrix.

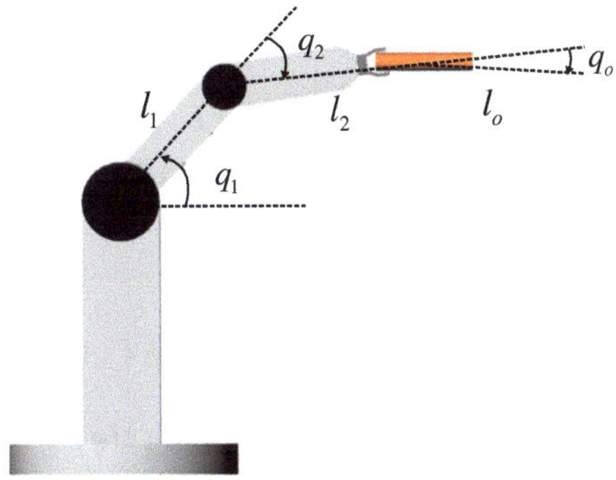

Fig. 3.3 A two-link planar robot holding an unknown tool

From Eqs. (3.16) and (3.19), we have

$$\dot{q}s = J^{-1}(q)\dot{x}, \tag{3.20}$$

$$\ddot{q} = J^{-1}(q)\ddot{x} - J^{-1}(q)\dot{J}(q)J^{-1}(q)\dot{x}. \tag{3.21}$$

For the two-link robot, the robot dynamics is described as

$$M(q)\ddot{q} + \left(\tfrac{1}{2}\dot{M}(q) + S(q,\dot{q})\right)\dot{q} + g(q) = \tau, \tag{3.22}$$

where

$$M(q) = \begin{bmatrix} M_{11} & M_{12} \\ M_{21} & M_{22} \end{bmatrix}, \tag{3.23}$$

$$\tfrac{1}{2}\dot{M}(q) + S(q,\dot{q}) = \begin{bmatrix} -m_2\sin(q_2)l_1l_{c2}\dot{q}_2 & -m_2\sin(q_2)l_1l_{c2}(\dot{q}_1+\dot{q}_2) \\ m_2\sin(q_2)l_1l_{c2}\dot{q}_1 & 0 \end{bmatrix}, \tag{3.24}$$

$$g(q) = \begin{bmatrix} g_1 \\ g_2 \end{bmatrix}, \tag{3.25}$$

where

$$\begin{aligned}
M_{11} &= m_1l_{c1}^2 + I_1 + m_2[l_1^2 + l_{c2}^2 + 2\cos(q_2)l_1l_{c2}] + I_2, \\
M_{22} &= m_2l_{c2}^2 + I_2, \\
M_{12} &= M_{21} = m_2\cos(q_2)l_1l_{c2} + m_2l_{c2}^2 + I_2, \\
g_1 &= m_1g\cos(q_1)l_{c1} + m_2g[\cos(q_1+q_2)l_{c2} + \cos(q_1)l_1], \\
g_2 &= m_2g\cos(q_1+q_2)l_{c2}.
\end{aligned} \tag{3.26}$$

Using Eq. (3.21), the dynamics equation is obtained as

$$\begin{aligned}
&M(q)J^{-1}(q)\ddot{x} + \big(-M(q)J^{-1}(q)\dot{J}(q) \\
&+\tfrac{1}{2}\dot{M}(q) + S(q,\dot{q})\big)J^{-1}(q)\dot{x} + g(q) = \tau.
\end{aligned} \tag{3.27}$$

Multiplying Eq. (3.28) by $\boldsymbol{J}^{-T}(\boldsymbol{q})$ yields

$$\begin{aligned}&\boldsymbol{J}^{-T}(\boldsymbol{q})\boldsymbol{M}(\boldsymbol{q})\boldsymbol{J}^{-1}(\boldsymbol{q})\ddot{\boldsymbol{x}}+\boldsymbol{J}^{-T}(\boldsymbol{q})\big(\tfrac{1}{2}\dot{\boldsymbol{M}}(\boldsymbol{q})+\boldsymbol{S}(\boldsymbol{q},\dot{\boldsymbol{q}})\\&-\boldsymbol{M}(\boldsymbol{q})\boldsymbol{J}^{-1}(\boldsymbol{q})\dot{\boldsymbol{J}}(\boldsymbol{q})\big)\boldsymbol{J}^{-1}(\boldsymbol{q})\dot{\boldsymbol{x}}+\boldsymbol{J}^{-T}(\boldsymbol{q})\boldsymbol{g}(\boldsymbol{q})=\boldsymbol{J}^{-T}(\boldsymbol{q})\boldsymbol{\tau},\end{aligned} \tag{3.28}$$

which is written as

$$\boldsymbol{M}_x(\boldsymbol{q})\ddot{\boldsymbol{x}}+\boldsymbol{C}_x(\boldsymbol{q},\dot{\boldsymbol{q}})\dot{\boldsymbol{x}}+\boldsymbol{g}_x(\boldsymbol{q})=\boldsymbol{F}, \tag{3.29}$$

where $\boldsymbol{M}_x(\boldsymbol{q})=\boldsymbol{J}^{-T}(\boldsymbol{q})\boldsymbol{M}(\boldsymbol{q})\boldsymbol{J}^{-1}(\boldsymbol{q})$, $\boldsymbol{g}_x(\boldsymbol{q})=\boldsymbol{J}^{-T}(\boldsymbol{q})\boldsymbol{g}(\boldsymbol{q})$, $\boldsymbol{C}_x(\boldsymbol{q},\dot{\boldsymbol{q}})=\boldsymbol{J}^{-T}(\boldsymbol{q})(\frac{1}{2}\dot{\boldsymbol{M}}(\boldsymbol{q})+\boldsymbol{S}(\boldsymbol{q},\dot{\boldsymbol{q}})-\boldsymbol{M}(\boldsymbol{q})\boldsymbol{J}^{-1}(\boldsymbol{q})\dot{\boldsymbol{J}}(\boldsymbol{q}))\boldsymbol{J}^{-1}(\boldsymbol{q})$, and $\boldsymbol{F}=\boldsymbol{J}^{-T}(\boldsymbol{q})\boldsymbol{\tau}$.

Therefore, Eq. (3.29) cannot be expressed in a form as in *Property* 1.3, because $\boldsymbol{J}^{-1}(\boldsymbol{q})$ is not linear in the unknown kinematic parameters. For example, the gravitational force $\boldsymbol{g}_x(\boldsymbol{q})$ for the two-link robot in task space is specified as

$$\boldsymbol{g}_x(\boldsymbol{q})=\boldsymbol{J}^{-T}(\boldsymbol{q})\boldsymbol{g}(\boldsymbol{q})=\begin{bmatrix}\frac{(l_2c_{12}+l_oc_{12o})g_1+(l_2s_{12}+l_os_{12o})g_2}{l_1l_2s_2+l_1l_os_{2o}}\\ \frac{(-(l_1c_1+l_2c_{12}+l_oc_{12o})g_1-(l_1s_1+l_2s_{12}+l_os_{12o})g_2}{l_1l_2s_2+l_1l_os_{2o}}\end{bmatrix}, \tag{3.30}$$

which cannot be parameterized as the product of a regressor and unknown kinematic parameters.

In general, in the presence of uncertainty in kinematics, the relationship between the control force $\boldsymbol{F}$ in Cartesian space and the torque $\boldsymbol{\tau}$ in joint space is also uncertain.

Therefore, the standard adaptive control method as discussed in Sect. 3.1 cannot be applied directly to overcome the uncertainty in both kinematics and dynamics. Hence, in the presence of kinematic uncertainty, it results in tracking errors or even an unstable response in the end effector's motion.

We are now ready to introduce the concepts of adaptive Jacobian control for a robot manipulator with uncertain kinematics. From Eq. (1.39), *Property* 1.5, Eq. (3.16) can be expressed as

$$\dot{\boldsymbol{x}}=\boldsymbol{J}(\boldsymbol{q})\dot{\boldsymbol{q}}=\boldsymbol{Y}_k(\boldsymbol{q},\dot{\boldsymbol{q}})\boldsymbol{\theta}_k, \tag{3.31}$$

where $\boldsymbol{Y}_k(\boldsymbol{q},\dot{\boldsymbol{q}})\in\Re^{n\times n_k}$ is called the *kinematic regressor matrix.*

To design the adaptive control in sensory space, a reference vector $\dot{\boldsymbol{x}}_r\in\Re^n$ is defined in sensory space as

$$\dot{\boldsymbol{x}}_r=\dot{\boldsymbol{x}}_d-\alpha(\boldsymbol{x}-\boldsymbol{x}_d), \tag{3.32}$$

where $\boldsymbol{x}$ or $\boldsymbol{x}-\boldsymbol{x}_d$ is measured from a position sensor, $\boldsymbol{x}_d\in\Re^n$ is a desired trajectory, and $\dot{\boldsymbol{x}}_d=\frac{\mathrm{d}\boldsymbol{x}_d}{\mathrm{d}t}\in\Re^n$ is the desired velocity specified in sensory task space.

Differentiating Eq. (3.32) with respect to time, we have

$$\ddot{x}_r = \ddot{x}_d - \alpha(\dot{x} - \dot{x}_d), \tag{3.33}$$

where $\dot{x} = \frac{dx}{dt}$ is the task-space velocity and $\ddot{x}_d = \frac{d\dot{x}_d}{dt} \in \Re^n$ is the desired acceleration in sensory task space.

In the presence of kinematic uncertainty, the parameters of the Jacobian matrix are uncertain. Using *Property* 1.5, the estimated sensory-space velocity can be defined by using the uncertain Jacobian matrix as

$$\dot{\hat{x}} = \hat{J}(q, \hat{\theta}_k)\dot{q} = Y_k(q, \dot{q})\hat{\theta}_k, \tag{3.34}$$

where $\dot{\hat{x}} \in \Re^n$ denotes the estimated sensory-space velocity, $\hat{J}(q, \hat{\theta}_k) \in \Re^{n\times n}$ is the adaptive Jacobian matrix, and $\hat{\theta}_k \in \Re^{n_k}$ are estimated kinematic parameters.

Next, we define an adaptive task-space sliding vector using Eqs. (3.34) and (3.32) as

$$\hat{s}_x = \dot{\hat{x}} - \dot{x}_r = \hat{J}(q, \hat{\theta}_k)\dot{q} - \dot{x}_r. \tag{3.35}$$

The above vector is adaptive in the sense that the parameters of the adaptive Jacobian matrix will be updated by a kinematic adaptation law. In contrast to the case of the standard sliding vector, the convergence of $\hat{s}_x$ to zero does not imply directly the convergence of the tracking error $x - x_d$ to zero. A kinematic update law has to be designed to ensure the convergence of the tracking error.

Differentiating Eq. (3.35) with respect to time, we have

$$\dot{\hat{s}}_x = \ddot{\hat{x}} - \ddot{x}_r = \hat{J}(q, \hat{\theta}_k)\ddot{q} + \dot{\hat{J}}(q, \hat{\theta}_k)\dot{q} - \ddot{x}_r, \tag{3.36}$$

where $\ddot{\hat{x}}$ denotes the derivative of $\dot{\hat{x}}$.

Next, a reference vector is defined in joint space as

$$\dot{q}_r = \hat{J}^{-1}(q, \hat{\theta}_k)\dot{x}_r, \tag{3.37}$$

where $\hat{J}^{-1}(q, \hat{\theta}_k)$ is the inverse of the approximate Jacobian matrix $\hat{J}(q, \hat{\theta}_k)$. It is assumed that the robot is operating in a finite task space such that the adaptive Jacobian matrix is of full rank. A suitable projection algorithm can be used to keep the estimated parameters within a specified bound to avoid singularities due to the estimated parameters.

Note that $\hat{\theta}_k$ is just part of the state variables of the adaptive control system and therefore can be used in the control variables even if it is nonlinear in the variables. This is the key idea in designing the adaptive Jacobian controller.

From Eq. (3.37), we have

$$\ddot{q}_r = \hat{J}^{-1}(q, \hat{\theta}_k)\ddot{x}_r + \dot{\hat{J}}^{-1}(q, \hat{\theta}_k)\dot{x}_r, \tag{3.38}$$

where $\dot{\hat{J}}^{-1}(q, \hat{\theta}_k) = -\hat{J}^{-1}(q, \hat{\theta}_k)\dot{\hat{J}}(q, \hat{\theta}_k)\hat{J}^{-1}(q, \hat{\theta}_k)$.

Hence, an adaptive sliding vector is defined in joint space as

$$s = \dot{q} - \dot{q}_r = \hat{J}^{-1}(q, \hat{\theta}_k)((\dot{\hat{x}} - \dot{x}_d) + \alpha(x - x_d)) = \hat{J}^{-1}(q, \hat{\theta}_k)\hat{s}_x, \tag{3.39}$$

and the time derivative of the joint-space sliding vector is obtained as

$$\dot{s} = \ddot{q} - \ddot{q}_r = \hat{J}^{-1}(q, \hat{\theta}_k)\dot{\hat{s}}_x + \dot{\hat{J}}^{-1}(q, \hat{\theta}_k)\hat{s}_x. \tag{3.40}$$

Therefore, the joint-space velocity and its time derivative can be specified as

$$\dot{q} = s + \dot{q}_r, \tag{3.41}$$

$$\ddot{q} = \dot{s} + \ddot{q}_r. \tag{3.42}$$

Substituting Eqs. (3.41) and (3.42) into Eq. (3.4), the equations of motion can be expressed as

$$\begin{aligned} &M(q)\dot{s} + \left(\tfrac{1}{2}\dot{M}(q) + S(q, \dot{q})\right)s \\ &+M(q)\ddot{q}_r + \left(\tfrac{1}{2}\dot{M}(q) + S(q, \dot{q})\right)\dot{q}_r + g(q) = \tau. \end{aligned} \tag{3.43}$$

From *Property* 1.3, the last four terms of Eq. (3.43) are linear in a set of dynamics parameters $\theta_d \in \Re^{n_d}$ and hence can be expressed as

$$M(q)\ddot{q}_r + \left(\tfrac{1}{2}\dot{M}(q) + S(q, \dot{q})\right)\dot{q}_r + g(q) = Y_d(q, \dot{q}, \dot{q}_r, \ddot{q}_r)\theta_d, \tag{3.44}$$

where $Y_d(q, \dot{q}, \dot{q}_r, \ddot{q}_r) \in \Re^{n \times n_d}$ is the dynamic regressor matrix.

From Eq. (3.44), the dynamic equation (3.43) can be expressed as

$$M(q)\dot{s} + \left(\tfrac{1}{2}\dot{M}(q) + S(q, \dot{q})\right)s + Y_d(q, \dot{q}, \dot{q}_r, \ddot{q}_r)\theta_d = \tau. \tag{3.45}$$

The adaptive Jacobian controller is designed as

$$\tau = -\hat{J}^T(q, \hat{\theta}_k)(K_v\Delta\dot{x} + K_p\Delta x) - \hat{J}^T(q, \hat{\theta}_k)K_s\hat{s}_x + Y_d(q, \dot{q}, \dot{q}_r, \ddot{q}_r)\hat{\theta}_d, \tag{3.46}$$

where $\Delta\dot{x} = \dot{x} - \dot{x}_d$, $\Delta x = x - x_d$, $K_v \in \Re^{n \times n}$, $K_p \in \Re^{n \times n}$, and $K_s \in \Re^{n \times n}$ are positive definite matrices. In the adaptive tracking controller (3.46), the first term is an adaptive Jacobian transpose feedback law of the task-space velocity and position errors, the next term is an adaptive Jacobian transpose feedback law of the adaptive sliding vector (3.35), and the last term is an estimated control input based on Eq. (3.44). The estimated kinematic parameters $\hat{\boldsymbol{\theta}}_k$ of the approximate Jacobian matrix $\hat{\boldsymbol{J}}(\boldsymbol{q}, \hat{\boldsymbol{\theta}}_k)$ are updated by

$$\dot{\hat{\boldsymbol{\theta}}}_k = \boldsymbol{L}_k \boldsymbol{Y}_k^T(\boldsymbol{q}, \dot{\boldsymbol{q}})(\boldsymbol{K}_v \Delta\dot{\boldsymbol{x}} + \boldsymbol{K}_p \Delta\boldsymbol{x}), \tag{3.47}$$

and the estimated dynamic parameters $\hat{\boldsymbol{\theta}}_d$ of the dynamic model are updated by

$$\dot{\hat{\boldsymbol{\theta}}}_d = -\boldsymbol{L}_d \boldsymbol{Y}_d^T(\boldsymbol{q}, \dot{\boldsymbol{q}}, \dot{\boldsymbol{q}}_r, \ddot{\boldsymbol{q}}_r)\boldsymbol{s}. \tag{3.48}$$

where $\boldsymbol{L}_k \in \Re^{q \times q}$ and $\boldsymbol{L}_d \in \Re^{n_d \times n_d}$ are positive definite matrices. A block diagram of the adaptive Jacobian control system is illustrated in Fig. 3.4.

Although some kinematic parameters appear in $\hat{\boldsymbol{\theta}}_d$, they are updated separately in $\hat{\boldsymbol{\theta}}_k$ to preserve linearity. The linear parameterization of the kinematic parameters is obtained from Eq. (3.31). The estimated kinematic parameters $\hat{\boldsymbol{\theta}}_k$ of the approximate Jacobian matrix $\hat{\boldsymbol{J}}(\boldsymbol{q}, \hat{\boldsymbol{\theta}}_k)$ are updated by the parameter update Eq. (3.47). The estimated parameters $\hat{\boldsymbol{\theta}}_k$ are then used in the inverse Jacobian matrix $\hat{\boldsymbol{J}}^{-1}(\boldsymbol{q}, \hat{\boldsymbol{\theta}}_k)$, and hence $\dot{\boldsymbol{q}}_r$ and $\ddot{\boldsymbol{q}}_r$ in the dynamic regressor matrix. Since $\hat{\boldsymbol{J}}(\boldsymbol{q}, \hat{\boldsymbol{\theta}}_k)$ and its inverse $\hat{\boldsymbol{J}}^{-1}(\boldsymbol{q}, \hat{\boldsymbol{\theta}}_k)$ are updated by $\boldsymbol{q}$ and $\hat{\boldsymbol{\theta}}_k$, $\dot{\hat{\boldsymbol{J}}}(\boldsymbol{q}, \hat{\boldsymbol{\theta}}_k)$ and $\dot{\hat{\boldsymbol{J}}}^{-1}(\boldsymbol{q}, \hat{\boldsymbol{\theta}}_k) = -\hat{\boldsymbol{J}}^{-1}(\boldsymbol{q}, \hat{\boldsymbol{\theta}}_k)\dot{\hat{\boldsymbol{J}}}(\boldsymbol{q}, \hat{\boldsymbol{\theta}}_k)\hat{\boldsymbol{J}}^{-1}(\boldsymbol{q}, \hat{\boldsymbol{\theta}}_k)$ are functions of $\boldsymbol{q}$, $\dot{\boldsymbol{q}}$, $\hat{\boldsymbol{\theta}}_k$, $\Delta\boldsymbol{x}$, and $\Delta\dot{\boldsymbol{x}}$, because $\dot{\hat{\boldsymbol{\theta}}}_k$ is described by Eq. (3.47).

The closed-loop equation is obtained by substituting Eq. (3.46) into Eq. (3.45) to give

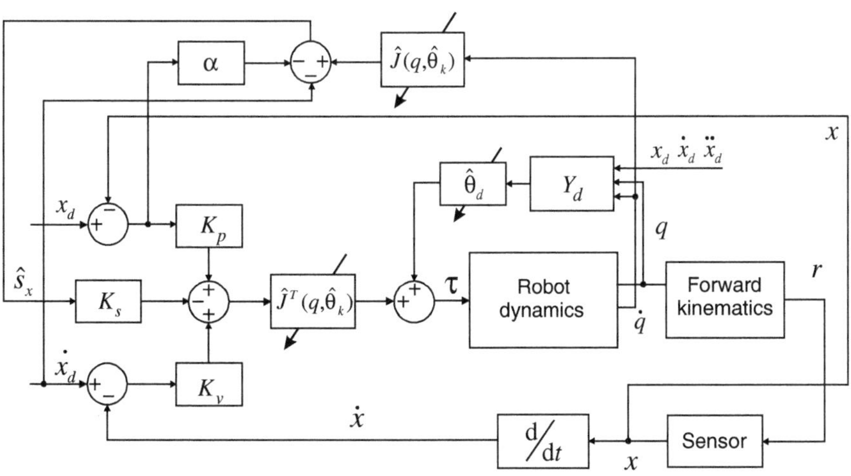

Fig. 3.4 A block diagram of the adaptive Jacobian control system

$$\begin{aligned} &\boldsymbol{M}(\boldsymbol{q})\dot{\boldsymbol{s}} + \left(\tfrac{1}{2}\dot{\boldsymbol{M}}(\boldsymbol{q}) + \boldsymbol{S}(\boldsymbol{q},\dot{\boldsymbol{q}})\right)\boldsymbol{s} + \boldsymbol{Y}_d(\boldsymbol{q},\dot{\boldsymbol{q}},\dot{\boldsymbol{q}}_r,\ddot{\boldsymbol{q}}_r)\Delta\boldsymbol{\theta}_d \\ &+\hat{\boldsymbol{J}}^T(\boldsymbol{q},\hat{\boldsymbol{\theta}}_k)(\boldsymbol{K}_v\Delta\dot{\boldsymbol{x}} + \boldsymbol{K}_p\Delta\boldsymbol{x}) + \hat{\boldsymbol{J}}^T(\boldsymbol{q},\hat{\boldsymbol{\theta}}_k)\boldsymbol{K}_s\hat{\boldsymbol{s}}_x = \boldsymbol{0}, \end{aligned} \tag{3.49}$$

where $\Delta\boldsymbol{\theta}_d = \boldsymbol{\theta}_d - \hat{\boldsymbol{\theta}}_d$.

A Lyapunov-like function candidate is defined as

$$\begin{aligned} V &= \tfrac{1}{2}\boldsymbol{s}^T\boldsymbol{M}(\boldsymbol{q})\boldsymbol{s} + \tfrac{1}{2}\Delta\boldsymbol{\theta}_d^T\boldsymbol{L}_d^{-1}\Delta\boldsymbol{\theta}_d \\ &+\tfrac{1}{2}\Delta\boldsymbol{x}^T(\boldsymbol{K}_p + \alpha\boldsymbol{K}_v)\Delta\boldsymbol{x} + \tfrac{1}{2}\Delta\boldsymbol{\theta}_k^T\boldsymbol{L}_k^{-1}\Delta\boldsymbol{\theta}_k, \end{aligned} \tag{3.50}$$

where $\Delta\boldsymbol{\theta}_k = \boldsymbol{\theta}_k - \hat{\boldsymbol{\theta}}_k$. Differentiating Eq. (3.50) with respect to time and using *Property* 1.1, we have

$$\begin{aligned} \dot{V} &= \boldsymbol{s}^T\boldsymbol{M}(\boldsymbol{q})\dot{\boldsymbol{s}} + \tfrac{1}{2}\boldsymbol{s}^T\dot{\boldsymbol{M}}(\boldsymbol{q})\boldsymbol{s} - \Delta\boldsymbol{\theta}_d^T\boldsymbol{L}_d^{-1}\dot{\hat{\boldsymbol{\theta}}}_d, \\ &+\Delta\boldsymbol{x}^T(\boldsymbol{K}_p + \alpha\boldsymbol{K}_v)\Delta\dot{\boldsymbol{x}} - \Delta\boldsymbol{\theta}_k^T\boldsymbol{L}_k^{-1}\dot{\hat{\boldsymbol{\theta}}}_k. \end{aligned} \tag{3.51}$$

Substituting $\boldsymbol{M}(\boldsymbol{q})\dot{\boldsymbol{s}}$ from Eq. (3.49), $\dot{\hat{\boldsymbol{\theta}}}_k$ from Eq. (3.47), and $\dot{\hat{\boldsymbol{\theta}}}_d$ from Eq. (3.48) into the above equation, using *Property* 1.2 and Eq. (3.39) yields

$$\begin{aligned} \dot{V} &= -\boldsymbol{s}^T\hat{\boldsymbol{J}}^T(\boldsymbol{q},\hat{\boldsymbol{\theta}}_k)\boldsymbol{K}_s\hat{\boldsymbol{s}}_x - \boldsymbol{s}^T\hat{\boldsymbol{J}}^T(\boldsymbol{q},\hat{\boldsymbol{\theta}}_k)(\boldsymbol{K}_v\Delta\dot{\boldsymbol{x}} + \boldsymbol{K}_p\Delta\boldsymbol{x}) \\ &\quad + \Delta\boldsymbol{x}^T(\boldsymbol{K}_p + \alpha\boldsymbol{K}_v)\Delta\dot{\boldsymbol{x}} - \Delta\boldsymbol{\theta}_k^T\boldsymbol{Y}_k^T(\boldsymbol{q},\dot{\boldsymbol{q}})(\boldsymbol{K}_v\Delta\dot{\boldsymbol{x}} + \boldsymbol{K}_p\Delta\boldsymbol{x}) \\ &= -\hat{\boldsymbol{s}}_x^T\boldsymbol{K}_s\hat{\boldsymbol{s}}_x - \hat{\boldsymbol{s}}_x^T(\boldsymbol{K}_v\Delta\dot{\boldsymbol{x}} + \boldsymbol{K}_p\Delta\boldsymbol{x}) + \Delta\boldsymbol{x}^T(\boldsymbol{K}_p + \alpha\boldsymbol{K}_v)\Delta\dot{\boldsymbol{x}} \\ &\quad - \Delta\boldsymbol{\theta}_k^T\boldsymbol{Y}_k^T(\boldsymbol{q},\dot{\boldsymbol{q}})(\boldsymbol{K}_v\Delta\dot{\boldsymbol{x}} + \boldsymbol{K}_p\Delta\boldsymbol{x}). \end{aligned} \tag{3.52}$$

From Eqs. (3.36), (3.31) and (3.32), we have

$$\hat{\boldsymbol{s}}_x = \Delta\dot{\boldsymbol{x}} + \alpha\Delta\boldsymbol{x} - \boldsymbol{Y}_k(\boldsymbol{q},\dot{\boldsymbol{q}})\Delta\boldsymbol{\theta}_k, \tag{3.53}$$

where

$$\boldsymbol{Y}_k(\boldsymbol{q},\dot{\boldsymbol{q}})\Delta\boldsymbol{\theta}_k = \boldsymbol{J}(\boldsymbol{q})\dot{\boldsymbol{q}} - \hat{\boldsymbol{J}}(\boldsymbol{q},\hat{\boldsymbol{\theta}}_k)\dot{\boldsymbol{q}} = \dot{\boldsymbol{x}} - \dot{\hat{\boldsymbol{x}}}. \tag{3.54}$$

Substituting Eq. (3.53) into Eq. (3.52) yields

$$\begin{aligned} \dot{V} &= -(\Delta\dot{\boldsymbol{x}} + \alpha\Delta\boldsymbol{x} - \boldsymbol{Y}_k(\boldsymbol{q},\dot{\boldsymbol{q}})\Delta\boldsymbol{\theta}_k)^T\boldsymbol{K}_s(\Delta\dot{\boldsymbol{x}} + \alpha\Delta\boldsymbol{x} - \boldsymbol{Y}_k(\boldsymbol{q},\dot{\boldsymbol{q}})\Delta\boldsymbol{\theta}_k) \\ &\quad - \Delta\dot{\boldsymbol{x}}^T\boldsymbol{K}_v\Delta\dot{\boldsymbol{x}} - \alpha\Delta\boldsymbol{x}^T\boldsymbol{K}_p\Delta\boldsymbol{x} \le 0. \end{aligned} \tag{3.55}$$

We are now in a position to state the following Theorem:

Theorem 3.1 *For a finite task space such that the adaptive Jacobian matrix is non-singular, the adaptive Jacobian control law* (3.46) *and the parameter update laws*

(3.47) *and* (3.48) *for the robot system* (3.4) *result in the convergence of position and velocity tracking errors. That is,* $\boldsymbol{x} - \boldsymbol{x}_d \rightarrow \boldsymbol{0}$ *and* $\dot{\boldsymbol{x}} - \dot{\boldsymbol{x}}_d \rightarrow \boldsymbol{0}$*, as* $t \rightarrow \infty$.

In addition, the estimated task-space velocity converges to the actual task-space velocity, i.e., $\dot{\hat{\boldsymbol{x}}} \rightarrow \dot{\boldsymbol{x}}, t \rightarrow \infty$.

Proof Since $\boldsymbol{M}(\boldsymbol{q})$ is uniformly positive definite, V in Eq. (3.50) is positive definite in $\boldsymbol{s}$, $\Delta\boldsymbol{x}$, $\Delta\boldsymbol{\theta}_k$, and $\Delta\boldsymbol{\theta}_d$. Since $\dot{V} \leq 0$, V is also bounded, and therefore $\boldsymbol{s}$, $\Delta\boldsymbol{x}$, $\Delta\boldsymbol{\theta}_k$, and $\Delta\boldsymbol{\theta}_d$ are bounded vectors. This implies that $\hat{\boldsymbol{\theta}}_k$, $\hat{\boldsymbol{\theta}}_d$ are bounded, $\boldsymbol{x}$ is bounded if $\boldsymbol{x}_d$ is bounded, and $\hat{s}_x = \hat{\boldsymbol{J}}(\boldsymbol{q}, \hat{\boldsymbol{\theta}}_k)\boldsymbol{s}$ is bounded, as seen from Eq. (3.39). Since $\Delta\boldsymbol{x}$ is bounded, $\dot{\boldsymbol{x}}_r$ in Eq. (3.32) is also bounded if $\dot{\boldsymbol{x}}_d$ is bounded. Therefore, $\dot{\boldsymbol{q}}_r$ in Eq. (3.37) is also bounded if the approximate Jacobian matrix is nonsingular. From Eq. (3.39), $\dot{\boldsymbol{q}}$ is bounded, and the boundedness of $\dot{\boldsymbol{q}}$ means that $\dot{\boldsymbol{x}}$ is bounded, since the Jacobian matrix is bounded. Hence, $\Delta\dot{\boldsymbol{x}}$ is bounded and $\ddot{\boldsymbol{x}}_r$ in Eq. (3.33) is also bounded if $\ddot{\boldsymbol{x}}_d$ is bounded. From Eq. (3.47), $\dot{\hat{\boldsymbol{\theta}}}_k$ is therefore bounded, since $\Delta\boldsymbol{x}$, $\Delta\dot{\boldsymbol{x}}$, $\dot{\boldsymbol{q}}$ are bounded and $\boldsymbol{Y}_k(\cdot)$ is a trigonometric function of $\boldsymbol{q}$. Therefore, $\ddot{\boldsymbol{q}}_r$ in Eq. (3.38) is bounded. From the closed-loop equation (3.49), we can conclude that $\dot{\boldsymbol{s}}$ is bounded. The boundedness of $\dot{\boldsymbol{s}}$ implies the boundedness of $\ddot{\boldsymbol{q}}$, as seen from Eq. (3.40). From Eq. (3.36), $\dot{\hat{s}}_x$ is therefore bounded. Finally, differentiating equation (3.53) with respect to time and rearranging yields

$$\Delta\ddot{\boldsymbol{x}} + \alpha\Delta\dot{\boldsymbol{x}} = \dot{\hat{s}}_x + \dot{\boldsymbol{Y}}_k(\boldsymbol{q}, \dot{\boldsymbol{q}}, \ddot{\boldsymbol{q}})\Delta\boldsymbol{\theta}_k - \boldsymbol{Y}_k(\boldsymbol{q}, \dot{\boldsymbol{q}})\dot{\hat{\boldsymbol{\theta}}}_k, \tag{3.56}$$

which means that $\Delta\ddot{\boldsymbol{x}} = \ddot{\boldsymbol{x}} - \ddot{\boldsymbol{x}}_d$ is also bounded.

Next, let us check the uniform continuity of $\dot{V}$. Differentiating Eq. (3.55) with respect to time gives

$$\ddot{V} = -2\hat{s}_x^T \boldsymbol{K}_s \dot{\hat{s}}_x - 2\Delta\dot{\boldsymbol{x}}^T \boldsymbol{K}_v \Delta\ddot{\boldsymbol{x}} - 2\alpha\Delta\boldsymbol{x}^T \boldsymbol{K}_p \Delta\dot{\boldsymbol{x}}. \tag{3.57}$$

This shows that $\ddot{V}$ is bounded, since $\hat{s}_x$, $\dot{\hat{s}}_x$ and $\Delta\boldsymbol{x}$, $\Delta\dot{\boldsymbol{x}}$, $\Delta\ddot{\boldsymbol{x}}$ are all bounded. Hence, $\dot{V}$ is uniformly continuous. Using Barbalat's Lemma or the Lyapunov-Like Lemma, we have $\Delta\boldsymbol{x} = \boldsymbol{x} - \boldsymbol{x}_d \rightarrow 0$, $\Delta\dot{\boldsymbol{x}} = \dot{\boldsymbol{x}} - \dot{\boldsymbol{x}}_d \rightarrow \boldsymbol{0}$, and $\hat{s}_x \rightarrow \boldsymbol{0}$ as $t \rightarrow \infty$. Finally, from Eq. (3.54), the convergence of $\Delta\boldsymbol{x}$, $\Delta\dot{\boldsymbol{x}}$, and $\hat{s}_x$ implies that $\dot{\boldsymbol{x}} - \dot{\hat{\boldsymbol{x}}} \rightarrow \boldsymbol{0}$ as $t \rightarrow \infty$. $\triangle\triangle\triangle$

3.3 Adaptive Jacobian Tracking Control Without Task-Space Velocity

The adaptive tracking controller in Eq. (3.46) requires the task-space velocity $\dot{\boldsymbol{x}}$, which is obtained from numerical differentiation of the task-space position. In this section, an adaptive Jacobian tracking control problem without using the task-space velocity is considered. To avoid the need for numerical differentiation of position, which is usually noisy, a known signal $\boldsymbol{y}$ based on filtered differentiation of the measured position $\boldsymbol{x}$ is introduced as

$$\dot{\boldsymbol{y}} + \lambda \boldsymbol{y} = \lambda \dot{\boldsymbol{x}}. \tag{3.58}$$

The signal $\boldsymbol{y}$ can be computed by measuring $\boldsymbol{x}$ alone. With the Laplace variable p, $\boldsymbol{y}$ can be written from Eqs. (3.31) and (3.58) as (Fig. 3.5)

$$\boldsymbol{y} = \frac{\lambda p}{p+\lambda}\boldsymbol{x} = \boldsymbol{W}_k(t)\boldsymbol{\theta}_k, \tag{3.59}$$

where

$$\boldsymbol{W}_k(t) = \frac{\lambda}{p+\lambda}\boldsymbol{Y}_k(\boldsymbol{q}, \dot{\boldsymbol{q}}), \tag{3.60}$$

with $\boldsymbol{y}(0) = \boldsymbol{0}$ and $\boldsymbol{W}_k(0) = \boldsymbol{0}$, since the robot usually starts from a rest position. Other linear filters may also be used based on noise or vibration models.

Using the filtered signals, the adaptive Jacobian tracking controller without using task-space velocity is designed as

$$\boldsymbol{\tau} = -\hat{\boldsymbol{J}}^T(\boldsymbol{q}, \hat{\boldsymbol{\theta}}_k)(\boldsymbol{K}_v \Delta\dot{\hat{\boldsymbol{x}}} + \boldsymbol{K}_p \Delta \boldsymbol{x}) + \bar{\boldsymbol{Y}}_d(\boldsymbol{q}, \dot{\boldsymbol{q}}, \dot{\boldsymbol{q}}_r, \ddot{\bar{\boldsymbol{q}}}_r, \hat{\boldsymbol{\theta}}_k)\hat{\boldsymbol{\theta}}_d, \tag{3.61}$$

where $\Delta\dot{\hat{\boldsymbol{x}}} = \dot{\hat{\boldsymbol{x}}} - \dot{\boldsymbol{x}}_d$, $\bar{\boldsymbol{Y}}_d(\boldsymbol{q}, \dot{\boldsymbol{q}}, \dot{\boldsymbol{q}}_r, \ddot{\bar{\boldsymbol{q}}}_r, \hat{\boldsymbol{\theta}}_k)$ is a dynamic regressor matrix as detailed later, and $\dot{\boldsymbol{q}}_r$ and $\dot{\hat{\boldsymbol{q}}}_r$ are defined based on the adaptive sliding vector as detailed later. The estimated dynamic parameters $\hat{\boldsymbol{\theta}}_d$ are updated by a dynamic adaptation law,

$$\dot{\hat{\boldsymbol{\theta}}}_d = -\boldsymbol{L}_d \bar{\boldsymbol{Y}}_d(\boldsymbol{q}, \dot{\boldsymbol{q}}, \dot{\boldsymbol{q}}_r, \ddot{\bar{\boldsymbol{q}}}_r, \hat{\boldsymbol{\theta}}_k)\boldsymbol{s}, \tag{3.62}$$

and the estimated kinematic parameters $\hat{\boldsymbol{\theta}}_k$ are updated by a kinematic adaptation law,

$$\dot{\hat{\boldsymbol{\theta}}}_k = -\boldsymbol{L}_k \boldsymbol{W}_k^T(t)\boldsymbol{K}_v(\boldsymbol{W}_k(t)\hat{\boldsymbol{\theta}}_k - \boldsymbol{y}) + \boldsymbol{L}_k \boldsymbol{Y}_k^T(\boldsymbol{q}, \dot{\boldsymbol{q}})(\boldsymbol{K}_p + \alpha \boldsymbol{K}_v)\Delta \boldsymbol{x}. \tag{3.63}$$

Similarly, all gain matrices are symmetric positive definite. In the adaptive control law (3.61), the first term is an approximate Jacobian transpose feedback law of the task-space velocity and position errors, and the last term is an estimated dynamic compensation term based on Eq. (3.78), which will be detailed later. A block diagram of the control system is illustrated in Fig. 3.6.

Similarly, a reference vector is first defined in sensory space as

$$\dot{\boldsymbol{x}}_r = \dot{\boldsymbol{x}}_d - \alpha \Delta \boldsymbol{x}, \tag{3.64}$$

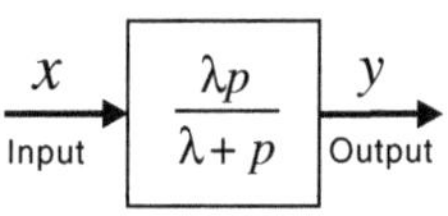

Fig. 3.5 The signal $\boldsymbol{y}$ is based on filtered differentiation of the measured position $\boldsymbol{x}$

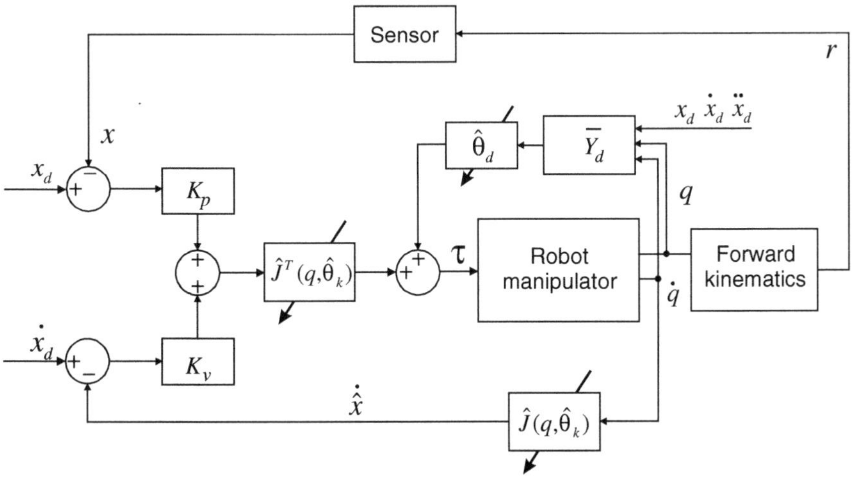

Fig. 3.6 A block diagram of the adaptive Jacobian control system without using task-space velocity

and differentiating it with respect to time yields

$$\ddot{\boldsymbol{x}}_r = \ddot{\boldsymbol{x}}_d - \alpha\Delta\dot{\boldsymbol{x}}. \tag{3.65}$$

The adaptive task-space sliding vector is defined as,

$$\hat{\boldsymbol{s}}_x = \dot{\hat{\boldsymbol{x}}} - \dot{\boldsymbol{x}}_r = \hat{\boldsymbol{J}}(\boldsymbol{q}, \hat{\boldsymbol{\theta}}_k)\dot{\boldsymbol{q}} - \dot{\boldsymbol{x}}_r, \tag{3.66}$$

and differentiating it with respect to time yields

$$\dot{\hat{\boldsymbol{s}}}_x = \ddot{\hat{\boldsymbol{x}}} - \ddot{\boldsymbol{x}}_r = \hat{\boldsymbol{J}}(\boldsymbol{q}, \hat{\boldsymbol{\theta}}_k)\ddot{\boldsymbol{q}} + \dot{\hat{\boldsymbol{J}}}(\boldsymbol{q}, \hat{\boldsymbol{\theta}}_k)\dot{\boldsymbol{q}} - \ddot{\boldsymbol{x}}_r, \tag{3.67}$$

where $\ddot{\hat{\boldsymbol{x}}}$ denotes the derivative of $\dot{\hat{\boldsymbol{x}}}$. Next, let

$$\dot{\boldsymbol{q}}_r = \hat{\boldsymbol{J}}^{-1}(\boldsymbol{q}, \hat{\boldsymbol{\theta}}_k)\dot{\boldsymbol{x}}_r. \tag{3.68}$$

From Eq. (3.68), we have

$$\ddot{\boldsymbol{q}}_r = \hat{\boldsymbol{J}}^{-1}(\boldsymbol{q}, \hat{\boldsymbol{\theta}}_k)\ddot{\boldsymbol{x}}_r + \dot{\hat{\boldsymbol{J}}}^{-1}(\boldsymbol{q}, \hat{\boldsymbol{\theta}}_k)\dot{\boldsymbol{x}}_r, \tag{3.69}$$

where $\dot{\hat{\boldsymbol{J}}}^{-1}(\boldsymbol{q}, \hat{\boldsymbol{\theta}}_k) = -\hat{\boldsymbol{J}}^{-1}(\boldsymbol{q}, \hat{\boldsymbol{\theta}}_k)\dot{\hat{\boldsymbol{J}}}(\boldsymbol{q}, \hat{\boldsymbol{\theta}}_k)\hat{\boldsymbol{J}}^{-1}(\boldsymbol{q}, \hat{\boldsymbol{\theta}}_k)$.

From Eq. (3.65), note that $\ddot{\boldsymbol{x}}_r$ in Eq. (3.69) contains the task-space velocity. To eliminate the need for the task-space velocity, we define

$$\ddot{\hat{\boldsymbol{x}}}_r = \ddot{\boldsymbol{x}}_d - \alpha\Delta\dot{\hat{\boldsymbol{x}}}, \tag{3.70}$$

where $\Delta\dot{\hat{x}} = \dot{\hat{x}} - \dot{x}_d$. Using Eq. (3.70), we define

$$\ddot{\hat{q}}_r = \hat{J}^{-1}(q, \hat{\theta}_k)\ddot{\hat{x}}_r + \dot{\hat{J}}^{-1}(q, \hat{\theta}_k)\dot{x}_r. \tag{3.71}$$

From Eqs. (3.70) and (3.65), we have

$$\ddot{\hat{x}}_r = \ddot{x}_d - \alpha\Delta\dot{x} + \alpha(\dot{x} - \dot{\hat{x}}) = \ddot{x}_r + \alpha(\dot{x} - \dot{\hat{x}}). \tag{3.72}$$

Substituting Eq. (3.72) into Eq. (3.71) and using Eq. (3.69) yields

$$\ddot{\hat{q}}_r = \ddot{q}_r + \alpha\hat{J}^{-1}(q, \hat{\theta}_k)(\dot{x} - \dot{\hat{x}}) = \ddot{q}_r - \alpha\dot{q} + \alpha\hat{J}^{-1}(q, \hat{\theta}_k)J(q)\dot{q}. \tag{3.73}$$

Next, an adaptive sliding vector is defined in joint space as,

$$s = \dot{q} - \dot{q}_r = \hat{J}^{-1}(q, \hat{\theta}_k)((\dot{\hat{x}} - \dot{x}_d) + \alpha(x - x_d)) = \hat{J}^{-1}(q, \hat{\theta}_k)\hat{s}_x \tag{3.74}$$

and

$$\dot{s} = \ddot{q} - \ddot{q}_r. \tag{3.75}$$

Substituting $\ddot{q}_r$ from Eq. (3.73) into Eq. (3.75) yields

$$\dot{s} = \ddot{q} - (\ddot{\hat{q}}_r + \alpha\dot{q}) + \alpha\hat{J}^{-1}(q, \hat{\theta}_k)J(q)\dot{q}. \tag{3.76}$$

Substituting Eqs. (3.74) and (3.76) into Eq. (3.4), the robot dynamics can be expressed as,

$$\begin{aligned} &M(q)\dot{s} + \big(\tfrac{1}{2}\dot{M}(q) + S(q, \dot{q})\big)s + M(q)\ddot{\hat{q}}_r + \big(\tfrac{1}{2}\dot{M}(q) + S(q, \dot{q})\big)\dot{q}_r + g(q) \\ &\quad + \alpha M(q)\dot{q} - \alpha M(q)\hat{J}^{-1}(q, \hat{\theta}_k)J(q)\dot{q} = \tau. \end{aligned} \tag{3.77}$$

The last six terms of Eq. (3.77) are linear in a set of dynamics parameters $\bar{\theta}_d$ and hence can be expressed as

$$\begin{aligned} &M(q)\ddot{\hat{q}}_r + \big(\tfrac{1}{2}\dot{M}(q) + S(q, \dot{q})\big)\dot{q}_r + g(q) + \alpha M(q)\dot{q} - \alpha M(q)\hat{J}^{-1}(q, \hat{\theta}_k)J(q)\dot{q} \\ &\quad = \bar{Y}_d(q, \dot{q}, \dot{q}_r, \ddot{\hat{q}}_r, \hat{\theta}_k)\bar{\theta}_d, \end{aligned} \tag{3.78}$$

Therefore, the dynamic equation (3.77) can be written

$$M(q)\dot{s} + \big(\tfrac{1}{2}\dot{M}(q) + S(q, \dot{q})\big)s + \bar{Y}_d(q, \dot{q}, \dot{q}_r, \ddot{\hat{q}}_r, \hat{\theta}_k)\bar{\theta}_d = \tau. \tag{3.79}$$

The closed-loop dynamics is obtained by substituting (3.61) into (3.79),

$$\begin{aligned} \boldsymbol{M}(\boldsymbol{q})\dot{\boldsymbol{s}} + \left(\tfrac{1}{2}\dot{\boldsymbol{M}}(\boldsymbol{q}) + \boldsymbol{S}(\boldsymbol{q},\dot{\boldsymbol{q}})\right)\boldsymbol{s} + \bar{\boldsymbol{Y}}_d(\boldsymbol{q},\dot{\boldsymbol{q}},\dot{\boldsymbol{q}}_r,\hat{\ddot{\boldsymbol{q}}}_r,\hat{\boldsymbol{\theta}}_k)\Delta\boldsymbol{\theta}_d \\ +\hat{\boldsymbol{J}}^T(\boldsymbol{q},\hat{\boldsymbol{\theta}}_k)(\boldsymbol{K}_v\Delta\dot{\hat{\boldsymbol{x}}} + \boldsymbol{K}_p\Delta\boldsymbol{x}) = \boldsymbol{0}, \end{aligned} \tag{3.80}$$

where $\Delta\boldsymbol{\theta}_d = \bar{\boldsymbol{\theta}}_d - \hat{\boldsymbol{\theta}}_d$. Since $\hat{\boldsymbol{J}}(\boldsymbol{q},\hat{\boldsymbol{\theta}}_k)$ and its inverse $\hat{\boldsymbol{J}}^{-1}(\boldsymbol{q},\hat{\boldsymbol{\theta}}_k)$ are updated by $\boldsymbol{q}$ and $\hat{\boldsymbol{\theta}}_k$, $\dot{\hat{\boldsymbol{J}}}(\boldsymbol{q},\hat{\boldsymbol{\theta}}_k)$ and $\dot{\hat{\boldsymbol{J}}}^{-1}(\boldsymbol{q},\hat{\boldsymbol{\theta}}_k) = -\hat{\boldsymbol{J}}^{-1}(\boldsymbol{q},\hat{\boldsymbol{\theta}}_k)\dot{\hat{\boldsymbol{J}}}(\boldsymbol{q},\hat{\boldsymbol{\theta}}_k)\hat{\boldsymbol{J}}^{-1}(\boldsymbol{q},\hat{\boldsymbol{\theta}}_k)$ are functions of $\boldsymbol{q}$, $\dot{\boldsymbol{q}}$, $\Delta\boldsymbol{\theta}_k$, and $\Delta\boldsymbol{x}$, because $\dot{\hat{\boldsymbol{\theta}}}_k$ is described by Eq. (3.63).

To prove stability, a Lyapunov-like function candidate is defined as

$$V = \tfrac{1}{2}\boldsymbol{s}^T\boldsymbol{M}(\boldsymbol{q})\boldsymbol{s} + \tfrac{1}{2}\Delta\boldsymbol{\theta}_d^T\boldsymbol{L}_d^{-1}\Delta\boldsymbol{\theta}_d + \tfrac{1}{2}\Delta\boldsymbol{\theta}_k^T\boldsymbol{L}_k^{-1}\Delta\boldsymbol{\theta}_k + \tfrac{1}{2}\Delta\boldsymbol{x}^T(\boldsymbol{K}_p + \alpha\boldsymbol{K}_v)\Delta\boldsymbol{x}, \tag{3.81}$$

where $\Delta\boldsymbol{\theta}_k = \boldsymbol{\theta}_k - \hat{\boldsymbol{\theta}}_k$. Differentiating with respect to time and using *Property* 1.1, we have

$$\dot{V} = \boldsymbol{s}^T\boldsymbol{M}(\boldsymbol{q})\dot{\boldsymbol{s}} + \tfrac{1}{2}\boldsymbol{s}^T\dot{\boldsymbol{M}}(\boldsymbol{q})\boldsymbol{s} - \Delta\boldsymbol{\theta}_d^T\boldsymbol{L}_d^{-1}\dot{\hat{\boldsymbol{\theta}}}_d - \Delta\boldsymbol{\theta}_k^T\boldsymbol{L}_k^{-1}\dot{\hat{\boldsymbol{\theta}}}_k + \Delta\boldsymbol{x}^T(\boldsymbol{K}_p + \alpha\boldsymbol{K}_v)\Delta\dot{\boldsymbol{x}}. \tag{3.82}$$

Substituting $\boldsymbol{M}(\boldsymbol{q})\dot{\boldsymbol{s}}$ from Eq. (3.80), $\dot{\hat{\boldsymbol{\theta}}}_k$ from Eq. (3.63), and $\dot{\hat{\boldsymbol{\theta}}}_d$ from Eq. (3.62) into the above equation, using *Property* 1.2, Eqs. (3.74) and (3.59), we have

$$\begin{aligned} \dot{V} &= -\hat{\boldsymbol{s}}_x^T\boldsymbol{K}_v\Delta\dot{\hat{\boldsymbol{x}}} - \hat{\boldsymbol{s}}_x^T\boldsymbol{K}_p\Delta\boldsymbol{x} + \Delta\boldsymbol{x}^T(\boldsymbol{K}_p + \alpha\boldsymbol{K}_v)\Delta\dot{\boldsymbol{x}} \\ &\quad - \Delta\boldsymbol{\theta}_k^T\boldsymbol{W}_k^T(t)\boldsymbol{K}_v\boldsymbol{W}_k(t)\Delta\boldsymbol{\theta}_k - \Delta\boldsymbol{\theta}_k^T\boldsymbol{Y}_k^T(\boldsymbol{q},\dot{\boldsymbol{q}})(\boldsymbol{K}_p + \alpha\boldsymbol{K}_v)\Delta\boldsymbol{x}. \end{aligned} \tag{3.83}$$

From Eqs. (3.66), (3.31) and (3.64), we have

$$\hat{\boldsymbol{s}}_x = \Delta\dot{\hat{\boldsymbol{x}}} + \alpha\Delta\boldsymbol{x} = \Delta\dot{\boldsymbol{x}} + \alpha\Delta\boldsymbol{x} - \boldsymbol{Y}_k(\boldsymbol{q},\dot{\boldsymbol{q}})\Delta\boldsymbol{\theta}_k, \tag{3.84}$$

where

$$\boldsymbol{Y}_k(\boldsymbol{q},\dot{\boldsymbol{q}})\Delta\boldsymbol{\theta}_k = \boldsymbol{J}(\boldsymbol{q})\dot{\boldsymbol{q}} - \hat{\boldsymbol{J}}(\boldsymbol{q},\hat{\boldsymbol{\theta}}_k)\dot{\boldsymbol{q}} = \dot{\boldsymbol{x}} - \dot{\hat{\boldsymbol{x}}}. \tag{3.85}$$

Substituting $\Delta\dot{\hat{\boldsymbol{x}}} = \Delta\dot{\boldsymbol{x}} - \boldsymbol{Y}_k(\boldsymbol{q},\dot{\boldsymbol{q}})\Delta\boldsymbol{\theta}_k$ and $\hat{\boldsymbol{s}}_x = \Delta\dot{\boldsymbol{x}} + \alpha\Delta\boldsymbol{x} - \boldsymbol{Y}_k(\boldsymbol{q},\dot{\boldsymbol{q}})\Delta\boldsymbol{\theta}_k$ into Eq. (3.83), we have

$$\begin{aligned} \dot{V} &= -\Delta\dot{\boldsymbol{x}}^T\boldsymbol{K}_v\Delta\dot{\boldsymbol{x}} + 2\Delta\dot{\boldsymbol{x}}^T\boldsymbol{K}_v\boldsymbol{Y}_k(\boldsymbol{q},\dot{\boldsymbol{q}})\Delta\boldsymbol{\theta}_k - \Delta\boldsymbol{\theta}_k^T\boldsymbol{Y}_k^T(\boldsymbol{q},\dot{\boldsymbol{q}})\boldsymbol{K}_v\boldsymbol{Y}_k(\boldsymbol{q},\dot{\boldsymbol{q}})\Delta\boldsymbol{\theta}_k \\ &\quad - \alpha\Delta\boldsymbol{x}^T\boldsymbol{K}_p\Delta\boldsymbol{x} - \Delta\boldsymbol{\theta}_k^T\boldsymbol{W}_k^T(t)\boldsymbol{K}_v\boldsymbol{W}_k(t)\Delta\boldsymbol{\theta}_k. \end{aligned} \tag{3.86}$$

Since $\Delta\dot{\hat{x}} = \Delta\dot{x} - Y_k(q, \dot{q})\Delta\theta_k$, the above equation can be simplified to

$$\dot{V} = -\Delta\dot{\hat{x}}^T K_v \Delta\dot{\hat{x}} - \alpha\Delta x^T K_p \Delta x - \Delta\theta_k^T W_k^T(t) K_v W_k(t)\Delta\theta_k. \quad (3.87)$$

We are now in a position to state the following theorem:

Theorem 3.2 *For a finite task space such that the approximate Jacobian matrix is nonsingular, the adaptive Jacobian control law* (3.61) *and the parameter update laws* (3.63) *and* (3.62) *for the robot system* (3.4) *result in the convergence of position and velocity tracking errors. That is,* $x - x_d \to 0$ *and* $\dot{x} - \dot{x}_d \to 0$*, as* $t \to \infty$*. In addition,* $W_k(t)\Delta\theta_k \to 0$ *as* $t \to \infty$.

Proof Since $M(q)$ is uniformly positive definite, V in Eq. (3.81) is positive definite in s, Δx, $\Delta\theta_k$, and $\Delta\theta_d$. Since $\dot{V} \le 0$, V is also bounded, and therefore s, Δx, $\Delta\theta_k$, and $\Delta\theta_d$ are bounded vectors. This implies that $\hat{\theta}_k$, $\hat{\theta}_d$ are bounded, x is bounded if x_d is bounded, and $\hat{s}_x = \hat{J}(q, \hat{\theta}_k)s$ is bounded, as seen from Eq. (3.74). Then we can conclude that $\Delta\dot{\hat{x}}$ is also bounded. Since Δx is bounded, $\dot{x}_r$ in Eq. (3.64) is also bounded if $\dot{x}_d$ is bounded. Therefore, $\dot{q}_r$ in Eq. (3.68) is also bounded if the inverse approximate Jacobian matrix is bounded. From Eq. (3.74), $\dot{q}$ is bounded, and the boundedness of $\dot{q}$ means that $\dot{x}$ is bounded, since the Jacobian matrix is bounded. Hence, $\Delta\dot{x}$ is bounded, and $\ddot{x}_r$ in Eq. (3.65) is also bounded if $\ddot{x}_d$ is bounded. In addition, $\ddot{\hat{x}}_r$ in Eq. (3.70) is bounded, since $\Delta\dot{\hat{x}}$ is bounded. From Eq. (3.63), $\dot{\hat{\theta}}_k$ is bounded, since Δx, $\Delta\theta_k$, $\dot{q}$ are bounded and $Y_k(\cdot)$ is a trigonometric function of q. Therefore, $\ddot{\hat{q}}_r$ in Eq. (3.71) is bounded. From the closed-loop Eq. (3.80), we can conclude that $\dot{s}$ is bounded. The boundedness of $\dot{s}$ implies the boundedness of $\ddot{q}$, as seen from Eq. (3.76). From Eq. (3.67), $\dot{\hat{s}}_x$ is therefore bounded. Differentiating Eq. (3.84) with respect to time and rearranging yields

$$\Delta\ddot{\hat{x}} + \alpha\Delta\dot{x} = \dot{\hat{s}}_x, \quad (3.88)$$

which means that $\Delta\ddot{\hat{x}} = \ddot{\hat{x}} - \ddot{x}_d$ is also bounded.

Next, let us check the uniform continuity of $\dot{V}$. Differentiating Eq. (3.87) with respect to time gives

$$\ddot{V} = -2\Delta\dot{\hat{x}}^T K_v \Delta\ddot{\hat{x}} - 2\alpha\Delta x^T K_p \Delta\dot{x} - 2\Delta\theta_k^T W_k^T(t) K_v (\dot{W}_k(t)\Delta\theta_k - W_k(t)\dot{\hat{\theta}}_k), \quad (3.89)$$

where $W(t)$ and $\dot{W}(t)$ are bounded, since $\dot{q}$, $\ddot{q}$ are bounded. This shows that $\ddot{V}$ is bounded, since Δx, $\Delta\dot{x}$, $\Delta\dot{\hat{x}}$, $\Delta\ddot{\hat{x}}$, $\Delta\theta_k$, $\dot{\hat{\theta}}_k$ are all bounded. Hence, $\dot{V}$ is uniformly continuous. Using Barbalat's Lemma or the Lyapunov-Like Lemma, we have $\Delta x = x - x_d \to 0$, $\Delta\dot{\hat{x}} = \dot{\hat{x}} - \dot{x}_d \to 0$, and $W_k(q)\Delta\theta_k \to 0$ as $t \to \infty$. Finally, differentiating Eq. (3.66) with respect to time and rearranging yields

$$\Delta\ddot{x} + \alpha\Delta\dot{x} = \dot{\hat{s}}_x + \dot{Y}_k(q, \dot{q}, \ddot{q})\Delta\theta_k - Y_k(q, \dot{q})\dot{\hat{\theta}}_k, \quad (3.90)$$

which means that $\Delta\ddot{x} = \ddot{x} - \ddot{x}_d$ is also bounded. Since Δx and $\Delta\ddot{x}$ are bounded, we have $\Delta\dot{x} \to \mathbf{0}$ as $t \to \infty$. △△△

3.4 Adaptive Vision-Based Tracking Control with Uncertain Depth

In this section, we consider adaptive Jacobian tracking control for a robot with uncertain kinematics, dynamics, and camera depth information. The extension to the case of a redundant robot is also shown in this section. When cameras are used to monitor the position of the end effector, the task coordinates are defined as image coordinates. From Eq. (2.62) in Chap. 2, the velocity of the image feature is related to the joint-space velocity by

$$\dot{\boldsymbol{x}} = \boldsymbol{Z}^{-1}(\boldsymbol{q})\boldsymbol{J}(\boldsymbol{q})\dot{\boldsymbol{q}}, \tag{3.91}$$

where $\boldsymbol{J}(\boldsymbol{q}) = \boldsymbol{L}(\boldsymbol{x})\boldsymbol{J}_m(\boldsymbol{q})$ denotes the Jacobian matrix, which includes kinematic parameters and camera parameters, and

$$\boldsymbol{Z}^{-1}(\boldsymbol{q}) = \begin{bmatrix} \frac{1}{z_1(\boldsymbol{q})}\boldsymbol{I}_2, & 0 & \cdots & 0 \\ 0 & \frac{1}{z_2(\boldsymbol{q})}\boldsymbol{I}_2 & \cdots & 0 \\ \vdots & \vdots & \ddots & \vdots \\ 0 & 0 & \cdots & \frac{1}{z_m(\boldsymbol{q})}\boldsymbol{I}_2 \end{bmatrix} \tag{3.92}$$

is a diagonal matrix that contains the depth information of the feature points with respect to the camera image frame.

Note from Eqs. (3.91) and (3.92) that the overall Jacobian matrix $\boldsymbol{Z}^{-1}(\boldsymbol{q})\boldsymbol{J}(\boldsymbol{q})$ is not linearly parameterizable if the depth is not constant, because it is inversely proportional to the depth.

Example 3.1 Consider a 3-DOF anthropomorphic manipulator with a camera installed a distance away from the robot, as illustrated in Fig. 3.7.

The equation of forward kinematics for the robot in Fig. 3.7 is obtained as

$$\begin{aligned} r_1 &= l_2\cos(q_1)\cos(q_2) + l_3\cos(q_1)\cos(q_2+q_3), \\ r_2 &= l_1 + l_2\sin(q_2) + l_3\sin(q_2+q_3), \\ r_3 &= -l_2\sin(q_1)\cos(q_2) - l_3\sin(q_1)\cos(q_2+q_3), \end{aligned} \tag{3.93}$$

where l_1, l_2, l_3 represent the lengths of links, and q_1, q_2, q_3 are joint angles. Based on the forward kinematics, the Jacobian matrix from joint space to Cartesian space is specified as

Fig. 3.7 Visual servoing of a 3-DOF robot

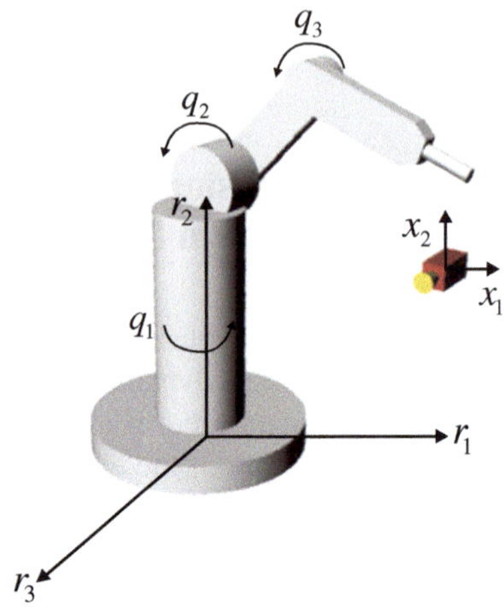

$$J_m(q) = \begin{bmatrix} -s_1(l_2c_2 + l_3c_{23}) & -s_2c_1l_2 - s_{23}l_3c_1 & -s_{23}l_3c_1 \\ 0 & c_2l_2 + c_{23}l_3 & c_{23}l_3 \\ -c_1(l_2c_2 + l_3c_{23}) & s_2s_1l_2 + s_{23}l_3s_1 & s_{23}l_3s_1 \end{bmatrix}, \quad (3.94)$$

where $s_1 = \sin(q_1)$, $c_1 = \cos(q_1)$, $s_2 = \sin(q_2)$, $c_2 = \cos(q_2)$, $s_{23} = \sin(q_2 + q_3)$, and $c_{23} = \cos(q_2 + q_3)$.

The position of the robot end effector in image space is measured with the camera. Based on the pinhole camera model, the image-space position of the end effector is related to the Cartesian-space position of the end effector as follows:

$$\begin{aligned} x_1 &= \tfrac{\beta r_1}{z} = \tfrac{\beta r_1}{d+r_3}, \\ x_2 &= \tfrac{\beta r_2}{z} = \tfrac{\beta r_2}{d+r_3}, \end{aligned} \quad (3.95)$$

where β is a scaling vector, d denotes the distance from the camera frame to the robot base frame, and z describes the depth from the camera frame to the robot end effector, which is defined as

$$z = d + r_3 = d - l_2\sin(q_1)\cos(q_2) - l_3\sin(q_1)\cos(q_2 + q_3). \quad (3.96)$$

It can be seen from Eq. (3.96) that the depth is not constant.

The image Jacobian matrix is specified as

$$J_s(r) = \begin{bmatrix} \frac{\beta}{z} & 0 & -\frac{x_1}{z} \\ 0 & \frac{\beta}{z} & -\frac{x_2}{z} \end{bmatrix}. \quad (3.97)$$

The overall Jacobian matrix that represents the mapping from joint space to image space is specified as

$$\boldsymbol{J}_s(\boldsymbol{r})\boldsymbol{J}_m(\boldsymbol{q}) = \boldsymbol{Z}^{-1}(\boldsymbol{q})\boldsymbol{J}(\boldsymbol{q}) = \boldsymbol{Z}^{-1}(\boldsymbol{q})\boldsymbol{L}(\boldsymbol{x})\boldsymbol{J}_m(\boldsymbol{q})$$
$$= \underbrace{\begin{bmatrix} \frac{1}{z} & 0 \\ 0 & \frac{1}{z} \end{bmatrix}}_{\boldsymbol{Z}^{-1}(\boldsymbol{q})} \underbrace{\begin{bmatrix} \beta & 0 & -x_1 \\ 0 & \beta & -x_2 \end{bmatrix}}_{\boldsymbol{L}(\boldsymbol{x})} \underbrace{\begin{bmatrix} -s_1(l_2c_2+l_3c_{23}) & -s_2c_1l_2-s_{23}l_3c_1 & -s_{23}l_3c_1 \\ 0 & c_2l_2+c_{23}l_3 & c_{23}l_3 \\ -c_1(l_2c_2+l_3c_{23}) & s_2s_1l_2+s_{23}l_3s_1 & s_{23}l_3s_1 \end{bmatrix}}_{\boldsymbol{J}_m(\boldsymbol{q})}, \tag{3.98}$$

which cannot be linearly parameterized, since it is inversely proportional to z. □

Hence the kinematic parameters in the overall Jacobian matrix cannot be extracted to form a lumped kinematic parameter vector $\boldsymbol{\theta}$ that includes the unknown depth parameters, i.e., $\boldsymbol{Z}^{-1}(\boldsymbol{q})\boldsymbol{J}(\boldsymbol{q})\dot{\boldsymbol{q}} \neq \boldsymbol{Y}(\boldsymbol{q}, \dot{\boldsymbol{q}})\boldsymbol{\theta}$. Therefore, the adaptive controllers presented in Sects. 3.2 and 3.3 cannot be used directly unless the depth information is constant or slowly time-varying.

However, both $\boldsymbol{Z}(\boldsymbol{q})\dot{\boldsymbol{x}}$ and $\boldsymbol{J}(\boldsymbol{q})\dot{\boldsymbol{q}}$ are linear in sets of kinematic parameters such as robot link lengths and camera intrinsic and extrinsic parameters. First, from *Property* 1.5, $\boldsymbol{J}(\boldsymbol{q})\dot{\boldsymbol{q}}$ can be expressed as

$$\boldsymbol{J}(\boldsymbol{q})\dot{\boldsymbol{q}} = \boldsymbol{Y}_k(\boldsymbol{q}, \dot{\boldsymbol{q}})\boldsymbol{\theta}_k, \tag{3.99}$$

where $\boldsymbol{Y}_k(\boldsymbol{q}, \dot{\boldsymbol{q}}) \in \Re^{2m\times n_k}$ is the kinematic regressor matrix, and $\boldsymbol{\theta}_k = (\theta_{k1}, \ldots, \theta_{kn_k})^T \in \Re^{n_k}$ represent the unknown kinematic parameters.

Next, we have the following property:

Property 3.1 *The product* $\boldsymbol{Z}(\boldsymbol{q})\dot{\boldsymbol{x}}$ *can be linearly parameterized as*

$$\boldsymbol{Z}(\boldsymbol{q})\dot{\boldsymbol{x}} = \boldsymbol{Y}_z(\boldsymbol{q}, \dot{\boldsymbol{x}})\boldsymbol{\theta}_z, \tag{3.100}$$

where $\boldsymbol{Y}_z(\boldsymbol{q}, \dot{\boldsymbol{x}}) \in \Re^{2m\times n_z}$ *is called the* depth regressor matrix, *and* $\boldsymbol{\theta}_z = (\theta_{z1}, \ldots, \theta_{zn_z})^T \in \Re^{n_z}$ *represents the unknown depth parameters.* □

In this section, the above properties in Eqs. (3.100) and (3.99) will be explored to introduce an adaptive Jacobian tracking controller with adaptation to the uncertain parameters of the depth information even if it appears nonlinearly in the overall Jacobian matrix.

Example 3.2 Consider the 3-DOF manipulator in Example 3.1. The term $\boldsymbol{Z}(\boldsymbol{q})\dot{\boldsymbol{x}}$ can be written as

$$\boldsymbol{Z}(\boldsymbol{q})\dot{\boldsymbol{x}} = \begin{bmatrix} z\dot{x}_1 \\ z\dot{x}_2 \end{bmatrix} = \begin{bmatrix} (d + l_2\sin(q_1)\cos(q_2) + l_3\sin(q_1)\cos(q_2+q_3))\dot{x}_1 \\ (d + l_2\sin(q_1)\cos(q_2) + l_3\sin(q_1)\cos(q_2+q_3))\dot{x}_2 \end{bmatrix}$$
$$= \boldsymbol{Y}_z(\boldsymbol{q}, \dot{\boldsymbol{x}})\boldsymbol{\theta}_z, \tag{3.101}$$

where

$$Y_z(\boldsymbol{q}, \dot{\boldsymbol{x}}) = \begin{bmatrix} \dot{x}_1 & s_1c_2\dot{x}_1 & s_1c_{23}\dot{x}_1 \\ \dot{x}_2 & s_1c_2\dot{x}_2 & s_1c_{23}\dot{x}_2 \end{bmatrix}, \tag{3.102}$$

$$\boldsymbol{\theta}_z = [d, l_2, l_3]^T, \tag{3.103}$$

and the term $\boldsymbol{J}(\boldsymbol{q})\dot{\boldsymbol{q}}$ can be written as $\boldsymbol{J}(\boldsymbol{q})\dot{\boldsymbol{q}} = \boldsymbol{Y}_k(\boldsymbol{q}, \dot{\boldsymbol{q}})\boldsymbol{\theta}_k$, where

$$\boldsymbol{Y}_k(\boldsymbol{q}, \dot{\boldsymbol{q}}) =$$
$$\begin{bmatrix} -s_1c_2\dot{q}_1 - c_1s_2\dot{q}_2 & -s_1c_{23}\dot{q}_1 - c_1s_{23}(\dot{q}_2+\dot{q}_3) & c_1c_2x_1\dot{q}_1 - s_1s_2x_1\dot{q}_2 & c_1c_{23}x_1\dot{q}_1 - s_1s_{23}x_1(\dot{q}_2+\dot{q}_3) \\ c_2\dot{q}_2 & c_{23}(\dot{q}_2+\dot{q}_3) & c_1c_2x_2\dot{q}_1 - s_1s_2x_2\dot{q}_2 & c_1c_{23}x_2\dot{q}_1 - s_1s_{23}x_2(\dot{q}_2+\dot{q}_3) \end{bmatrix},$$
$$\boldsymbol{\theta}_k = [\beta l_2, \beta l_3, l_2, l_3]^T. \tag{3.104}$$

□

In the presence of kinematic and camera parameter uncertainties, the estimated image-space velocities vector is obtained as

$$\dot{\hat{\boldsymbol{x}}} = \hat{\boldsymbol{Z}}^{-1}(\boldsymbol{q}, \hat{\boldsymbol{\theta}}_z)\hat{\boldsymbol{J}}(\boldsymbol{q}, \hat{\boldsymbol{\theta}}_k)\dot{\boldsymbol{q}}, \tag{3.105}$$

where $\dot{\hat{\boldsymbol{x}}} = [\dot{\hat{\boldsymbol{x}}}_1^T, \dot{\hat{\boldsymbol{x}}}_2^T, \ldots, \dot{\hat{\boldsymbol{x}}}_m^T]^T \in \Re^{2m}$, $\dot{\hat{\boldsymbol{x}}}_i$ denotes the approximate image velocity of the ith feature point, and m is the total number of feature points. The matrices $\hat{\boldsymbol{Z}}(\boldsymbol{q}, \hat{\boldsymbol{\theta}}_z) \in \Re^{2m\times 2m}$ and $\hat{\boldsymbol{J}}(\boldsymbol{q}, \hat{\boldsymbol{\theta}}_k) \in \Re^{2m\times n}$ are the approximate matrices for $\boldsymbol{Z}(\boldsymbol{q})$ and $\boldsymbol{J}(\boldsymbol{q})$ respectively, which are defined as

$$\hat{\boldsymbol{Z}}^{-1}(\boldsymbol{q}, \hat{\boldsymbol{\theta}}_z) = \begin{bmatrix} \frac{1}{\hat{z}_1(\boldsymbol{q}, \hat{\boldsymbol{\theta}}_{z1})}\boldsymbol{I}_2, & 0 & \cdots & 0 \\ 0 & \frac{1}{\hat{z}_2(\boldsymbol{q}, \hat{\boldsymbol{\theta}}_{z2})}\boldsymbol{I}_2 & \cdots & 0 \\ \vdots & \vdots & \ddots & \vdots \\ 0 & 0 & \cdots & \frac{1}{\hat{z}_m(\boldsymbol{q}, \hat{\boldsymbol{\theta}}_{zm})}\boldsymbol{I}_2 \end{bmatrix}, \tag{3.106}$$

where $\boldsymbol{I}_2 \in \Re^{2\times 2}$ represents the identity matrix, and

$$\hat{\boldsymbol{J}}(\boldsymbol{q}, \hat{\boldsymbol{\theta}}_k) = \begin{bmatrix} \hat{\boldsymbol{J}}_1(\boldsymbol{q}, \hat{\boldsymbol{\theta}}_{k1}) & \hat{\boldsymbol{J}}_2(\boldsymbol{q}, \hat{\boldsymbol{\theta}}_{k2}) & \cdots & \hat{\boldsymbol{J}}_m(\boldsymbol{q}, \hat{\boldsymbol{\theta}}_{km}) \end{bmatrix}^T, \tag{3.107}$$

where $\hat{\boldsymbol{J}}_i(\boldsymbol{q}, \hat{\boldsymbol{\theta}}_{ki})$ is the approximate Jacobian matrix corresponding to the ith feature point.

From Eqs. (3.100) and (3.99), we can obtain the following properties of linear parameterization as

$$\hat{\boldsymbol{Z}}(\boldsymbol{q}, \hat{\boldsymbol{\theta}}_z)\dot{\boldsymbol{x}} = \boldsymbol{Y}_z(\boldsymbol{q}, \dot{\boldsymbol{x}})\hat{\boldsymbol{\theta}}_z, \tag{3.108}$$

$$\hat{\boldsymbol{J}}(\boldsymbol{q}, \hat{\boldsymbol{\theta}}_k)\dot{\boldsymbol{q}} = \boldsymbol{Y}_k(\boldsymbol{q}, \dot{\boldsymbol{q}})\hat{\boldsymbol{\theta}}_k, \tag{3.109}$$

where $\hat{\boldsymbol{\theta}}_k$ denote the estimated parameters of $\boldsymbol{J}(\boldsymbol{q})$, and $\hat{\boldsymbol{\theta}}_z$ represent the estimated depth parameters of $\boldsymbol{Z}(\boldsymbol{q})$.

To construct the adaptive Jacobian tracking controller, an adaptive image-space sliding vector is first defined as

$$\hat{\boldsymbol{s}}_x = \dot{\hat{\boldsymbol{x}}} - \dot{\boldsymbol{x}}_r, \tag{3.110}$$

where

$$\dot{\boldsymbol{x}}_r = \dot{\boldsymbol{x}}_d - \alpha(\boldsymbol{x} - \boldsymbol{x}_d) \tag{3.111}$$

is a reference vector defined in image space. Next, let

$$\dot{\boldsymbol{q}}_r = \hat{\boldsymbol{J}}^+(\boldsymbol{q}, \hat{\boldsymbol{\theta}}_k)\hat{\boldsymbol{Z}}(\boldsymbol{q}, \hat{\boldsymbol{\theta}}_z)\dot{\boldsymbol{x}}_r + (\boldsymbol{I}_n - \hat{\boldsymbol{J}}^+(\boldsymbol{q}, \hat{\boldsymbol{\theta}}_k)\hat{\boldsymbol{J}}(\boldsymbol{q}, \hat{\boldsymbol{\theta}}_k))\boldsymbol{\psi}, \tag{3.112}$$

where $\boldsymbol{\psi} \in \Re^n$ is minus the gradient of the convex function to be optimized, and $\boldsymbol{I}_n \in \Re^{n\times n}$ represents the identity matrix.

Next, an adaptive sliding vector in joint space is defined as

$$\boldsymbol{s} = \dot{\boldsymbol{q}} - \dot{\boldsymbol{q}}_r. \tag{3.113}$$

From Eqs. (3.112) and (3.113), the sliding vector in joint space $\boldsymbol{s}$ can be related to the image-space sliding vector $\hat{\boldsymbol{s}}_x$ by

$$\hat{\boldsymbol{Z}}^{-1}(\boldsymbol{q}, \hat{\boldsymbol{\theta}}_z)\hat{\boldsymbol{J}}(\boldsymbol{q}, \hat{\boldsymbol{\theta}}_k)\boldsymbol{s} = \hat{\boldsymbol{Z}}^{-1}(\boldsymbol{q}, \hat{\boldsymbol{\theta}}_z)\hat{\boldsymbol{J}}(\boldsymbol{q}, \hat{\boldsymbol{\theta}}_k)\dot{\boldsymbol{q}} - \dot{\boldsymbol{x}}_r = \dot{\hat{\boldsymbol{x}}} - \dot{\boldsymbol{x}}_r = \hat{\boldsymbol{s}}_x. \tag{3.114}$$

Differentiating Eq. (3.113) with respect to time, the time derivative of $\boldsymbol{s}$ is specified as

$$\dot{\boldsymbol{s}} = \ddot{\boldsymbol{q}} - \ddot{\boldsymbol{q}}_r. \tag{3.115}$$

The dynamic equation can be written using the sliding vector $\boldsymbol{s}$ and its time derivative as

$$\boldsymbol{M}(\boldsymbol{q})\dot{\boldsymbol{s}} + \left(\tfrac{1}{2}\dot{\boldsymbol{M}}(\boldsymbol{q}) + \boldsymbol{S}(\boldsymbol{q}, \dot{\boldsymbol{q}})\right)\boldsymbol{s} + \boldsymbol{Y}_d(\boldsymbol{q}, \dot{\boldsymbol{q}}, \dot{\boldsymbol{q}}_r, \ddot{\boldsymbol{q}}_r)\boldsymbol{\theta}_d = \boldsymbol{\tau}. \tag{3.116}$$

We can now design the adaptive Jacobian tracking controller with uncertain kinematics and depth information, which consists of
(i) a control law based on the approximate Jacobian matrices and approximate depth information as

$$\boldsymbol{\tau} = -\hat{\boldsymbol{J}}^T(\boldsymbol{q}, \hat{\boldsymbol{\theta}}_k)\hat{\boldsymbol{Z}}^{-1}(\boldsymbol{q}, \hat{\boldsymbol{\theta}}_z)(\boldsymbol{K}_v\Delta\dot{\boldsymbol{x}} + \boldsymbol{K}_p\Delta\boldsymbol{x}) + \boldsymbol{Y}_d(\boldsymbol{q}, \dot{\boldsymbol{q}}, \dot{\boldsymbol{q}}_r, \ddot{\boldsymbol{q}}_r)\hat{\boldsymbol{\theta}}_d, \tag{3.117}$$

(ii) a dynamic adaptation law

$$\dot{\hat{\theta}}_d = -L_d Y_d^T(q, \dot{q}, \dot{q}_r, \ddot{q}_r)s, \tag{3.118}$$

(iii) a kinematic adaptation law

$$\dot{\hat{\theta}}_k = L_k Y_k^T(q, \dot{q})\hat{Z}^{-1}(q, \hat{\theta}_z)(K_v \Delta\dot{x} + K_p \Delta x), \tag{3.119}$$

(iv) a depth adaptation law

$$\dot{\hat{\theta}}_z = -L_z Y_z^T(q, \dot{x})\hat{Z}^{-1}(q, \hat{\theta}_z)(K_v \Delta\dot{x} + K_p \Delta x), \tag{3.120}$$

where L_k, L_d, and L_z are symmetric positive definite matrices. A block diagram of the controller in Eq. (3.117) is illustrated in Fig. 3.8.

Comparing with the adaptive Jacobian controller in Sect. 3.2, we note that an additional update law (3.120) is now added to update the estimated depth parameters $\hat{\theta}_z$, but the main structure of the controller is still preserved. The design concept is thus a modular approach for analysis of adaptive control systems with multiple uncertain parameters. The linear parameterization is obtained from the system model described by Eq. (3.99), and the depth regressor is used to design the depth adaptation law (3.120) to update the estimated depth parameters $\hat{\theta}_z$ separately but concurrently with the other estimated parameters. Though $\hat{\theta}_z$ is nonlinear in the control variables, it is a state variable that can be used in the control law, since linear parameterization can be obtained elsewhere from the system model.

The closed-loop dynamics is obtained by substituting (3.117) into Eq. (3.116) to give

$$\begin{aligned} M(q)\dot{s} + \left(\tfrac{1}{2}\dot{M}(q) + S(q, \dot{q})\right)s + Y_d(q, \dot{q}, \dot{q}_r, \ddot{q}_r)\Delta\theta_d \\ + \hat{J}^T(q, \hat{\theta}_k)\hat{Z}^{-1}(q, \hat{\theta}_z)(K_v \Delta\dot{x} + K_p \Delta x) = \mathbf{0}. \end{aligned} \tag{3.121}$$

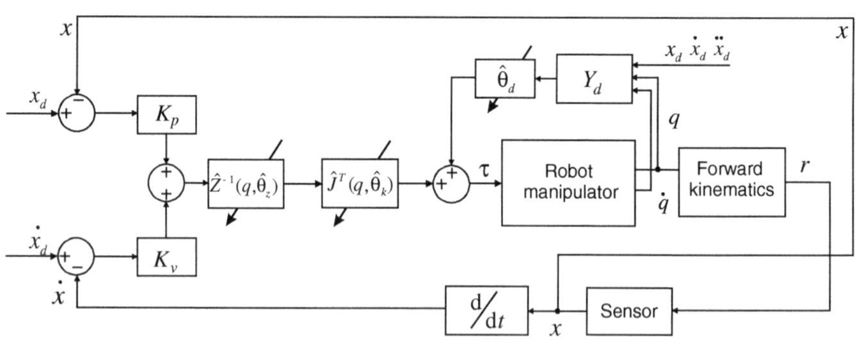

Fig. 3.8 Block diagram of an adaptive Jacobian control system with uncertain depth information

To prove stability, a Lyapunov-like function candidate is designed as

$$
\begin{aligned}
V = &\tfrac{1}{2}\boldsymbol{s}^T \boldsymbol{M}(\boldsymbol{q})\boldsymbol{s} + \tfrac{1}{2}\Delta\boldsymbol{\theta}_d^T \boldsymbol{L}_d^{-1}\Delta\boldsymbol{\theta}_d + \tfrac{1}{2}\Delta\boldsymbol{\theta}_k^T \boldsymbol{L}_k^{-1}\Delta\boldsymbol{\theta}_k \\
&+\tfrac{1}{2}\Delta\boldsymbol{\theta}_z^T \boldsymbol{L}_z^{-1}\Delta\boldsymbol{\theta}_z + \tfrac{1}{2}\Delta\boldsymbol{x}^T(\boldsymbol{K}_p + \alpha\boldsymbol{K}_v)\Delta\boldsymbol{x},
\end{aligned}
\tag{3.122}
$$

Differentiating with respect to time, substituting Eqs. (3.114), (3.118), (3.119), (3.120) and (3.121) yields,

$$
\begin{aligned}
\dot{V} = &-\hat{\boldsymbol{s}}_x^T(\boldsymbol{K}_v\Delta\dot{\boldsymbol{x}} + \boldsymbol{K}_p\Delta\boldsymbol{x}) - \boldsymbol{s}^T \boldsymbol{Y}_d(\boldsymbol{q}, \dot{\boldsymbol{q}}, \dot{\boldsymbol{q}}_r, \ddot{\boldsymbol{q}}_r)\Delta\boldsymbol{\theta}_d \\
&- \Delta\boldsymbol{\theta}_d^T \boldsymbol{L}_d^{-1}\dot{\hat{\boldsymbol{\theta}}}_d - \Delta\boldsymbol{\theta}_k^T \boldsymbol{L}_k^{-1}\dot{\hat{\boldsymbol{\theta}}}_k - \Delta\boldsymbol{\theta}_z^T \boldsymbol{L}_z^{-1}\dot{\hat{\boldsymbol{\theta}}}_z + \Delta\boldsymbol{x}^T(\alpha\boldsymbol{K}_v + \boldsymbol{K}_p)\Delta\dot{\boldsymbol{x}}.
\end{aligned}
\tag{3.123}
$$

From Eqs. (3.111), (3.105), (3.110) and (3.100), we have

$$
\begin{aligned}
\hat{\boldsymbol{s}}_x &= \hat{\boldsymbol{Z}}^{-1}(\boldsymbol{q}, \hat{\boldsymbol{\theta}}_z)\hat{\boldsymbol{J}}(\boldsymbol{q}, \hat{\boldsymbol{\theta}}_k)\dot{\boldsymbol{q}} - \dot{\boldsymbol{x}}_d + \alpha\Delta\boldsymbol{x} \\
&= \hat{\boldsymbol{Z}}^{-1}(\boldsymbol{q}, \hat{\boldsymbol{\theta}}_z)[-\boldsymbol{Y}_k(\boldsymbol{q}, \dot{\boldsymbol{q}})\Delta\boldsymbol{\theta}_k + \boldsymbol{Y}_z(\boldsymbol{q}, \dot{\boldsymbol{x}})\Delta\boldsymbol{\theta}_z + \hat{\boldsymbol{Z}}(\boldsymbol{q}, \hat{\boldsymbol{\theta}}_z)(\Delta\dot{\boldsymbol{x}} + \alpha\Delta\boldsymbol{x})].
\end{aligned}
\tag{3.124}
$$

Substituting parameter update laws (3.118)–(3.120), and Eq. (3.124) into Eq. (3.123) yields,

$$
\dot{V} \leq -\Delta\dot{\boldsymbol{x}}^T \boldsymbol{K}_v\Delta\dot{\boldsymbol{x}} - \alpha\Delta\boldsymbol{x}^T \boldsymbol{K}_p\Delta\boldsymbol{x} \leq 0. \tag{3.125}
$$

Theorem 3.3 *The adaptive Jacobian tracking control law* (3.117) *and the parameter update laws* (3.118)–(3.120) *guarantee the stability of the adaptive control system, and the position and velocity tracking errors of the m feature points converge to* 0 as $t \to \infty$.

Proof Since $\boldsymbol{M}(\boldsymbol{q})$ is uniformly positive definite, V in Eq. (3.122) is positive definite in $\boldsymbol{s}$, $\Delta\boldsymbol{x}$, $\Delta\boldsymbol{\theta}_k$, $\Delta\boldsymbol{\theta}_d$, and $\Delta\boldsymbol{\theta}_z$. Since $\dot{V} \leq 0$, V is bounded, and therefore $\boldsymbol{s}$, $\Delta\boldsymbol{x}$, $\Delta\boldsymbol{\theta}_k$, $\Delta\boldsymbol{\theta}_d$, and $\Delta\boldsymbol{\theta}_z$ are bounded vectors. This implies that $\hat{\boldsymbol{\theta}}_k$, $\hat{\boldsymbol{\theta}}_d$, and $\hat{\boldsymbol{\theta}}_z$ are bounded, $\boldsymbol{x}$ is bounded if $\boldsymbol{x}_d$ is bounded, and $\hat{\boldsymbol{s}}_x = \hat{\boldsymbol{J}}(\boldsymbol{q}, \hat{\boldsymbol{\theta}}_k)\boldsymbol{s}$ is also bounded. Since $\Delta\boldsymbol{x}$ is bounded, $\dot{\boldsymbol{x}}_r$ in Eq. (3.111) is bounded if $\dot{\boldsymbol{x}}_d$ is bounded. Therefore, $\dot{\boldsymbol{q}}_r$ in Eq. (3.112) is also bounded if the approximate Jacobian matrix is of full rank. From Eq. (3.113), $\dot{\boldsymbol{q}}$ is bounded, and the boundedness of $\dot{\boldsymbol{q}}$ means that $\dot{\boldsymbol{x}}$ is bounded, since the Jacobian matrix is bounded. Hence, $\Delta\dot{\boldsymbol{x}}$ is bounded, and $\ddot{\boldsymbol{x}}_r$ is also bounded if $\ddot{\boldsymbol{x}}_d$ is bounded. From Eq. (3.119), $\dot{\hat{\boldsymbol{\theta}}}_k$ is therefore bounded, since $\Delta\boldsymbol{x}$, $\Delta\dot{\boldsymbol{x}}$, $\dot{\boldsymbol{q}}$ are bounded and $\boldsymbol{Y}_k(\cdot)$ is a trigonometric function of $\boldsymbol{q}$. Similarly, $\dot{\hat{\boldsymbol{\theta}}}_z$ is bounded from Eq. (3.120), and $\boldsymbol{Y}_z(\boldsymbol{q})$ is a trigonometric function of $\boldsymbol{q}$. Therefore, $\ddot{\boldsymbol{q}}_r$ is bounded. From the closed-loop equation (3.121), we can conclude that $\dot{\boldsymbol{s}}$ is bounded. The boundedness of $\dot{\boldsymbol{s}}$ implies the boundedness of $\ddot{\boldsymbol{q}}$, as can be seen from Eq. (3.115). Then $\ddot{\boldsymbol{x}} = \dot{\boldsymbol{J}}(\boldsymbol{q})\dot{\boldsymbol{q}} + \boldsymbol{J}(\boldsymbol{q})\ddot{\boldsymbol{q}}$ is bounded, which means that $\Delta\ddot{\boldsymbol{x}} = \ddot{\boldsymbol{x}} - \ddot{\boldsymbol{x}}_d$ is also bounded if $\ddot{\boldsymbol{x}}_d$ is bounded. From the boundedness of $\Delta\ddot{x}$ and $\Delta\dot{x}$, we can conclude that $\Delta\dot{x}$ and Δx are uniformly continuous. Since V is bounded, it is clear from inequality (3.125) that Δx and $\Delta\dot{x} \in L_2(0, \infty)$. Then, we have $\Delta x = x - x_d \rightarrow 0$ and $\Delta\dot{x} = \dot{x} - \dot{x}_d \rightarrow 0$ as $t \rightarrow \infty$. $\triangle\triangle\triangle$

3.5 Adaptive Jacobian Tracking Control with Uncertain Actuator Model

In the previous section, we have seen that the update law for uncertain depth parameters can be added and updated concurrently while the structure and properties of the original adaptive control systems are preserved and the stability of the overall system is ensured. In this section, we show that the design concept can be applied to vision-based control with an uncertain actuator model. We consider the adaptive Jacobian tracking control problem for a robot with uncertain kinematics, dynamics, camera depth information, and actuator model and show that an additional adaptation law can be constructed to update the estimated actuator parameters while preserving the structure of the adaptive vision-based control system in the previous section.

If a DC motor driven by an amplifier is used as an actuator at each joint of the robot, the dynamics of the robot with n degrees of freedom can be expressed as

$$M(q)\ddot{q} + \left(\tfrac{1}{2}\dot{M}(q) + S(q, \dot{q})\right)\dot{q} + g(q) = Ku, \tag{3.126}$$

where $u \in \Re^n$ is either a voltage or current input to the amplifiers, and $K \in \Re^{n\times n}$ is a diagonal transmission matrix that relates the actuator input u to the control torque τ. That is, $\tau = Ku$.

If the transmission matrix K is known exactly, the controller can be derived using the inverse transmission matrix such that $u = K^{-1}\tau$, where the control torque τ can be specified as Eq. (3.117), to deal with the uncertainties in dynamics, kinematics, and depth information.

Using $\tau = Ku$, Eq. (3.116) can be expressed as

$$M(q)\dot{s} + \left(\tfrac{1}{2}\dot{M}(q) + S(q, \dot{q})\right)s + Y_d(q, \dot{q}, \dot{q}_r, \ddot{q}_r)\theta_d = Ku, \tag{3.127}$$

where the sliding vector s is defined similarly as in Eq. (3.113).

In the presence of uncertainty in the actuator model, the transmission matrix is estimated as $\hat{K}$. We now develop the adaptive Jacobian tracking control scheme using the uncertain transmission matrix $\hat{K}$, which is composed of

(i) a control law based on an approximate transmission matrix $\hat{\boldsymbol{K}} \in \Re^{n \times n}$ as

$$\begin{aligned}\boldsymbol{u} = & -\hat{\boldsymbol{K}}^{-1}\hat{\boldsymbol{J}}^T(\boldsymbol{q}, \hat{\boldsymbol{\theta}}_k)\hat{\boldsymbol{Z}}^{-1}(\boldsymbol{q}, \hat{\boldsymbol{\theta}}_z)(\boldsymbol{K}_v\Delta\dot{\boldsymbol{x}} + \boldsymbol{K}_p\Delta\boldsymbol{x}) \\ & + \hat{\boldsymbol{K}}^{-1}\boldsymbol{Y}_d(\boldsymbol{q}, \dot{\boldsymbol{q}}, \dot{\boldsymbol{q}}_r, \ddot{\boldsymbol{q}}_r)\hat{\boldsymbol{\theta}}_d + \hat{\boldsymbol{K}}^{-1}\boldsymbol{Y}_a(\boldsymbol{\tau}_o)\hat{\boldsymbol{\theta}}_a,\end{aligned} \tag{3.128}$$

(ii) a dynamic adaptation law

$$\dot{\hat{\boldsymbol{\theta}}}_d = -\boldsymbol{L}_d\boldsymbol{Y}_d^T(\boldsymbol{q}, \dot{\boldsymbol{q}}, \dot{\boldsymbol{q}}_r, \ddot{\boldsymbol{q}}_r)\boldsymbol{s}, \tag{3.129}$$

(iii) a kinematic adaptation law

$$\dot{\hat{\boldsymbol{\theta}}}_k = \boldsymbol{L}_k\boldsymbol{Y}_k^T(\boldsymbol{q}, \dot{\boldsymbol{q}})(\boldsymbol{K}_v\Delta\dot{\boldsymbol{x}} + \boldsymbol{K}_p\Delta\boldsymbol{x}), \tag{3.130}$$

(iv) a depth adaptation law

$$\dot{\hat{\boldsymbol{\theta}}}_z = -\boldsymbol{\Phi}\boldsymbol{L}_z\boldsymbol{Y}_z^T(\boldsymbol{q}, \dot{\boldsymbol{x}})\hat{\boldsymbol{Z}}^{-1}(\boldsymbol{q}, \hat{\boldsymbol{\theta}}_z)(\boldsymbol{K}_v\Delta\dot{\boldsymbol{x}} + \boldsymbol{K}_p\Delta\boldsymbol{x}), \tag{3.131}$$

(v) an actuator adaptation law

$$\dot{\hat{\boldsymbol{\theta}}}_a = -\boldsymbol{L}_a\boldsymbol{Y}_a(\boldsymbol{\tau}_o)\boldsymbol{s}, \tag{3.132}$$

where $\boldsymbol{L}_a \in \Re^{n \times n}$ is a positive definite matrix, $\boldsymbol{Y}_a(\boldsymbol{\tau}_o) = diag\{-\tau_{o1}, -\tau_{o2}, \ldots, -\tau_{on}\}$, and τ_{oi} denotes the ith element of the vector $\boldsymbol{\tau}_o$, which is defined as

$$\boldsymbol{\tau}_o = \hat{\boldsymbol{J}}^T(\boldsymbol{q}, \hat{\boldsymbol{\theta}}_k)\hat{\boldsymbol{Z}}^{-1}(\boldsymbol{q}, \hat{\boldsymbol{\theta}}_z)(\boldsymbol{K}_v\Delta\dot{\boldsymbol{x}} + \boldsymbol{K}_p\Delta\boldsymbol{x}) - \boldsymbol{Y}_d(\boldsymbol{q}, \dot{\boldsymbol{q}}, \dot{\boldsymbol{q}}_r, \ddot{\boldsymbol{q}}_r)\hat{\boldsymbol{\theta}}_d. \tag{3.133}$$

In the adaptive control law (3.128), a constant $\hat{\boldsymbol{K}}^{-1}$ is used to transform the control torque to an approximate actuator input. It can be seen that an additional update law for the uncertain actuator parameters is now added and updated concurrently, while still preserving the structure of the adaptive vision-based control system in the previous section. A block diagram of the control system is illustrated in Fig. 3.9.

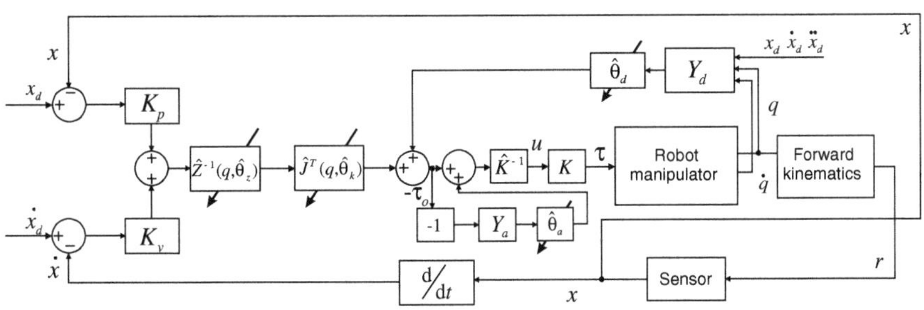

Fig. 3.9 Block diagram of an adaptive Jacobian control system with uncertain actuator model

The closed-loop dynamics is obtained by substituting (3.128) into Eq. (3.127) to give

$$\begin{aligned} &\boldsymbol{M}(\boldsymbol{q})\dot{\boldsymbol{s}} + \left(\tfrac{1}{2}\dot{\boldsymbol{M}}(\boldsymbol{q}) + \boldsymbol{S}(\boldsymbol{q},\dot{\boldsymbol{q}})\right)\boldsymbol{s} + \boldsymbol{Y}_d(\boldsymbol{q},\dot{\boldsymbol{q}},\dot{\boldsymbol{q}}_r,\ddot{\boldsymbol{q}}_r)\Delta\boldsymbol{\theta}_d \\ &+\hat{\boldsymbol{J}}^T(\boldsymbol{q},\hat{\boldsymbol{\theta}}_k)\hat{\boldsymbol{Z}}^{-1}(\boldsymbol{q},\hat{\boldsymbol{\theta}}_z)(\boldsymbol{K}_v\Delta\dot{\boldsymbol{x}} + \boldsymbol{K}_p\Delta\boldsymbol{x}) \\ &+(\boldsymbol{K}\hat{\boldsymbol{K}}^{-1} - \boldsymbol{I}_n)\boldsymbol{\tau}_o - \boldsymbol{K}\hat{\boldsymbol{K}}^{-1}\boldsymbol{Y}_a(\boldsymbol{\tau}_o)\hat{\boldsymbol{\theta}}_a = \boldsymbol{0}, \end{aligned} \tag{3.134}$$

where $\boldsymbol{\tau}_o$ is defined in Eq. (3.133).

Since $\boldsymbol{K}$, $\hat{\boldsymbol{K}}$, and $\boldsymbol{Y}_a(\boldsymbol{\tau}_o)$ are diagonal matrices, the last two terms of Eq. (3.134) can be expressed as

$$(\boldsymbol{K}\hat{\boldsymbol{K}}^{-1} - \boldsymbol{I}_n)\boldsymbol{\tau}_o - \boldsymbol{K}\hat{\boldsymbol{K}}^{-1}\boldsymbol{Y}_a(\boldsymbol{\tau}_o)\hat{\boldsymbol{\theta}}_a = \boldsymbol{Y}_a(\boldsymbol{\tau}_o)(\bar{\boldsymbol{\theta}}_a - \boldsymbol{K}\hat{\boldsymbol{K}}^{-1}\hat{\boldsymbol{\theta}}_a), \tag{3.135}$$

where $\bar{\theta}_{ai} = 1 - \frac{k_i}{\hat{k}_i}$ and $k_i, \hat{k}_i$ are the ith diagonal elements of $\boldsymbol{K}, \hat{\boldsymbol{K}}$ respectively.

Substituting Eq. (3.135) into (3.134), we have

$$\begin{aligned} &\boldsymbol{M}(\boldsymbol{q})\dot{\boldsymbol{s}} + \left(\tfrac{1}{2}\dot{\boldsymbol{M}}(\boldsymbol{q}) + \boldsymbol{S}(\boldsymbol{q},\dot{\boldsymbol{q}})\right)\boldsymbol{s} + \boldsymbol{Y}_d(\boldsymbol{q},\dot{\boldsymbol{q}},\dot{\boldsymbol{q}}_r,\ddot{\boldsymbol{q}}_r)\Delta\boldsymbol{\theta}_d \\ &+\hat{\boldsymbol{J}}^T(\boldsymbol{q},\hat{\boldsymbol{\theta}}_k)\hat{\boldsymbol{Z}}^{-1}(\boldsymbol{q},\hat{\boldsymbol{\theta}}_z)(\boldsymbol{K}_v\Delta\dot{\boldsymbol{x}} + \boldsymbol{K}_p\Delta\boldsymbol{x}) + \boldsymbol{Y}_a(\boldsymbol{\tau}_o)\Delta\bar{\boldsymbol{\theta}}_a = \boldsymbol{0}, \end{aligned} \tag{3.136}$$

where $\Delta\bar{\boldsymbol{\theta}}_a = \bar{\boldsymbol{\theta}}_a - \boldsymbol{K}\hat{\boldsymbol{K}}^{-1}\hat{\boldsymbol{\theta}}_a$ and hence $\Delta\dot{\bar{\boldsymbol{\theta}}}_a = -\boldsymbol{K}\hat{\boldsymbol{K}}^{-1}\dot{\hat{\boldsymbol{\theta}}}_a$.

To prove stability, a Lyapunov-like function candidate given as

$$\begin{aligned} V = &\tfrac{1}{2}\boldsymbol{s}^T\boldsymbol{M}(\boldsymbol{q})\boldsymbol{s} + \tfrac{1}{2}\Delta\boldsymbol{\theta}_d^T\boldsymbol{L}_d^{-1}\Delta\boldsymbol{\theta}_d + \tfrac{1}{2}\Delta\boldsymbol{\theta}_k^T\boldsymbol{L}_k^{-1}\Delta\boldsymbol{\theta}_k + \tfrac{1}{2}\Delta\boldsymbol{\theta}_z^T\boldsymbol{L}_z^{-1}\Delta\boldsymbol{\theta}_z \\ &+\tfrac{1}{2}\Delta\bar{\boldsymbol{\theta}}_a^T\boldsymbol{L}_a^{-1}\hat{\boldsymbol{K}}\boldsymbol{K}^{-1}\Delta\bar{\boldsymbol{\theta}}_a + \tfrac{1}{2}\Delta\boldsymbol{x}^T(\boldsymbol{K}_p + \alpha\boldsymbol{K}_v)\Delta\boldsymbol{x}, \end{aligned} \tag{3.137}$$

Differentiating Eq. (3.137) with respect to time and using *Property* 1.1, we have

$$\begin{aligned} \dot{V} = &\boldsymbol{s}^T\boldsymbol{M}(\boldsymbol{q})\dot{\boldsymbol{s}} + \tfrac{1}{2}\boldsymbol{s}^T\dot{\boldsymbol{M}}(\boldsymbol{q})\boldsymbol{s} - \Delta\boldsymbol{\theta}_d^T\boldsymbol{L}_d^{-1}\dot{\hat{\boldsymbol{\theta}}}_d - \Delta\boldsymbol{\theta}_k^T\boldsymbol{L}_k^{-1}\dot{\hat{\boldsymbol{\theta}}}_k - \Delta\boldsymbol{\theta}_z^T\boldsymbol{L}_z^{-1}\dot{\hat{\boldsymbol{\theta}}}_z \\ &-\Delta\bar{\boldsymbol{\theta}}_a^T\boldsymbol{L}_a^{-1}\dot{\hat{\boldsymbol{\theta}}}_a + \Delta\boldsymbol{x}^T(\boldsymbol{K}_p + \alpha\boldsymbol{K}_v)\Delta\dot{\boldsymbol{x}}. \end{aligned} \tag{3.138}$$

Substituting $\boldsymbol{M}(\boldsymbol{q})\dot{\boldsymbol{s}}$ from Eq. (3.136), $\dot{\hat{\boldsymbol{\theta}}}_k$ from Eq. (3.130), $\dot{\hat{\boldsymbol{\theta}}}_z$ from Eq. (3.131), $\dot{\hat{\boldsymbol{\theta}}}_d$ from Eq. (3.129), $\dot{\hat{\boldsymbol{\theta}}}_a$ from Eq. (3.132), and the task-space sliding vector defined in Eq. (3.124) into the above equation and using *Property* 1.2 yields

$$\dot{V} = -\Delta\dot{\boldsymbol{x}}^T\boldsymbol{K}_v\Delta\dot{\boldsymbol{x}} - \alpha\Delta\boldsymbol{x}^T\boldsymbol{K}_p\Delta\boldsymbol{x} \le 0. \tag{3.139}$$

We are now in a position to state the following theorem:

Theorem 3.4 *The adaptive Jacobian tracking control law* (3.128) *and the parameter update laws* (3.130), (3.129), *and* (3.132) *for the robot system* (3.126) *result in the convergence of position and velocity tracking errors. That is,* $\boldsymbol{x} - \boldsymbol{x}_d \to \boldsymbol{0}$ *and* $\dot{\boldsymbol{x}} - \dot{\boldsymbol{x}}_d \to \boldsymbol{0}$*, as* $t \to \infty$.

Proof Since $\boldsymbol{M}(\boldsymbol{q})$ is uniformly positive definite, V in Eq. (3.137) is positive definite in $\boldsymbol{s}$, $\Delta\boldsymbol{x}$, $\Delta\boldsymbol{\theta}_k$, $\Delta\boldsymbol{\theta}_d$, and $\Delta\bar{\boldsymbol{\theta}}_a$. Since $\dot{V} \leq 0$, V is also bounded, and therefore $\boldsymbol{s}$, $\Delta\boldsymbol{x}$, $\Delta\boldsymbol{\theta}_k$, $\Delta\boldsymbol{\theta}_d$, and $\Delta\bar{\boldsymbol{\theta}}_a$ are bounded vectors. It can be proved similarly that $\Delta\boldsymbol{x}$, $\Delta\dot{\boldsymbol{x}}$, $\Delta\ddot{\boldsymbol{x}}$ are all bounded, which ensures the boundedness of $\ddot{V}$. Therefore, $\dot{V}$ is uniformly continuous. Using Barbalat's Lemma or the Lyapunov-Like Lemma, we have $\Delta\boldsymbol{x} = \boldsymbol{x} - \boldsymbol{x}_d \to \boldsymbol{0}$, $\Delta\dot{\boldsymbol{x}} = \dot{\boldsymbol{x}} - \dot{\boldsymbol{x}}_d \to \boldsymbol{0}$, and $\boldsymbol{s} \to \boldsymbol{0}$ as $t \to \infty$. △△△

3.6 Summary and Notes

To deal with dynamic uncertainty, many robot adaptive controllers have been proposed for tracking control of a robot manipulator (Arimoto 1996; Craig et al. 1987; Craig 1988; Slotine and Li 1987a, b, 1988; Middleton and Goodwin 1988; Koditschek 1987; Wen and Bayard 1988; Paden and Panja 1988; Kelly and Carelli 1988; Ortega and Spong 1989; Sadegh and Horowitz 1990; Niemeyer and Slotine 1991; Berghuis et al. 1993; Lewis et al. 1993; Lee and Khalil 1997; Tomei 2000). A key point in adaptive control is that the tracking error will converge regardless of whether the trajectory is persistently exciting (Slotine and Li 1987a). That is, one does not need parameter convergence for task convergence.

In most adaptive controllers, the kinematics of the robot is assumed to be known exactly. The adaptive Jacobian tracking controller (3.46) for robots with uncertainties in kinematics and dynamics was developed in Cheah et al. (2004). The proposed controller requires the differentiation of the task-space position, which is noisy. An adaptive Jacobian tracking control law based on filtered differentiation of the measured task-space position was developed in Cheah et al. (2006a). Observer design techniques can also be used to estimate the velocities (Liang et al. 2010). To avoid singularities associated with the Euler angle representation, an adaptive Jacobian tracking controller based on unit quaternion was developed in Braganza et al. (2005). An adaptive Jacobian controller based on prediction error was proposed in Wang and Xie (2009). A sliding PID tracking control scheme with uncertain Jacobian was proposed in Garcia-Rodriguez and Parra-Vega (2012).

In visual task-space control, it is also difficult to obtain the exact image Jacobian matrix, due to the calibration errors of the camera's intrinsic parameters and the uncertainty existing in the depth information. In addition, in the image Jacobian matrix, the depth information is inversely proportional to the velocity of image features, and hence the overall Jacobian matrix is not linearly parameterizable. Thus the adaptive Jacobian controllers are effective only in cases in which the depth information is constant or slowly time-varying. Vision-based tracking controllers with uncertain depth were proposed in Wang et al. (2007), but the uncertainty of robot kinematics and dynamics was not considered, and the depth information was not updated online. By exploring *Property* 1.5 and 3.1, an adaptive Jacobian tracking control with uncertain depth information was proposed in Cheah et al. (2007, 2010), where the uncertain depth, kinematic, and dynamic parameters are updated concurrently.

Chapter 4
Region Control

In most human reaching movements, the desired targets are regions with arbitrary shapes rather than points. Examples are placing chess pieces on a chess board, putting a spoon into a cup, typing on a keyboard, inserting a thermometer into a patient's mouth, placing an object on a table. Other examples include keeping a vehicle within a travel lane, playing a game of darts, throwing a basketball into a net. This chapter introduces a control concept called region control. In region control, the desired objective can be specified as a region instead of a point or a trajectory. Since the desired region can be specified to be arbitrarily small, the region control method is also a generalization of the setpoint control problem and trajectory tracking control problem. The desired region can also be specified as a performance bound to ensure transient and steady-state response of the system. Since a region is specified instead of a point or a trajectory, less control effort or attention is required, saving energy in the reaching tasks. Moreover, region control can also be used as a secondary objective to set constraints on the motion control tasks so that a robot operates in a finite region within the sensing zone and singularities are avoided.

4.1 Definition of the Region Function

4.1.1 Single Objective Function

In the conventional setpoint control problem, the desired position of the end effector is specified as a point, as illustrated in Fig. 4.1.

In this chapter, we consider a region control problem whereby the desired target is specified as a region. For illustration, we first consider a region-reaching control problem whereby the desired region is specified in task space by an inequality function as follows:

$$f(\boldsymbol{x}) \leq 0, \tag{4.1}$$

C.C. Cheah and X. Li, *Task-Space Sensory Feedback Control of Robot Manipulators*, Intelligent Systems, Control and Automation: Science and Engineering 73,
DOI 10.1007/978-981-287-062-9_4

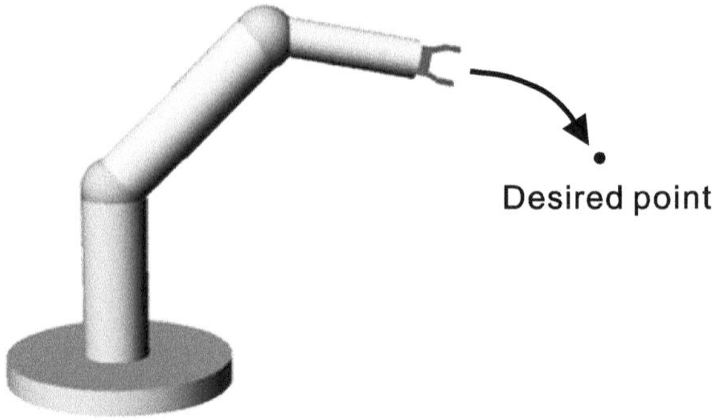

Fig. 4.1 The objective is specified as a point

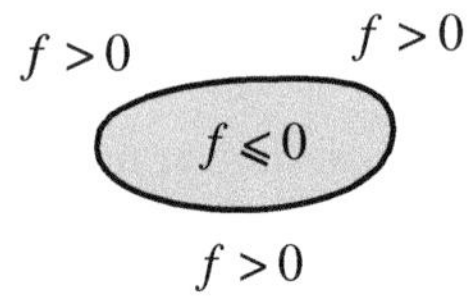

Fig. 4.2 Illustration of the objective function

where $f(\boldsymbol{x}) \in \Re$ is a continuous scalar function with continuous partial derivative. The region function $f(\boldsymbol{x})$ is defined such that when the position of the end effector is outside the desired region, it is positive. Once the end effector enters the region, $f(\boldsymbol{x})$ is nonpositive. An illustration of the objective function is shown in Fig. 4.2.

The control objective is to move the end effector to the desired region, as illustrated in Fig. 4.3.

Example 4.1 A desired circular region in 2D task space is specified as follows:

$$f(\boldsymbol{x}) = (x_1 - x_{o1})^2 + (x_2 - x_{o2})^2 - r^2 \leq 0, \tag{4.2}$$

where $\boldsymbol{x} = [x_1, x_2]^T \in \Re^2$ denotes the position of the end effector in task space, r is the radius of the circle, $\boldsymbol{x}_o = [x_{o1}, x_{o2}]^T$ is the center of a circle as shown in Fig. 4.4. From Eq. (4.2), note that $f(\boldsymbol{x})$ is positive when the position of the end effector

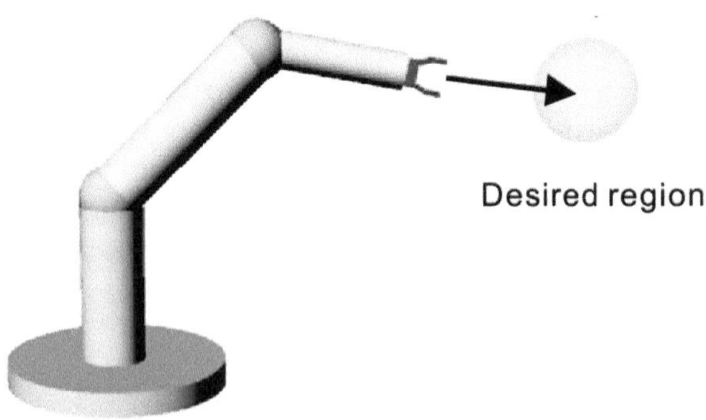

Fig. 4.3 The objective is specified as a region

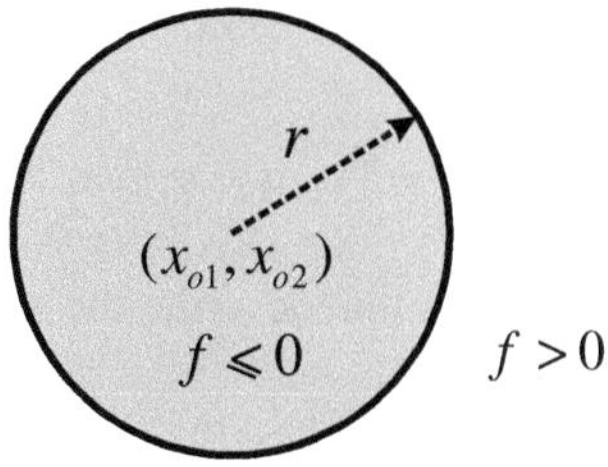

Fig. 4.4 Desired region specified as a circle

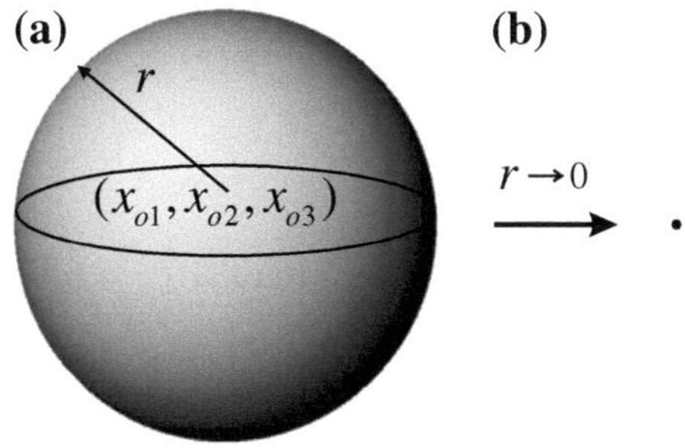

Fig. 4.5 **a** Desired region is specified as a sphere in 3D space. **b** The sphere reduces to a point when the radius is specified as arbitrarily small

is outside the circle, and it becomes nonpositive when the end effector enters the circle. □

Example 4.2 The desired region can also be specified as a 3D sphere, as shown in Fig. 4.5. The objective function of the region is specified as

$$f(\boldsymbol{x}) = (x_1 - x_{o1})^2 + (x_2 - x_{o2})^2 + (x_3 - x_{o3})^2 - r^2 \leq 0, \tag{4.3}$$

where $\boldsymbol{x} = [x_1, x_2, x_3]^T \in \Re^3$ denotes the position of the end effector in task space, r is the radius of the sphere, and $\boldsymbol{x}_o = [x_{o1}, x_{o2}, x_{o3}]^T$ is the center of the sphere. Note that the sphere reduces to a point when the radius is specified to be arbitrarily small, as illustrated in Fig. 4.5. Thus region control is also a generalization of conventional robot control. □

Example 4.3 Another 3D region is shown in Fig. 4.6, where the objective function is specified as

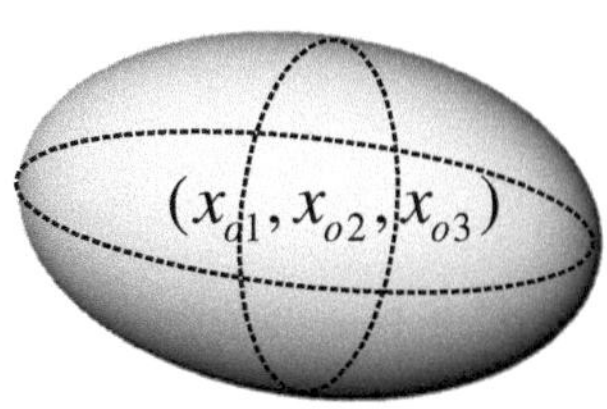

Fig. 4.6 Desired region specified as a 3D region

$$f(\boldsymbol{x}) = \frac{(x_1-x_{o1})^2}{c_1^2} + \frac{(x_2-x_{o2})^2}{c_2^2} + \frac{(x_3-x_{o3})^2}{c_3^2} - 1 \leq 0, \tag{4.4}$$

where $\boldsymbol{x} = [x_1, x_2, x_3]^T \in \Re^3$ denotes the position of the end effector in task space, $\boldsymbol{x}_o = [x_{o1}, x_{o2}, x_{o3}]^T$ is the center of the region, and c_1, c_2, and c_3 are positive constants that describe the size of the region. □

The potential energy function for the region-reaching controller is specified in task space as

$$P(\boldsymbol{x}) = \frac{k_p}{N}[\max(0, f(\boldsymbol{x}))]^N. \tag{4.5}$$

That is,

$$P(\boldsymbol{x}) = \begin{cases} 0, & f(\boldsymbol{x}) \leq 0, \\ \frac{k_p}{N} f^N(\boldsymbol{x}), & f(\boldsymbol{x}) > 0, \end{cases} \tag{4.6}$$

where k_p is a positive constant, and N is a positive integer specified so that the potential energy and its derivative are continuous. A graphical illustration of the potential energy function with a circular desired region is shown in Fig. 4.7. Differentiating the potential energy (4.6) with respect to $\boldsymbol{x}$, we have

$$\left(\frac{\partial P(\boldsymbol{x})}{\partial \boldsymbol{x}}\right)^T = \begin{cases} \boldsymbol{0}, & f(\boldsymbol{x}) \leq 0, \\ k_p[f(\boldsymbol{x})]^{N-1}\left(\frac{\partial f(\boldsymbol{x})}{\partial \boldsymbol{x}}\right)^T, & f(\boldsymbol{x}) > 0, \end{cases} \tag{4.7}$$

which can be written as

$$\left(\frac{\partial P(\boldsymbol{x})}{\partial \boldsymbol{x}}\right)^T = k_p[\max(0, f(\boldsymbol{x}))]^{N-1}\left(\frac{\partial f(\boldsymbol{x})}{\partial \boldsymbol{x}}\right)^T. \tag{4.8}$$

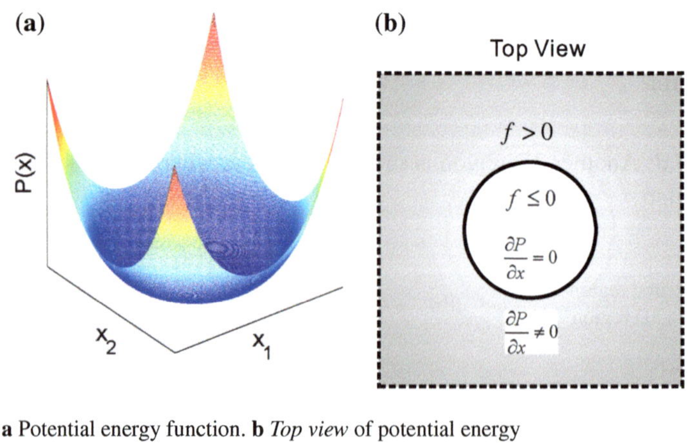

Fig. 4.7 **a** Potential energy function. **b** *Top view* of potential energy

The desired region is defined such that the boundedness of the potential energy guarantees the boundedness of the gradient of the potential energy.

For example, consider the circular region defined in Eq. (4.2). If the corresponding potential energy in Eq. (4.5) is bounded, the region function $f(\boldsymbol{x})$ is bounded. From Eq. (4.2), it can be seen that the boundedness of $f(\boldsymbol{x})$ ensures the boundedness of the position error $\Delta\boldsymbol{x} = [x_1 - x_{o1}, x_2 - x_{o2}]^T$. Differentiating the potential energy with respect to $\boldsymbol{x}$ yields the gradient of the potential energy:

$$\begin{aligned}\left(\frac{\partial P(\boldsymbol{x})}{\partial \boldsymbol{x}}\right)^T &= k_p[\max(0, f(\boldsymbol{x}))]^{N-1}\left(\frac{\partial f(\boldsymbol{x})}{\partial \boldsymbol{x}}\right)^T \\ &= 2k_p[\max(0, f(\boldsymbol{x}))]^{N-1}\begin{bmatrix}(x_1 - x_{o1}) \\ (x_2 - x_{o2})\end{bmatrix},\end{aligned} \tag{4.9}$$

which is bounded, since both $f(\boldsymbol{x})$ and $\Delta\boldsymbol{x}$ are bounded. The gradient of the potential energy will be employed in the region-reaching controller later, where the negative of the term $(\frac{\partial f(\boldsymbol{x})}{\partial \boldsymbol{x}})^T$ provides the directional information toward the desired region for the robot. Note that if $N = 2$, then $P(\boldsymbol{x}) \in C^1$, and if $N = 3$, then $P(\boldsymbol{x}) \in C^2$.

In summary, the region function is defined with the following properties:

Property 4.1 The functions $f(\boldsymbol{x})$ are continuous scalar functions with continuous partial derivatives that satisfy $|f(\boldsymbol{x})| \to \infty$ as $||\boldsymbol{x}|| \to \infty$. □

Property 4.2 The boundedness of $f(\boldsymbol{x})$ represents the boundedness of the gradient of the potential energy $\frac{\partial P(\boldsymbol{x})}{\partial \boldsymbol{x}} = k_p[\max(0, f(\boldsymbol{x}))]^{N-1}(\frac{\partial f(\boldsymbol{x})}{\partial \boldsymbol{x}})^T$. □

4.1.2 Multiple Objective Functions

The desired region can also be specified as multiple objective functions as follows:

$$\boldsymbol{f}(\boldsymbol{x}) = [f_1(\boldsymbol{x}), f_2(\boldsymbol{x}), \ldots, f_M(\boldsymbol{x})]^T \leq \boldsymbol{0}, \tag{4.10}$$

where $i = 1, 2, \ldots, M$, with M is the total number of objective functions. The vector inequality $\boldsymbol{f}(\boldsymbol{x}) \leq \boldsymbol{0}$ implies that $f_i(\boldsymbol{x}) \leq 0$ for all $i = 1, \ldots, M$. Similarly, $f_i(\boldsymbol{x})$ are specified to satisfy *Properties* 4.1 and 4.2.

Example 4.4 A desired region can be specified as the intersection of the two circles shown in Fig. 4.8a. In this case, the objective functions are specified as

$$\begin{aligned}f_1(\boldsymbol{x}) &= \left(x_1 - x_{1o_1}\right)^2 + \left(x_2 - x_{1o_2}\right)^2 - r_1^2 \leq 0, \\ f_2(\boldsymbol{x}) &= \left(x_1 - x_{2o_1}\right)^2 + \left(x_2 - x_{2o_2}\right)^2 - r_2^2 \leq 0,\end{aligned} \tag{4.11}$$

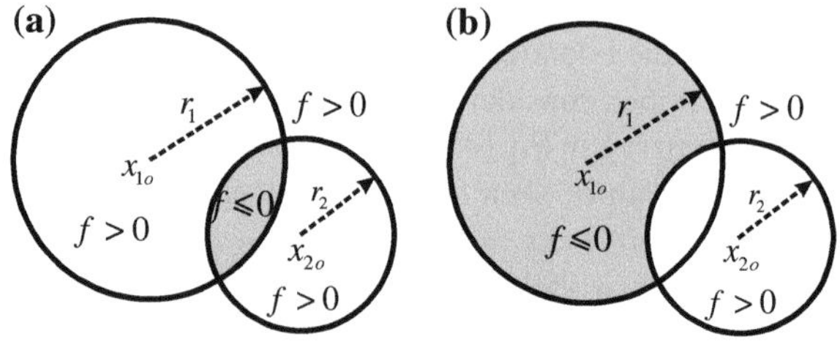

Fig. 4.8 Desired region as the intersection of two circles

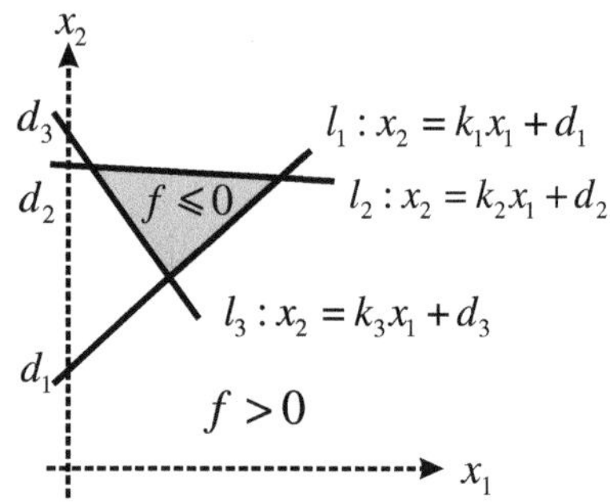

Fig. 4.9 Desired region as a triangle

where r_1, r_2 are the radii of the two circles, $\boldsymbol{x}_{1o} = [x_{1o_1}, x_{1o_2}]^T$ and $\boldsymbol{x}_{2o} = [x_{2o_1}, x_{2o_2}]^T$ represent the centers of the two circles. If $f_2(\boldsymbol{x})$ changes to $f_2(\boldsymbol{x}) = r_2^2 - (x_1 - x_{2o_1})^2 - (x_2 - x_{2o_2})^2 \leq 0$, then the desired region is as illustrated in Fig. 4.8b. □

Example 4.5 In Fig. 4.9, the desired region is a triangle specified by

$$\begin{aligned} f_1(\boldsymbol{x}) &= x_2 - (k_3x_1 + d_3) \leq 0, \\ f_2(\boldsymbol{x}) &= (k_2x_1 + d_2) - x_2 \leq 0, \\ f_3(\boldsymbol{x}) &= x_2 - (k_1x_1 + d_1) \leq 0, \end{aligned} \tag{4.12}$$

such that $k_1, d_1, k_2, d_2, k_3, d_3$ are constants that specify the lines 1, 2, 3. □

Example 4.6 In Fig. 4.10, the desired region is a cube in 3D space, defined by

$$\begin{aligned} f_1(\boldsymbol{x}) &= (x_1 - x_{o1})^2 - b_1^2 \leq 0, \\ f_2(\boldsymbol{x}) &= (x_2 - x_{o2})^2 - b_2^2 \leq 0, \\ f_3(\boldsymbol{x}) &= (x_3 - x_{o3})^2 - b_3^2 \leq 0, \end{aligned} \tag{4.13}$$

such that b_1, b_2, and b_3 are half the length, width, and height of the cube, while $\boldsymbol{x}_o = [x_{o1}, x_{o2}, x_{o3}]^T$ represent the coordinates of center of the cube. □

Example 4.7 A desired region for a robot with six degrees of freedom is shown as follows. Let $\boldsymbol{x} = [x_1, x_2, x_3, x_4, x_5, x_6]^T$, where $[x_1, x_2, x_3]^T$ represents the position

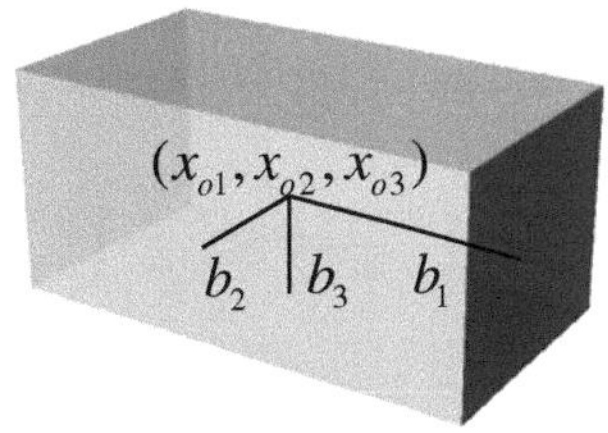

Fig. 4.10 Desired region as a block

and $[x_4, x_5, x_6]^T$ represents the orientation of the robot end effector. The desired regions of the position and orientation are specified as follows:

$$\begin{aligned} f_1(\boldsymbol{x}) &= \sqrt{(x_1 - x_{o1})^2 + (x_2 - x_{o2})^2 + (x_3 - x_{o3})^2} - r \le 0, \\ f_4(\boldsymbol{x}) &= \sqrt{(x_4 - x_{o4})^2} - b_2 \le 0, \\ f_5(\boldsymbol{x}) &= \sqrt{(x_5 - x_{o5})^2} - b_3 \le 0, \\ f_6(\boldsymbol{x}) &= \sqrt{(x_6 - x_{o6})^2} - b_4 \le 0, \end{aligned} \tag{4.14}$$

where $\boldsymbol{x}_o = [x_{o1}, x_{o2}, x_{o3}, x_{o4}, x_{o5}, x_{o6}]^T$ denotes the reference position and orientation, $f_1(\boldsymbol{x})$ is the desired region of the position, and $f_4(\boldsymbol{x})$, $f_5(\boldsymbol{x})$, and $f_6(\boldsymbol{x})$ are the desired regions of the orientation. □

The potential energy function for the region-reaching controller with multiple objective functions is specified as

$$P(\boldsymbol{x}) = \sum_{i=1}^{M} P_i(\boldsymbol{x}), \tag{4.15}$$

where

$$P_i(\boldsymbol{x}) = \frac{k_{pi}}{N}[\max(0, f_i(\boldsymbol{x}))]^N \tag{4.16}$$

and k_{pi} are positive constants. Note that $P(\boldsymbol{x}) = 0$ only if all the objective functions in (4.10) are satisfied. Similarly, the above Eq. (4.16) can be written as

$$P_i(\boldsymbol{x}) = \begin{cases} 0, & f_i(\boldsymbol{x}) \le 0, \\ \frac{k_{pi}}{N}[f_i(\boldsymbol{x})]^N, & f_i(\boldsymbol{x}) > 0. \end{cases} \tag{4.17}$$

Differentiating the potential energy function described by Eq. (4.17) with respect to $\boldsymbol{x}$, we have

$$\left(\frac{\partial P_i(\boldsymbol{x})}{\partial \boldsymbol{x}}\right)^T = k_{pi}[\max(0, f_i(\boldsymbol{x}))]^{N-1} \left(\frac{\partial f_i(\boldsymbol{x})}{\partial \boldsymbol{x}}\right)^T. \tag{4.18}$$

Differentiating the potential energy function described by Eq. (4.15) with respect to $\boldsymbol{x}$ and using Eq. (4.18), we have

$$\left(\frac{\partial P(\boldsymbol{x})}{\partial \boldsymbol{x}}\right)^T = \sum_{i=1}^{M} \left(\frac{\partial P_i(\boldsymbol{x})}{\partial \boldsymbol{x}}\right)^T = \sum_{i=1}^{M} k_{pi}[\max(0, f_i(\boldsymbol{x}))]^{N-1} \left(\frac{\partial f_i(\boldsymbol{x})}{\partial \boldsymbol{x}}\right)^T. \quad (4.19)$$

4.2 Region-Reaching Control

In this section, we consider a task-space region-reaching control problem whereby the objective of the robot end effector is specified as a static region described by Eqs. (4.1) or (4.10). We first consider the region-reaching control problem with exact Jacobian matrix. Sensory feedback region-reaching control with approximate Jacobian will be treated in the next section.

Based on Eq. (4.8), the task-space region-reaching controller is designed as

$$\boldsymbol{\tau} = -\boldsymbol{K}_v\dot{\boldsymbol{q}} - \boldsymbol{J}^T(\boldsymbol{q})k_p[\max(0, f(\boldsymbol{x}))]^{N-1} \left(\frac{\partial f(\boldsymbol{x})}{\partial \boldsymbol{x}}\right)^T + \boldsymbol{g}(\boldsymbol{q}), \quad (4.20)$$

where $\boldsymbol{K}_v \in \Re^{n\times n}$ is a positive definite velocity feedback gain matrix, $\boldsymbol{J}^T(\boldsymbol{q}) \in \Re^{n\times n}$ is the transpose of the Jacobian matrix. For the desired region described using multiple objective functions (4.10), the control term $k_p[\max(0, f(\boldsymbol{x}))]^{N-1}(\frac{\partial f(\boldsymbol{x})}{\partial \boldsymbol{x}})^T$ is replaced by $\sum_{i=1}^{M} k_{pi}[\max(0, f_i(\boldsymbol{x}_i))]^{N-1}(\frac{\partial f_i(\boldsymbol{x})}{\partial \boldsymbol{x}})^T$ in Eq. (4.19) accordingly.

From Eq. (1.5) in Chap. 1, the nonlinear robot dynamics is described as

$$\boldsymbol{M}(\boldsymbol{q})\ddot{\boldsymbol{q}} + \left(\tfrac{1}{2}\dot{\boldsymbol{M}}(\boldsymbol{q}) + \boldsymbol{S}(\boldsymbol{q}, \dot{\boldsymbol{q}})\right)\dot{\boldsymbol{q}} + \boldsymbol{g}(\boldsymbol{q}) = \boldsymbol{\tau}. \quad (4.21)$$

Substituting the control law (4.20) into the robot dynamics (4.21), the closed-loop equation is obtained as

$$\begin{aligned}&\boldsymbol{M}(\boldsymbol{q})\ddot{\boldsymbol{q}} + \left(\tfrac{1}{2}\dot{\boldsymbol{M}}(\boldsymbol{q}) + \boldsymbol{S}(\boldsymbol{q}, \dot{\boldsymbol{q}})\right)\dot{\boldsymbol{q}} + \boldsymbol{K}_v\dot{\boldsymbol{q}}\\ &+\boldsymbol{J}^T(\boldsymbol{q})k_p[\max(0, f(\boldsymbol{x}))]^{N-1} \left(\frac{\partial f(\boldsymbol{x})}{\partial \boldsymbol{x}}\right)^T = \boldsymbol{0}. \end{aligned} \quad (4.22)$$

To carry out the stability analysis, consider the following Lyapunov-like candidate:

$$V = \tfrac{1}{2}\dot{\boldsymbol{q}}^T\boldsymbol{M}(\boldsymbol{q})\dot{\boldsymbol{q}} + P(\boldsymbol{x}), \quad (4.23)$$

where $P(\boldsymbol{x})$ is defined in Eq. (4.5). Since $f(\Delta\boldsymbol{x})$ is continuous in $\boldsymbol{x}$, V is also a continuous scalar function with continuous partial derivatives.

Differentiating Eq. (4.23) with respect to time yields

$$\dot{V} = \dot{\boldsymbol{q}}^T \boldsymbol{M}(\boldsymbol{q})\ddot{\boldsymbol{q}} + \tfrac{1}{2}\dot{\boldsymbol{q}}^T \dot{\boldsymbol{M}}(\boldsymbol{q})\dot{\boldsymbol{q}} + \dot{\boldsymbol{x}}^T \left(\frac{\partial P(\boldsymbol{x})}{\partial \boldsymbol{x}}\right)^T. \tag{4.24}$$

Substituting Eqs. (4.8) and (4.22) into Eq. (4.24), we have

$$\dot{V} = -\dot{\boldsymbol{q}}^T \boldsymbol{S}(\boldsymbol{q}, \dot{\boldsymbol{q}})\dot{\boldsymbol{q}} - \dot{\boldsymbol{q}}^T \boldsymbol{K}_v \dot{\boldsymbol{q}} - k_p[\max(0, f(\boldsymbol{x}))]^{N-1}\dot{\boldsymbol{q}}^T \boldsymbol{J}^T(\boldsymbol{q}) \left(\frac{\partial f(\boldsymbol{x})}{\partial \boldsymbol{x}}\right)^T$$
$$+ k_p[\max(0, f(\boldsymbol{x}))]^{N-1}\dot{\boldsymbol{x}}^T \left(\frac{\partial f(\boldsymbol{x})}{\partial \boldsymbol{x}}\right)^T. \tag{4.25}$$

Since $\dot{\boldsymbol{x}} = \boldsymbol{J}(\boldsymbol{q})\dot{\boldsymbol{q}}$, if we simplify Eq. (4.25) and apply *Property* 1.2, $\dot{V}$ reduces to

$$\dot{V} = -\dot{\boldsymbol{q}}^T \boldsymbol{K}_v \dot{\boldsymbol{q}} \leq 0. \tag{4.26}$$

We consider a compact set

$$\Omega = \{(\boldsymbol{q}, \dot{\boldsymbol{q}}) : V \leq \gamma\}, \tag{4.27}$$

where γ is a positive constant, and the potential energy $P(\boldsymbol{x})$ has a unique minimum in the desired region. If $(\boldsymbol{q}(0), \dot{\boldsymbol{q}}(0)) \in \Omega$, it follows from (4.26) that $(\boldsymbol{q}, \dot{\boldsymbol{q}})$ remain in Ω.

We are now ready to state the following Theorem:

Theorem 4.1 *For the region-reaching controller described by Eq.* (4.20)*, the closed-loop system gives rise to the convergence of* $\boldsymbol{x}$ *to the desired region* $f(\boldsymbol{x}) \leq 0$ *and* $\dot{\boldsymbol{q}}$ *to* $\boldsymbol{0}$ *as* $t \rightarrow \infty$.

Proof Since V is positive definite in $\dot{\boldsymbol{q}}$ and $P(\boldsymbol{x})$, and $\dot{V}$ is negative definite in $\dot{\boldsymbol{q}}$, both $\dot{\boldsymbol{q}}$ and $P(\boldsymbol{x})$ are bounded. The boundedness of the potential energy $P(\boldsymbol{x})$ ensures the boundedness of the gradient $k_p[\max(0, f(\boldsymbol{x}))]^{N-1}(\frac{\partial f(\boldsymbol{x})}{\partial \boldsymbol{x}})^T$, as seen from *Properties* 4.1 and 4.2. From Eq. (4.22), it is seen that $\ddot{\boldsymbol{q}}$ is bounded, which also implies that $\dot{\boldsymbol{q}}$ is uniformly continuous. From Eq. (4.26), it is seen that $\dot{\boldsymbol{q}} \in L_2(0, \infty)$. Therefore, $\dot{\boldsymbol{q}} \rightarrow \boldsymbol{0}$. From the closed-loop Eq. (4.22), $\dot{\boldsymbol{q}} \rightarrow \boldsymbol{0}$ implies that $[\max(0, f(\boldsymbol{x}))]^{N-1}(\frac{\partial f(\boldsymbol{x})}{\partial \boldsymbol{x}})^T \rightarrow \boldsymbol{0}$ as $t \rightarrow \infty$, which is satisfied when $f(\boldsymbol{x}) < 0$. △△△

In general, the desired region can be defined so that $P(\boldsymbol{x})$ has a unique minimum that corresponds to the desired region. In cases in which a local minimum exists, there exists a compact set Ω such that $P(\boldsymbol{x})$ has a unique minimum.

The results of region-reaching control are now extended to the case of a redundant robot. The main idea is to specify a desired joint-space region for the redundant degree of freedom. Let us consider a redundant robot with n degrees of freedom, where the end effector position vector is given as $\boldsymbol{x} \in \Re^m$ $(m \leq n)$. Next, we partition the joint position vector $\boldsymbol{q}$ as $[\boldsymbol{q}_a, \boldsymbol{q}_b]$, where $\boldsymbol{q}_a \in \Re^{n-m}$ denotes the first n–m joints, and

$\boldsymbol{q}_b \in \Re^m$ denotes the last m joints. Similarly, the Jacobian matrix can be partitioned as $\boldsymbol{J}(\boldsymbol{q}) = [\boldsymbol{J}_1(\boldsymbol{q})\ \boldsymbol{J}_2(\boldsymbol{q})]$ such that $\dot{\boldsymbol{x}} = \boldsymbol{J}_1(\boldsymbol{q})\dot{\boldsymbol{q}}_a + \boldsymbol{J}_2(\boldsymbol{q})\dot{\boldsymbol{q}}_b$. The desired region for the end effector can be similarly specified as

$$\boldsymbol{f}(\boldsymbol{x}) = [f_1(\boldsymbol{x}), f_2(\boldsymbol{x}), \ldots, f_{N_1}(\boldsymbol{x})]^T \leq \boldsymbol{0}, \tag{4.28}$$

where $i = 1, 2, \ldots, N_1$, N_1 is the total number of primary objective functions, $f_i(\Delta\boldsymbol{x}) \in \Re$ are scalar functions with continuous first partial derivatives. The vector inequality $\boldsymbol{f}(\boldsymbol{x}) \leq \boldsymbol{0}$ implies that $f_i(\boldsymbol{x}) \leq 0$ for all $i = 1, \ldots, N_1$. In addition, we specify a desired region for $\boldsymbol{q}_a$ as

$$\boldsymbol{h}(\boldsymbol{q}_a) = [h_1(\boldsymbol{q}_a), h_2(\boldsymbol{q}_a), \ldots, h_{N_2}(\boldsymbol{q}_a)]^T \leq \boldsymbol{0}, \tag{4.29}$$

where $j = 1, 2, \ldots, N_2$, N_2 is the total number of secondary objective functions, $h_j(\boldsymbol{q}_a) \in \Re$ are scalar functions with continuous first partial derivatives. The vector inequality $\boldsymbol{h}(\boldsymbol{q}_a) \leq \boldsymbol{0}$ implies that $h_j(\boldsymbol{q}_a) \leq 0$ for all $j = 1, \ldots, N_2$. Similarly, $h_j(\boldsymbol{q}_a)$ are defined such that *Properties* 4.1 and 4.2 are satisfied.

For example, consider a 3-link planner robot as shown in Fig. 4.11. Suppose that the robot end effector is required to move to a primary region. Then a secondary region can be defined for the first joint to confine it to a desired region, as illustrated in Fig. 4.11. In the presence of an obstacle in the workspace, the specification of the secondary region may be useful in avoiding the obstacle. This secondary region can also be used to narrow down the many possible configurations in the task-space control of a redundant robot.

The potential energy functions for the desired regions described in inequalities (4.28) and (4.29) can be specified respectively as

Fig. 4.11 Extension to redundant robots

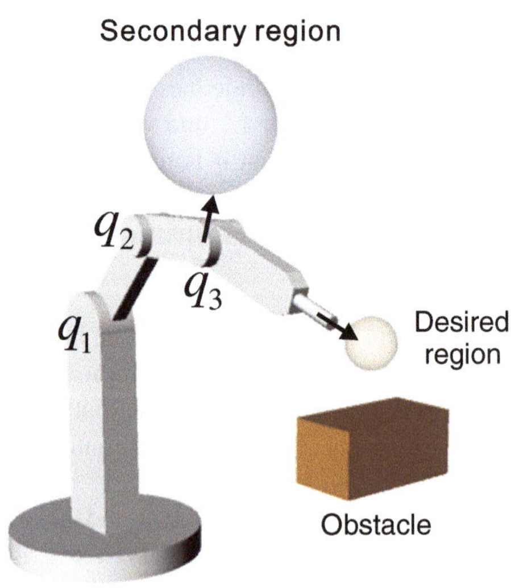

$$P_p(\boldsymbol{x}) = \sum_{i=1}^{N_1} \frac{k_{pi}}{N}[\max(0,\ f_i(\boldsymbol{x}))]^N, \tag{4.30}$$

$$P_s(\boldsymbol{q}_a) = \sum_{j=1}^{N_2} \frac{k_{qj}}{N}[\max(0,\ h_j(\boldsymbol{q}_a))]^N, \tag{4.31}$$

where k_{pi}, k_{qj} are positive constants.

Differentiating the potential energy function (4.30) with respect to $\boldsymbol{x}$ and using similar arguments to those in the previous section, we have

$$\left(\frac{\partial P_p(\boldsymbol{x})}{\partial \boldsymbol{x}}\right)^T = \sum_{i=1}^{N_1} k_{pi}[\max(0,\ f_i(\boldsymbol{x}))]^{N-1} \left(\frac{\partial f_i(\boldsymbol{x})}{\partial \boldsymbol{x}}\right)^T. \tag{4.32}$$

Differentiating the potential energy function (4.31) with respect to $\boldsymbol{q}_a$, we have

$$\left(\frac{\partial P_s(\boldsymbol{q}_a)}{\partial \boldsymbol{q}_a}\right)^T = \sum_{j=1}^{N_2} k_{qj}[\max(0,\ h_j(\boldsymbol{q}_a))]^{N-1} \left(\frac{\partial h_j(\boldsymbol{q}_a)}{\partial \boldsymbol{q}_a}\right)^T. \tag{4.33}$$

From Eqs. (4.32) and (4.33), the task-space region-reaching controller is designed for a redundant robot as

$$\begin{aligned} \boldsymbol{\tau} = &-\boldsymbol{K}_v\dot{\boldsymbol{q}} + \boldsymbol{g}(\boldsymbol{q}) - \boldsymbol{J}^T(\boldsymbol{q}) \sum_{i=1}^{N_1} k_{pi}[\max(0,\ f_i(\boldsymbol{x}))]^{N-1} \left(\frac{\partial f_i(\boldsymbol{x})}{\partial \boldsymbol{x}}\right)^T \\ &- \begin{bmatrix} \boldsymbol{I}_{n-m} \\ \boldsymbol{0} \end{bmatrix} \sum_{j=1}^{N_2} k_{qj}[\max(0,\ h_j(\boldsymbol{q}_a))]^{N-1} \left(\frac{\partial h_j(\boldsymbol{q}_a)}{\partial \boldsymbol{q}_a}\right)^T, \end{aligned} \tag{4.34}$$

where $\boldsymbol{I}_{n-m}$ is the $(n-m) \times (n-m)$ identity matrix, $\boldsymbol{0}$ is the $m \times (n-m)$ zero matrix.

Substituting Eq. (4.34) into Eq. (4.21), the closed-loop equation is obtained as

$$\begin{aligned} \boldsymbol{M}(\boldsymbol{q})\ddot{\boldsymbol{q}} &+ \left[\tfrac{1}{2}\dot{\boldsymbol{M}}(\boldsymbol{q}) + \boldsymbol{S}(\boldsymbol{q}, \dot{\boldsymbol{q}})\right]\dot{\boldsymbol{q}} + \boldsymbol{J}^T(\boldsymbol{q}) \sum_{i=1}^{N_1} k_{pi}[\max(0,\ f_i(\boldsymbol{x}))]^{N-1} \left(\frac{\partial f_i(\boldsymbol{x})}{\partial \boldsymbol{x}}\right)^T \\ &+\boldsymbol{K}_v\dot{\boldsymbol{q}} + \begin{bmatrix} \boldsymbol{I}_{n-m} \\ \boldsymbol{0} \end{bmatrix} \sum_{j=1}^{N_2} k_{qj}[\max(0,\ h_j(\boldsymbol{q}_a))]^{N-1} \left(\frac{\partial h_j(\boldsymbol{q}_a)}{\partial \boldsymbol{q}_a}\right)^T = \boldsymbol{0}. \end{aligned} \tag{4.35}$$

To prove stability, consider the Lyaponov-like candidate

$$V = \tfrac{1}{2}\dot{\boldsymbol{q}}^T \boldsymbol{M}(\boldsymbol{q})\dot{\boldsymbol{q}} + P_p(\boldsymbol{x}) + P_s(\boldsymbol{q}_a), \tag{4.36}$$

where $P_p(\boldsymbol{x})$ and $P_s(\boldsymbol{q}_a)$ are defined in Eqs. (4.30) and (4.31). Note that V is a scalar function with continuous partial derivatives, since $\boldsymbol{f}(\boldsymbol{x})$ in continuous in $\boldsymbol{x}$.

Differentiating equation (4.36) with respect to time yields

$$\dot{V} = \dot{\boldsymbol{q}}^T \boldsymbol{M}(\boldsymbol{q})\ddot{\boldsymbol{q}} + \tfrac{1}{2}\dot{\boldsymbol{q}}^T \dot{\boldsymbol{M}}(\boldsymbol{q})\dot{\boldsymbol{q}} + \dot{\boldsymbol{x}}^T (\tfrac{\partial P_p(\boldsymbol{x})}{\partial \boldsymbol{x}})^T + \dot{\boldsymbol{q}}_a^T (\tfrac{\partial P_s(\boldsymbol{q}_a)}{\partial \boldsymbol{q}_a})^T. \quad (4.37)$$

Substituting Eqs. (4.32), (4.33) and (4.35) into Eq. (4.37), we have

$$\begin{aligned}\dot{V} = &-\sum_{i=1}^{N_1} k_{pi}[\max(0, f_i(\boldsymbol{x}))]^{N-1}\dot{\boldsymbol{q}}^T \boldsymbol{J}^T(\boldsymbol{q}) \left(\frac{\partial f_i(\boldsymbol{x})}{\partial \boldsymbol{x}}\right)^T \\ &+\sum_{j=1}^{N_2} k_{qj}[\max(0, h_j(\boldsymbol{q}_a))]^{N-1}\dot{\boldsymbol{q}}_a^T \left(\frac{\partial h_j(\boldsymbol{q}_a)}{\partial \boldsymbol{q}_a}\right)^T \\ &+\sum_{i=1}^{N_1} k_{pi}[\max(0, f_i(\boldsymbol{x}))]^{N-1}\dot{\boldsymbol{x}}^T \left(\frac{\partial f_i(\boldsymbol{x})}{\partial \boldsymbol{x}}\right)^T \\ &-\sum_{j=1}^{N_2} k_{qj}[\max(0, h_j(\boldsymbol{q}_a))]^{N-1}\dot{\boldsymbol{q}}^T \begin{bmatrix} \boldsymbol{I}_{n-m} \\ \boldsymbol{0} \end{bmatrix} \left(\frac{\partial h_j(\boldsymbol{q}_a)}{\partial \boldsymbol{q}_a}\right)^T \\ &-\dot{\boldsymbol{q}}^T \boldsymbol{K}_v \dot{\boldsymbol{q}} - \dot{\boldsymbol{q}}^T \boldsymbol{S}(\boldsymbol{q}, \dot{\boldsymbol{q}})\dot{\boldsymbol{q}}, \end{aligned} \quad (4.38)$$

since $\dot{\boldsymbol{x}} = \boldsymbol{J}(\boldsymbol{q})\dot{\boldsymbol{q}}$ and $\boldsymbol{q}^T \begin{bmatrix} \boldsymbol{I}_{n-m} \\ \boldsymbol{0} \end{bmatrix} = \boldsymbol{q}_a^T$, simplifying Eq. (4.38) and applying *Property* 1.2, $\dot{V}$ reduces to:

$$\dot{V} = -\dot{\boldsymbol{q}}^T \boldsymbol{K}_v \dot{\boldsymbol{q}}. \quad (4.39)$$

Similarly, we consider the compact set

$$\Omega = \{(\boldsymbol{q}, \dot{\boldsymbol{q}}) : V \leq \gamma\}, \quad (4.40)$$

where γ is a positive constant, and the potential energy functions $P_p(\boldsymbol{x})$ and $P_s(\boldsymbol{q}_a)$ have a unique minimum at the desired region. If $(\boldsymbol{q}(0), \dot{\boldsymbol{q}}(0)) \in \Omega$, it follows from (4.39) that $(\boldsymbol{q}, \dot{\boldsymbol{q}})$ remain in Ω.

We are now in a position to state the following theorem:

Theorem 4.2 *The closed-loop system described by Eq.* (4.35) *gives rise to the convergence of* $\boldsymbol{x}$ *to the region* $\boldsymbol{f}(\boldsymbol{x}) = [f_1(\boldsymbol{x}) \cdots f_{N_1}(\boldsymbol{x})]^T \leq \boldsymbol{0}$, *and* $\boldsymbol{q}_a$ *to the region* $\boldsymbol{h}(\boldsymbol{q}_a) = [h_1(\boldsymbol{q}_a) \cdots h_{N_2}(\boldsymbol{q}_a)]^T \leq \boldsymbol{0}$ *and* $\dot{\boldsymbol{q}}$ *to* $\boldsymbol{0}$ *as* $t \to \infty$.

Proof Since V is positive definite in $\dot{\boldsymbol{q}}$, $P_p(\boldsymbol{x})$, and $P_s(\boldsymbol{q}_a)$, and $\dot{V}$ is negative definite in $\dot{\boldsymbol{q}}$, $\dot{\boldsymbol{q}}$, $P_p(\boldsymbol{x})$, and $P_s(\boldsymbol{q}_a)$ are all bounded. The boundedness of the potential energy $P_p(\boldsymbol{x})$ ensures the boundedness of the gradient $\sum_{i=1}^{N_1} k_{pi}[\max(0, f_i(\boldsymbol{x}))]^{N-1}(\frac{\partial f_i(\boldsymbol{x})}{\partial \boldsymbol{x}})^T$. Similarly, the gradient $\sum_{j=1}^{N_2} k_{qj}[\max(0, h_j(\boldsymbol{q}_a))]^{N-1}(\frac{\partial h_j(\boldsymbol{q}_a)}{\partial \boldsymbol{q}_a})^T$ is also bounded, since

$P_s(\boldsymbol{q}_a)$ is bounded. From Eq. (4.35), it is seen that $\ddot{\boldsymbol{q}}$ is bounded, which also implies that $\dot{\boldsymbol{q}}$ is uniformly continuous. From Eq. (4.39), it is seen that $\dot{\boldsymbol{q}} \in L_2(0, \infty)$. Therefore, $\dot{\boldsymbol{q}} \to \mathbf{0}$. From the closed-loop Eq. (4.22), $\dot{\boldsymbol{q}} \to \mathbf{0}$ implies that

$$\sum_{i=1}^{N_1} \boldsymbol{J}^T(\boldsymbol{q}) \max(0, f_i(\boldsymbol{x})) \left(\frac{\partial f_i(\boldsymbol{x})}{\partial \boldsymbol{x}}\right)^T + \sum_{j=1}^{N_2} \begin{bmatrix} \boldsymbol{I}_{n-m} \\ \mathbf{0} \end{bmatrix} \max(0, h_j(\boldsymbol{q}_a)) \left(\frac{\partial h_j(\boldsymbol{q}_a)}{\partial \boldsymbol{q}_a}\right)^T = \mathbf{0}. \quad (4.41)$$

The above equation can be written as

$$\begin{bmatrix} \boldsymbol{I}_{n-m} & \boldsymbol{J}_1^T(\boldsymbol{q}) \\ \mathbf{0} & \boldsymbol{J}_2^T(\boldsymbol{q}) \end{bmatrix} \begin{bmatrix} \sum_{j=1}^{N_2} \max(0, h_j(\boldsymbol{q}_a)) \left(\frac{\partial h_j(\boldsymbol{q}_a)}{\partial \boldsymbol{q}_a}\right)^T \\ \sum_{i=1}^{N_1} \max(0, f_i(\boldsymbol{x})) \left(\frac{\partial f_i(\boldsymbol{x})}{\partial \boldsymbol{x}}\right)^T \end{bmatrix} = \mathbf{0}, \quad (4.42)$$

where $\boldsymbol{J}(\boldsymbol{q}) = [\boldsymbol{J}_1(\boldsymbol{q}) \;\; \boldsymbol{J}_2(\boldsymbol{q})]$. Hence, $\sum_{i=1}^{N_1} \max(0, f_i(\boldsymbol{x}))(\frac{\partial f_i(\boldsymbol{x})}{\partial \boldsymbol{x}})^T = \mathbf{0}$, and $\sum_{j=1}^{N_2} \max(0, h_j(\boldsymbol{q}_a))(\frac{\partial h_j(\boldsymbol{q}_a)}{\partial \boldsymbol{q}_a})^T = \mathbf{0}$, and $\dot{\boldsymbol{q}} = \mathbf{0}$ if the square matrix $\begin{bmatrix} \boldsymbol{I}_{n-m} & \boldsymbol{J}_1^T(\boldsymbol{q}) \\ \mathbf{0} & \boldsymbol{J}_2^T(\boldsymbol{q}) \end{bmatrix}$ is nonsingular (i.e., if $\boldsymbol{J}_2^T(\boldsymbol{q}) \in \Re^{m \times n}$ is nonsingular). This finally implies that $\boldsymbol{f}(\boldsymbol{x}) = [f_1(\boldsymbol{x}), \ldots, f_{N_1}(\boldsymbol{x})]^T \leq \mathbf{0}$ and, $\boldsymbol{h}(\boldsymbol{q}_a) = [h_1(\boldsymbol{q}_a), \ldots, h_{N_2}(\boldsymbol{q}_a)]^T \leq \mathbf{0}$ as $t \to \infty$. $\triangle\triangle\triangle$

The additional regions can be specified either in joint space or task space to set constraints on the motion of the robot so that it operates within the nonsingular regions and/or sensing zones. As illustrated in Fig. 4.12, the additional region is specified such that the motion of the end effector is constrained to the region within the field of view.

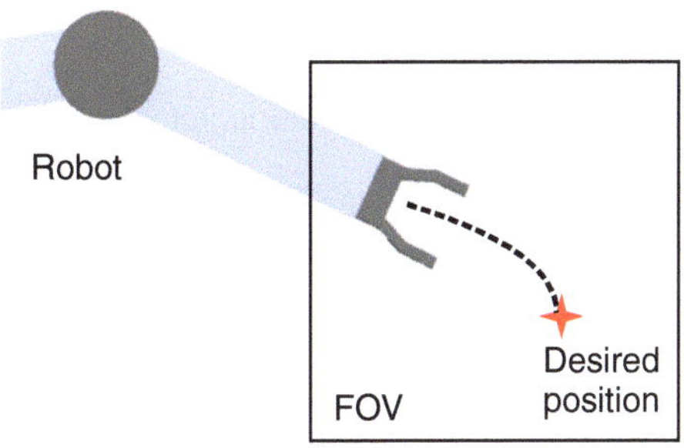

Fig. 4.12 The additional region is specified to match the field of view

4.3 Region-Reaching Control with Approximate Jacobian Matrix

We now consider the case of the sensory task-space region-reaching control problem with approximate Jacobian matrix. In the presence of uncertain kinematics, the Jacobian matrix is uncertain, and the region-reaching controller is designed as

$$\boldsymbol{\tau} = -\hat{\boldsymbol{J}}^T(\boldsymbol{q}) \sum_{i=1}^{M} k_{pi} \max(0, f_i(\boldsymbol{x})) \left(\frac{\partial f_i(\boldsymbol{x})}{\partial \boldsymbol{x}}\right)^T - \hat{\boldsymbol{J}}^T(\boldsymbol{q})\boldsymbol{K}_v \hat{\boldsymbol{J}}(\boldsymbol{q})\dot{\boldsymbol{q}} + \boldsymbol{g}(\boldsymbol{q}), \tag{4.43}$$

where the desired region is specified by multiple objective functions, as described by Eq. (4.10); the order of the potential energy is set as $N = 2$; and $\hat{\boldsymbol{J}}(\boldsymbol{q})$ denotes the approximate or estimated Jacobian matrix, which is chosen so that

$$\left\| \boldsymbol{J}^T(\boldsymbol{q}) - \hat{\boldsymbol{J}}^T(\boldsymbol{q}) \right\| \leq p, \tag{4.44}$$

and $\boldsymbol{K}_v = k_v \boldsymbol{I}_n$ is a velocity feedback gain matrix.

The closed-loop equation of the system is obtained by substituting (4.43) into (4.21):

$$\begin{aligned} &\boldsymbol{M}(\boldsymbol{q})\ddot{\boldsymbol{q}} + \left(\tfrac{1}{2}\dot{\boldsymbol{M}}(\boldsymbol{q}) + \boldsymbol{S}(\boldsymbol{q}, \dot{\boldsymbol{q}})\right)\dot{\boldsymbol{q}} + \hat{\boldsymbol{J}}^T(\boldsymbol{q}) k_v \hat{\boldsymbol{J}}(\boldsymbol{q})\dot{\boldsymbol{q}} \\ &+ \hat{\boldsymbol{J}}^T(\boldsymbol{q}) \sum_{i=1}^{M} k_{pi} \max(0, f_i(\boldsymbol{x})) \left(\tfrac{\partial f_i(\boldsymbol{x})}{\partial \boldsymbol{x}}\right)^T = \boldsymbol{0}. \end{aligned} \tag{4.45}$$

To prove stability, a Lyapunov-like candidate is designed as

$$\begin{aligned} V = &\tfrac{1}{2}\dot{\boldsymbol{q}}^T \boldsymbol{M}(\boldsymbol{q})\dot{\boldsymbol{q}} + \alpha \dot{\boldsymbol{q}}^T \boldsymbol{M}(\boldsymbol{q}) \hat{\boldsymbol{J}}^{-1}(\boldsymbol{q}) \sum_{i=1}^{M} k_{pi} \max(0, f_i(\boldsymbol{x})) \left(\tfrac{\partial f_i(\boldsymbol{x})}{\partial \boldsymbol{x}}\right)^T \\ &+ \tfrac{1}{2}(k_p + \alpha k_v) \sum_{i=1}^{M} a_i \left[\max(0, f_i(\boldsymbol{x}))\right]^2. \end{aligned} \tag{4.46}$$

For clarity of presentation, let

$$\sum_{i=1}^{M} k_{pi} \max(0, f_i(\boldsymbol{x})) \left(\tfrac{\partial f_i(\boldsymbol{x})}{\partial \boldsymbol{x}}\right)^T = k_p \sum_{i=1}^{M} \overline{\Delta \boldsymbol{x}_i} \triangleq k_p \overline{\Delta \boldsymbol{\xi}_o}, \tag{4.47}$$

where

$$\overline{\Delta \boldsymbol{x}_i} = a_i \max(0,\ f_i(\boldsymbol{x})) \left(\frac{\partial f_i(\boldsymbol{x})}{\partial \boldsymbol{x}}\right)^T, \tag{4.48}$$

$$\overline{\Delta \boldsymbol{\xi}_o} = \sum_{i=1}^{M} \overline{\Delta \boldsymbol{x}_i}, \tag{4.49}$$

and $a_i = \frac{k_{pi}}{k_p}$ is defined as the weight for each objective function.

From Eq. (4.48), note that

$$a_i^2 \left\| \frac{\partial f_i(\boldsymbol{x})}{\partial \boldsymbol{x}} \right\|^2 [\max(0,\ f_i(\boldsymbol{x}))]^2 = \overline{\Delta \boldsymbol{x}_i}^T \overline{\Delta \boldsymbol{x}_i}. \tag{4.50}$$

Taking the norm on (4.49) gives

$$\left\| \overline{\Delta \boldsymbol{\xi}_o} \right\| \le \left\| \overline{\Delta \boldsymbol{x}_1} \right\| + \cdots + \left\| \overline{\Delta \boldsymbol{x}_M} \right\|, \tag{4.51}$$

which implies that

$$\left\| \overline{\Delta \boldsymbol{\xi}_o} \right\|^2 \le [\left\| \overline{\Delta \boldsymbol{x}_1} \right\|, \ldots, \left\| \overline{\Delta \boldsymbol{x}_M} \right\|]\, \boldsymbol{\Lambda} \begin{bmatrix} \left\| \overline{\Delta \boldsymbol{x}_1} \right\| \\ \vdots \\ \left\| \overline{\Delta \boldsymbol{x}_M} \right\| \end{bmatrix} \le \lambda_{\max}[\boldsymbol{\Lambda}] \sum_{i=1}^{M} \left\| \overline{\Delta \boldsymbol{x}_i} \right\|^2, \tag{4.52}$$

where $\boldsymbol{\Lambda} = \begin{bmatrix} 1 & \cdots & 1 \\ \vdots & \ddots & \vdots \\ 1 & \cdots & 1 \end{bmatrix}$.

Substituting Eq. (4.50) into (4.52), we have

$$\begin{aligned} \left\| \overline{\Delta \boldsymbol{\xi}_o} \right\|^2 &\le \lambda_{\max}[\boldsymbol{\Lambda}] \sum_{i=1}^{M} a_i^2 \left\| \frac{\partial f_i(\boldsymbol{x})}{\partial \boldsymbol{x}} \right\|^2 [\max(0,\ f_i(\boldsymbol{x}))]^2 \\ &\le \lambda_{\max}[\boldsymbol{\Lambda}] b_2 \sum_{i=1}^{M} a_i\, [\max(0,\ f_i(\boldsymbol{x}))]^2, \end{aligned} \tag{4.53}$$

where $\left\| \frac{\partial f_i(\Delta \boldsymbol{x}_i)}{\partial \boldsymbol{x}} \right\|^2 \le b_i$ and $b_2 = \max_i a_i b_i$. Next, using (4.50) and (4.53), we note that

$$\frac{1}{4}\dot{\boldsymbol{q}}^T\boldsymbol{M}(\boldsymbol{q})\dot{\boldsymbol{q}} + \frac{1}{2}(k_p + \alpha k_v)\sum_{i=1}^{M} a_i\left[\max(0, f_i(\boldsymbol{x}))\right]^2$$

$$+\alpha\dot{\boldsymbol{q}}^T\boldsymbol{M}(\boldsymbol{q})\hat{\boldsymbol{J}}^{-1}(\boldsymbol{q})\sum_{i=1}^{M} k_{pi}\max(0, f_i(\boldsymbol{x})\left(\frac{\partial f_i(\boldsymbol{x})}{\partial \boldsymbol{x}}\right)^T$$

$$= \frac{1}{4}\left(\dot{\boldsymbol{q}} + 2\alpha\hat{\boldsymbol{J}}^{-1}(\boldsymbol{q})\overline{\Delta\boldsymbol{\xi}_o}\right)^T\boldsymbol{M}(\boldsymbol{q})\left(\dot{\boldsymbol{q}} + 2\alpha\hat{\boldsymbol{J}}^{-1}(\boldsymbol{q})\overline{\Delta\boldsymbol{\xi}_o}\right)$$

$$-\alpha^2\overline{\Delta\boldsymbol{\xi}_o}^T(\hat{\boldsymbol{J}}^{-1}(\boldsymbol{q}))^T\boldsymbol{M}(\boldsymbol{q})\hat{\boldsymbol{J}}^{-1}(\boldsymbol{q})\overline{\Delta\boldsymbol{\xi}_o} + \frac{1}{2}(k_p + \alpha k_v)\sum_{i=1}^{M} a_i\left[\max(0, f_i(\boldsymbol{x}))\right]^2$$

$$\geq \frac{1}{2}(k_p + \alpha k_v)\sum_{i=1}^{M} a_i\left[\max(0, f_i(\boldsymbol{x}))\right]^2 - \alpha^2\lambda_1\left\|\overline{\Delta\boldsymbol{\xi}_o}\right\|^2$$

$$\geq \left(\frac{1}{2}(k_p + \alpha k_v) - \alpha^2\lambda_1\lambda_{\max}[\boldsymbol{\Lambda}]b_2\right)\sum_{i=1}^{M} a_i\left[\max(0, f_i(\boldsymbol{x}))\right]^2. \tag{4.54}$$

Substituting the above inequality into the Lyapunov-like function (4.46) yields

$$V \geq \frac{1}{4}\dot{\boldsymbol{q}}^T\boldsymbol{M}(\boldsymbol{q})\dot{\boldsymbol{q}} + \left(\frac{1}{2}(k_p + \alpha k_v) - \alpha^2\lambda_1\lambda_{\max}[\boldsymbol{\Lambda}]b_2\right)\sum_{i=1}^{M} a_i\left[\max(0, f_i(\boldsymbol{x}))\right]^2. \tag{4.55}$$

Hence, k_p, k_v can be chosen sufficiently large that

$$\frac{1}{2}(k_p + \alpha k_v)b_2 - \alpha^2\lambda_1\lambda_{\max}[\boldsymbol{\Lambda}] > 0, \tag{4.56}$$

and hence V bounded from below by zero.

Similarly, differentiating (4.46) with respect to time and using (4.45) yields

$$\dot{V} = -W, \tag{4.57}$$

where

$$W = \dot{\boldsymbol{q}}^T\hat{\boldsymbol{J}}^T(\boldsymbol{q})k_v\hat{\boldsymbol{J}}(\boldsymbol{q})\dot{\boldsymbol{q}} + \alpha\overline{\Delta\boldsymbol{\xi}_o}^T k_p\overline{\Delta\boldsymbol{\xi}_o} + \dot{\boldsymbol{q}}^T\hat{\boldsymbol{J}}^T(\boldsymbol{q})k_p\overline{\Delta\boldsymbol{\xi}_o}$$

$$+\alpha\overline{\Delta\boldsymbol{\xi}_o}^T\boldsymbol{K}_v\hat{\boldsymbol{J}}(\boldsymbol{q})\dot{\boldsymbol{q}} - (k_p + \alpha k_v)\sum_{i=1}^{M} a_i\max(0, f_i(\boldsymbol{x}))\frac{\partial f_i(\boldsymbol{x})}{\partial \boldsymbol{x}}\dot{\boldsymbol{x}}$$

$$+\alpha\left\{\overline{\Delta\boldsymbol{\xi}_o}^T(\hat{\boldsymbol{J}}^{-1}(\boldsymbol{q}))^T\left[\frac{1}{2}\dot{\boldsymbol{M}}(\boldsymbol{q}) + \boldsymbol{S}(\boldsymbol{q}, \dot{\boldsymbol{q}})\right]\dot{\boldsymbol{q}} - \dot{\overline{\Delta\boldsymbol{\xi}_o}}^T(\hat{\boldsymbol{J}}^{-1}(\boldsymbol{q}))^T\boldsymbol{M}(\boldsymbol{q})\dot{\boldsymbol{q}}\right.$$

$$\left.-\overline{\Delta\boldsymbol{\xi}_o}^T(\dot{\hat{\boldsymbol{J}}}^{-1}(\boldsymbol{q}))^T\boldsymbol{M}(\boldsymbol{q})\dot{\boldsymbol{q}} - \overline{\Delta\boldsymbol{\xi}_o}^T(\hat{\boldsymbol{J}}^{-1}(\boldsymbol{q}))^T\dot{\boldsymbol{M}}(\boldsymbol{q})\dot{\boldsymbol{q}}\right\}. \tag{4.58}$$

Next, the time derivative of the Lyapunov-like function is shown to be negative definite. As seen from (4.57), this is equivalent to W being positive definite. From the last term on the right-hand side of (4.58), in analogy with the previous section, it is noted that there exists a constant $c_2 > 0$ such that

$$\begin{aligned}\alpha|\overline{\Delta\boldsymbol{\xi}_o}^T(\hat{\boldsymbol{J}}^{-1}(\boldsymbol{q}))^T\left[\tfrac{1}{2}\dot{\boldsymbol{M}}(\boldsymbol{q})+\boldsymbol{S}(\boldsymbol{q},\dot{\boldsymbol{q}})\right]\dot{\boldsymbol{q}}-\dot{\overline{\Delta\boldsymbol{\xi}_o}}^T(\hat{\boldsymbol{J}}^{-1}(\boldsymbol{q}))^T\boldsymbol{M}(\boldsymbol{q})\dot{\boldsymbol{q}}\\ -\overline{\Delta\boldsymbol{\xi}_o}^T\dot{\hat{\boldsymbol{J}}}^{+^T}(\boldsymbol{q})\boldsymbol{M}(\boldsymbol{q})\dot{\boldsymbol{q}}-\overline{\Delta\boldsymbol{\xi}_o}^T\hat{\boldsymbol{J}}^{+^T}(\boldsymbol{q})\dot{\boldsymbol{M}}(\boldsymbol{q})\dot{\boldsymbol{q}}|\geq-\alpha c_2\|\dot{\boldsymbol{q}}\|^2.\end{aligned} \tag{4.59}$$

Substituting (4.59) into (4.58) yields

$$\begin{aligned}W &\geq \dot{\boldsymbol{q}}^T(\hat{\boldsymbol{J}}^T(\boldsymbol{q})k_v\hat{\boldsymbol{J}}(\boldsymbol{q})-\alpha c_2\boldsymbol{I}_n)\dot{\boldsymbol{q}}+\alpha\overline{\Delta\boldsymbol{\xi}_o}^T k_p\overline{\Delta\boldsymbol{\xi}_o}\\ &\quad-\dot{\boldsymbol{q}}^T(\boldsymbol{J}^T(\boldsymbol{q})-\hat{\boldsymbol{J}}^T(\boldsymbol{q}))k_p\overline{\Delta\boldsymbol{\xi}_o}-\alpha\dot{\boldsymbol{q}}^T(\boldsymbol{J}^T(\boldsymbol{q})-\hat{\boldsymbol{J}}^T(\boldsymbol{q}))k_v\overline{\Delta\boldsymbol{\xi}_o}\\ &\geq\left(\lambda_{\min}[\hat{\boldsymbol{J}}^T(\boldsymbol{q})\hat{\boldsymbol{J}}(\boldsymbol{q})]k_v-\alpha c_2-\tfrac{1}{2}pk_p-\tfrac{1}{2}\alpha pk_v\right)\|\dot{\boldsymbol{q}}\|^2+\left(\alpha k_p-\tfrac{1}{2}pk_p-\tfrac{1}{2}\alpha pk_v\right)\|\overline{\Delta\boldsymbol{\xi}_o}\|^2.\end{aligned} \tag{4.60}$$

Equation (4.60) can be further simplified into the following form:

$$W \geq k_v\bar{l}_3\|\dot{\boldsymbol{q}}\|^2+k_v\bar{l}_4\|\overline{\Delta\boldsymbol{\xi}_o}\|^2, \tag{4.61}$$

where

$$\begin{aligned}\bar{l}_3 &= \lambda_{\min}\left[\hat{\boldsymbol{J}}^T(\boldsymbol{q})\hat{\boldsymbol{J}}(\boldsymbol{q})\right]-\alpha\tfrac{c_2}{k_v}-\tfrac{1}{2}p(a+\alpha),\\ \bar{l}_4 &= \alpha a-\tfrac{1}{2}p(a+\alpha).\end{aligned} \tag{4.62}$$

Hence if

$$\min\left\{\frac{2\left(\lambda_{\min}\left[\hat{\boldsymbol{J}}^T(\boldsymbol{q})\hat{\boldsymbol{J}}(\boldsymbol{q})\right]-\alpha\frac{c_2}{k_v}\right)}{a+\alpha},\ \frac{2\alpha a}{a+\alpha}\right\}>p, \tag{4.63}$$

then $\bar{l}_3 > 0$ and $\bar{l}_4 > 0$, and hence W is positive definite in $\dot{\boldsymbol{q}}$ and $\overline{\Delta\boldsymbol{\xi}_o}$.

Consider the compact set

$$\Omega=\{(\boldsymbol{q},\dot{\boldsymbol{q}}):V\leq\gamma\}, \tag{4.64}$$

where the potential energy for the region function $P(\boldsymbol{x})$ has a unique minimum at the desired region with a positive γ. If $(\dot{\boldsymbol{q}}(0),\boldsymbol{q}(0))\in\Omega$, then $(\boldsymbol{q},\dot{\boldsymbol{q}})\in\Omega$, since $\dot{V}\leq 0$.

Theorem 4.3 *The closed-loop system described by* (4.45) *gives rise to the convergence of* $\boldsymbol{x}$ *to the desired regions* $f(\boldsymbol{x})=[f_1(\boldsymbol{x})f_2(\boldsymbol{x})\ldots\ f_M(\boldsymbol{x})]^T\leq 0$ *and* $\dot{\boldsymbol{q}}$ *to* $\boldsymbol{0}$ *as* $t\to\infty$ *if the feedback gains* k_p *and* k_v *are chosen to satisfy conditions* (4.56) *and* (4.63).

Proof Since V is positive definite in $\dot{\boldsymbol{q}}$ and $\max(0, f_i(\boldsymbol{x}))$, and W is positive definite in $\overline{\Delta\boldsymbol{\xi}_o}$, both $\dot{\boldsymbol{q}}$ and $f_i(\boldsymbol{x})$ are bounded. From *Property* 4.2, the boundedness of $\max(0, f_i(\boldsymbol{x}))$ ensures the boundedness of $k_{pi}\max(0, f_i(\boldsymbol{x}))\left(\frac{\partial f_i(\boldsymbol{x})}{\partial \boldsymbol{x}}\right)^T$. From Eq. (4.45), it is seen that $\ddot{\boldsymbol{q}}$ is bounded, which also implies that $\dot{\boldsymbol{q}}$ is uniformly continuous. From Eqs. (4.57) and (4.61), it is seen that $\dot{\boldsymbol{q}} \in L_2(0, \infty)$. Therefore, $\dot{\boldsymbol{q}} \to \boldsymbol{0}$. From the closed-loop Eq. (4.45), $\dot{\boldsymbol{q}} \to \boldsymbol{0}$ implies that $\max(0, f_i(\boldsymbol{x}))\left(\frac{\partial f_i(\boldsymbol{x})}{\partial \boldsymbol{x}}\right)^T \to \boldsymbol{0}$ as $t \to \infty$, which is satisfied when $\boldsymbol{f}(\Delta\boldsymbol{x}) < \boldsymbol{0}$. Since the potential energy has a unique minimum at the desired region inside the compact set Ω, the position of the end effector converges to the desired region. $\triangle\triangle\triangle$

We now extend the controller described by Eq. (4.43) to a region-reaching controller with uncertain Jacobian matrix and gravitational force. The region-reaching controller is designed as

$$\boldsymbol{\tau} = -\hat{\boldsymbol{J}}^T(\boldsymbol{q})\sum_{i=1}^{M} k_{pi}\max(0, f_i(\boldsymbol{x}))\left(\frac{\partial f_i(\boldsymbol{x})}{\partial \boldsymbol{x}}\right)^T - \hat{\boldsymbol{J}}^T(\boldsymbol{q})\boldsymbol{K}_v\hat{\boldsymbol{J}}(\boldsymbol{q})\dot{\boldsymbol{q}} + \boldsymbol{Y}_g(\boldsymbol{q})\hat{\boldsymbol{\theta}}_g, \tag{4.65}$$

where the uncertain dynamic parameters for the gravitational force are updated by

$$\dot{\hat{\boldsymbol{\theta}}}_g = -\boldsymbol{L}_g\boldsymbol{Y}_g^T(\boldsymbol{q})\left(\dot{\boldsymbol{q}} + \alpha\hat{\boldsymbol{J}}^{-1}(\boldsymbol{q})\sum_{i=1}^{M} a_i\max(0, f_i(\boldsymbol{x}))\left(\frac{\partial f_i(\boldsymbol{x})}{\partial \boldsymbol{x}}\right)^T\right), \tag{4.66}$$

where $\boldsymbol{L}_g \in \Re^{n_g \times n_g}$ is a positive definite matrix, α is a positive constant.

The closed-loop equation of the system is obtained by substituting (4.65) into (4.21):

$$\begin{aligned}\boldsymbol{M}(\boldsymbol{q})\ddot{\boldsymbol{q}} + \left(\tfrac{1}{2}\dot{\boldsymbol{M}}(\boldsymbol{q}) + \boldsymbol{S}(\boldsymbol{q}, \dot{\boldsymbol{q}})\right)\dot{\boldsymbol{q}} + \hat{\boldsymbol{J}}^T(\boldsymbol{q})\boldsymbol{K}_v\hat{\boldsymbol{J}}(\boldsymbol{q})\dot{\boldsymbol{q}} + \boldsymbol{Y}_g(\boldsymbol{q})\Delta\boldsymbol{\theta}_g \\ + \hat{\boldsymbol{J}}^T(\boldsymbol{q})\sum_{i=1}^{M} k_{pi}\max(0, f_i(\boldsymbol{x}))\left(\frac{\partial f_i(\boldsymbol{x})}{\partial \boldsymbol{x}}\right)^T = \boldsymbol{0},\end{aligned} \tag{4.67}$$

where $\Delta\boldsymbol{\theta}_g = \boldsymbol{\theta}_g - \hat{\boldsymbol{\theta}}_g$.

We are now ready to prove the following theorem:

Theorem 4.4 *The closed-loop system described by Eqs.* (4.67) *and* (4.66) *gives rise to the convergence of* $\boldsymbol{x}$ *to the desired region* $\boldsymbol{f}(\Delta\boldsymbol{x}) \le \boldsymbol{0}$ *and* $\dot{\boldsymbol{q}}$ *to* $\boldsymbol{0}$ *as* $t \to \infty$ *if the feedback gains* k_p *and* k_v *are chosen to satisfy conditions* (4.56) *and* (4.63).

Proof The Lyapunov-like function is defined as

$$V + \tfrac{1}{2}\Delta\boldsymbol{\theta}_g^T\boldsymbol{L}_g^{-1}\Delta\boldsymbol{\theta}_g, \tag{4.68}$$

where V is as defined in Eq. (4.46).

Differentiating Eq. (4.68) with respect to time and using Eqs. (4.67) and (4.66), we have

$$\frac{\mathrm{d}}{\mathrm{d}t}(V + \frac{1}{2}\Delta\boldsymbol{\theta}_g^T \boldsymbol{L}_g^{-1}\Delta\boldsymbol{\theta}_g) = -W \leq 0, \tag{4.69}$$

where W is as defined in Eq. (4.58). It can be proved similarly that $\dot{\boldsymbol{q}} \rightarrow \mathbf{0}$ implies that $\max(0, f_i(\boldsymbol{x}))\left(\frac{\partial f_i(\boldsymbol{x})}{\partial \boldsymbol{x}}\right)^T \rightarrow \mathbf{0}$ as $t \rightarrow \infty$, which is satisfied when $\boldsymbol{f}(\boldsymbol{x}) < \mathbf{0}$. $\triangle\triangle\triangle$

When the desired region is specified to be arbitrarily small, it reduces to a point. Hence, the region control concept is a generalization of the setpoint control problem.

4.4 Region-Tracking Control with Unified Objective Function

In the previous sections, we considered region-reaching control problems whereby the desired objective can be specified as a static region. In this section, a region-tracking control problem with a dynamic desired region is considered. The dynamic region can be specified in various forms to suit different manipulation tasks, and thus it is a general objective function that unifies various control objectives. The dynamic region can be scaled and rotated to provide flexibility for the specification of robot tasks. When the region is specified as arbitrarily small, it reduces to the conventional trajectory-tracking task. It can also be specified as a performance bound to ensure transient and steady-state response of the system.

To illustrate the concept of region-tracking control, we first introduce a desired region as follows:

$$\boldsymbol{f}(\Delta\boldsymbol{x}) = [f_1(\Delta\boldsymbol{x}), \ldots, f_M(\Delta\boldsymbol{x})]^T \leq \mathbf{0}, \tag{4.70}$$

where M is the total number of objective functions, $\Delta\boldsymbol{x} = \boldsymbol{x} - \boldsymbol{x}_o$ represents the position error with respect to the reference position, where $\boldsymbol{x}_o(t) = [x_{o1}(t), \ldots, x_{om}(t)]^T \in \Re^m$ is a reference position that is time-varying. The vector inequality $\boldsymbol{f}(\Delta\boldsymbol{x}) \leq \mathbf{0}$ implies that $f_i(\Delta\boldsymbol{x}) \leq 0$ for all $i = 1, \ldots, M$. The functions $f_i(\Delta\boldsymbol{x})$ are scalar functions with continuous partial derivatives. An example of the moving region is shown in Fig. 4.13.

To define the general objective function, a variable $\Delta\boldsymbol{x}_T$ is defined as

$$\Delta\boldsymbol{x}_T = \bar{\boldsymbol{T}}(\boldsymbol{x} - \boldsymbol{x}_o) = \bar{\boldsymbol{T}}\Delta\boldsymbol{x}, \tag{4.71}$$

where $\bar{\boldsymbol{T}}$ is a transformation matrix. Using $\Delta\boldsymbol{x}_T$, the unified objective function is defined as follows:

$$\boldsymbol{f}(\Delta\boldsymbol{x}_T) = [f_1(\Delta\boldsymbol{x}_T), \ldots, f_M(\Delta\boldsymbol{x}_T)]^T \leq \mathbf{0}, \tag{4.72}$$

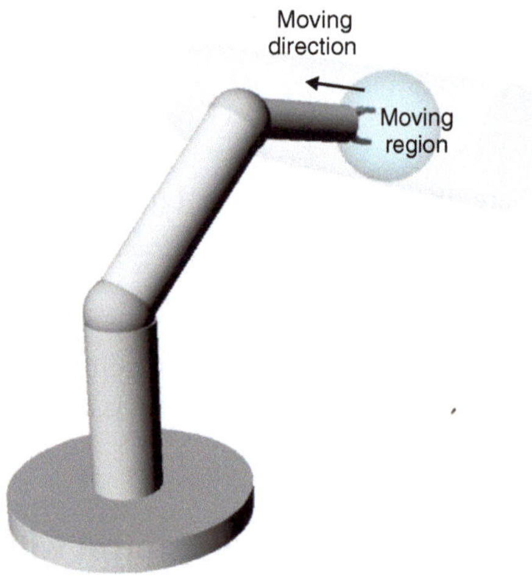

Fig. 4.13 The unified bound is specified as a moving region

where the vector inequality $\boldsymbol{f}(\Delta \boldsymbol{x}_T) \leq \boldsymbol{0}$ implies that $f_i(\Delta \boldsymbol{x}_T) \leq 0$ for all $i = 1, \ldots, M$. The functions $f_i(\Delta \boldsymbol{x}_T)$ are scalar functions with continuous partial derivatives.

The transformation matrix $\bar{\boldsymbol{T}}$ should be specified such that the inverse matrix $\bar{\boldsymbol{T}}^{-1}$ exists. As seen from Eq. (4.71), $\Delta \boldsymbol{x}_T$ is a product of the transformation matrix $\bar{\boldsymbol{T}}$ and the position error $\Delta \boldsymbol{x}$, which provides directional information toward the time-varying unified objective bound.

4.4.1 Moving Region and Dynamic Region

When the transformation matrix $\bar{\boldsymbol{T}}$ is specified as an identity matrix $\bar{\boldsymbol{T}} = \boldsymbol{I}_m$, the unified bound represents the moving region, as defined in Eq. (4.70).

When the unified bound is specified as a dynamic region, the desired region can be scaled and rotated to allow flexibility for robot movement by specifying the transformation matrix $\bar{\boldsymbol{T}}$ as a scaling matrix or a rotation matrix according to specific robot tasks. An example of a dynamic region is shown in Fig. 4.14.

Example 4.8 The transformation matrix $\bar{\boldsymbol{T}}$ can be specified as a scaling matrix in 3-D space as

$$\bar{\boldsymbol{T}} = \boldsymbol{S} = \begin{bmatrix} \frac{1}{S_1} & 0 & 0 \\ 0 & \frac{1}{S_2} & 0 \\ 0 & 0 & \frac{1}{S_3} \end{bmatrix}, \tag{4.73}$$

where $S_i(t)(i = 1, 2, 3)$ are positive scaling factors, and $\boldsymbol{S}$ is a symmetric nonsingular matrix. Therefore, we have

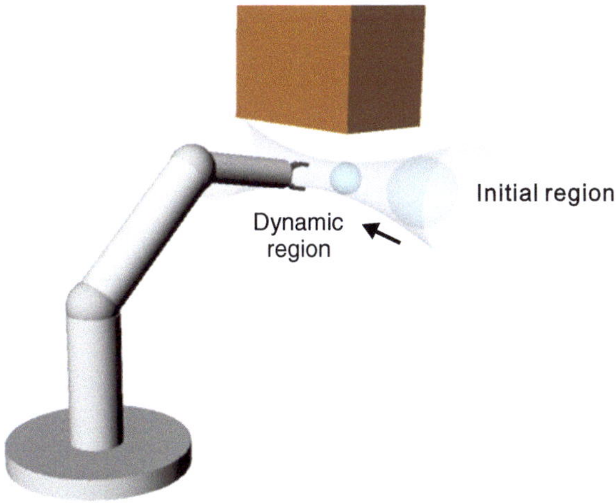

Fig. 4.14 The unified bound is specified as a dynamic region

$$\Delta \boldsymbol{x}_T = \bar{\boldsymbol{T}} \begin{bmatrix} x_1 - x_{o1} \\ x_2 - x_{o2} \\ x_3 - x_{o3} \end{bmatrix} = \begin{bmatrix} \frac{x_1 - x_{o1}}{S_1} \\ \frac{x_2 - x_{o2}}{S_2} \\ \frac{x_3 - x_{o3}}{S_3} \end{bmatrix}. \tag{4.74}$$

□

Various dynamic regions such as sphere, cube, and cylinder can be formed by choosing the appropriate functions and using the scaling matrix in Eq. (4.73).

Example 4.9 When the desired region is specified as a cube in 3-D space, the inequality functions in Eq. (4.72) can be specified as

$$\begin{aligned} f_1(\Delta \boldsymbol{x}_T) &= \frac{(x_1 - x_{o1})^2}{(S_1 b_1)^2} - 1 \leq 0, \\ f_2(\Delta \boldsymbol{x}_T) &= \frac{(x_2 - x_{o2})^2}{(S_2 b_2)^2} - 1 \leq 0, \\ f_3(\Delta \boldsymbol{x}_T) &= \frac{(x_3 - x_{o3})^2}{(S_3 b_3)^2} - 1 \leq 0, \end{aligned} \tag{4.75}$$

where $b_i (i = 1, 2, 3)$ are the individual regional bounds for each axis. □

Example 4.10 When the desired region is designated as a sphere, the inequality function in Eq. (4.72) can be specified as

$$f(\Delta \boldsymbol{x}_T) = \frac{(x_1 - x_{o1})^2}{(S_1 r)^2} + \frac{(x_2 - x_{o2})^2}{(S_2 r)^2} + \frac{(x_3 - x_{o3})^2}{(S_3 r)^2} - 1 \leq 0, \tag{4.76}$$

where Sr is the radius of the sphere, and $S_1 = S_2 = S_3 = S$. The size of the region increases as the factors $S_i (i = 1, 2, 3)$ increase, and conversely. □

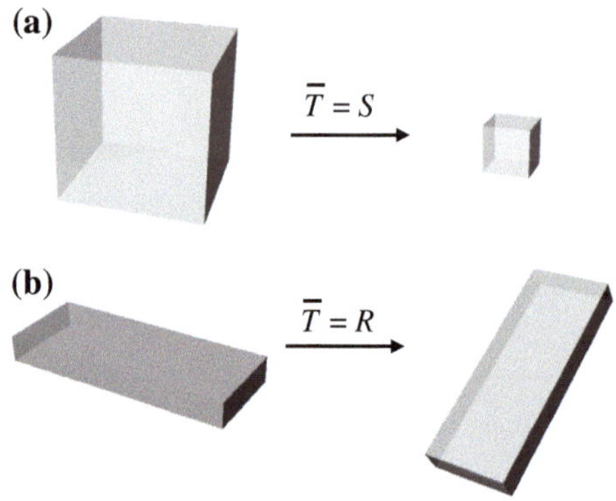

Fig. 4.15 Dynamic regions *before* and *after scaling* and rotation

Example 4.11 The transformation matrix $\bar{\boldsymbol{T}}$ can also be specified as a rotation matrix or a composite rotation matrix, such as

$$\bar{\boldsymbol{T}} = \boldsymbol{R} = \begin{bmatrix} \cos\phi & -\sin\phi & 0 \\ \sin\phi & \cos\phi & 0 \\ 0 & 0 & 1 \end{bmatrix} \begin{bmatrix} 1 & 0 & 0 \\ 0 & \cos\varphi & -\sin\varphi \\ 0 & \sin\varphi & \cos\varphi \end{bmatrix}, \tag{4.77}$$

where $\phi(t)$ and $\varphi(t)$ are rotation angles. □

In addition, the transformation matrix $\bar{\boldsymbol{T}}$ can be specified as the product of the scaling matrix and the rotation matrix, $\bar{\boldsymbol{T}} = \boldsymbol{R}\boldsymbol{S}$, so as to allow the desired region to be scaled and rotated simultaneously. Some examples of dynamic regions are illustrated in Fig. 4.15.

Therefore, the size of the dynamic region can be varied by specifying the transformation matrix. In an application in which high precision is required, it is possible to define the region to be arbitrarily small. When precision is not critical, the region could be scaled up to provide more flexibility in task specifications. Figure 4.16 illustrates a scenario of manipulation using the dynamic region. In Fig. 4.16, the

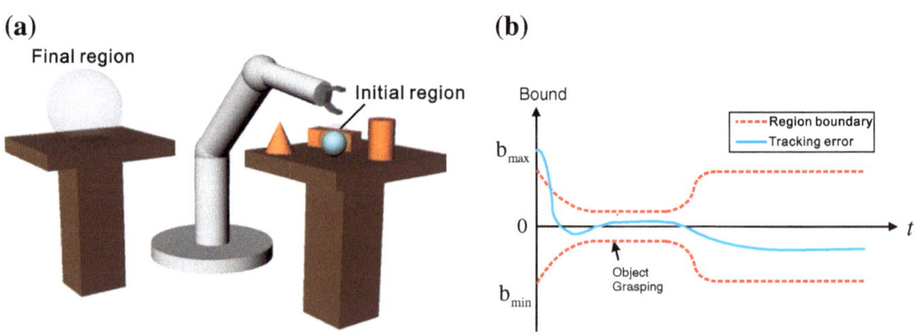

Fig. 4.16 A scenario of robotic manipulation using a scaling dynamic region. **a** The initial stage is small, while the final region is large. **b** Dynamic bound of tracking error

region is set small enough that the end effector can exactly grasp the desired object on a table; After the object has been grasped, the dynamic region is scaled up in such a way that the end effector can place the object at any position on another table.

4.4.2 Performance Bound

The unified objective bound can also be formulated as a performance bound, to specify the transient and steady-state response of the closed-loop system. When the unified objective bound is specified as the performance bound, the function $f_i(\Delta \boldsymbol{x}_T)$ denotes the bound in the ith coordinate. As the performance bound shrinks to a bound around the reference position $\boldsymbol{x}_o$, the tracking error $\Delta \boldsymbol{x} = \boldsymbol{x} - \boldsymbol{x}_o$ reduces correspondingly. Therefore, the variation of the tracking error is related to the variation of the performance bound, and the transformation matrix $\bar{\boldsymbol{T}}$ and the specific form of the function $f_i(\Delta \boldsymbol{x}_T)$ determine the transient performance of the system.

For example, when a transient response with an allowable maximum overshoot as illustrated in Fig. 4.17a is required, the region function $f_i(\Delta \boldsymbol{x}_T)$ can be specified as

$$f_i(\Delta \boldsymbol{x}_T) = \frac{\Delta x_{Ti}^2}{b_i^2} - 1 \leq 0, \tag{4.78}$$

where

$$b_i = \begin{cases} b_{ci}, & x_i \geq x_{oi}, \\ -M_p b_{ci}, & x_i < x_{oi}, \end{cases} \tag{4.79}$$

where $0 < M_p < 1$ is the allowable overshoot, the b_i denote the positions of the boundaries, and the b_{ci} are constants. The transformation matrix $\bar{\boldsymbol{T}}$ is specified as an $m \times m$ matrix as

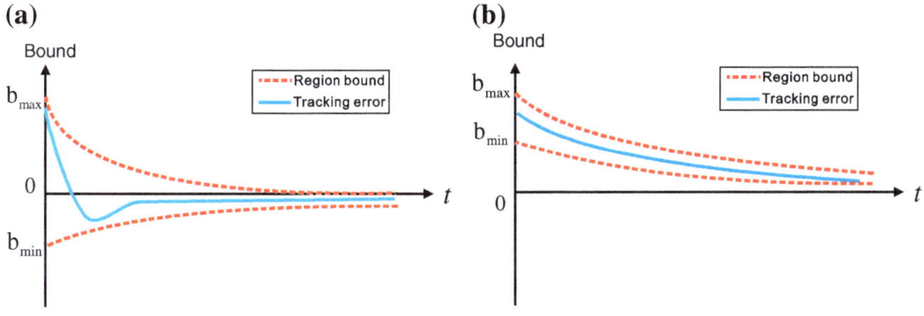

Fig. 4.17 An illustration of transient response: **a** with allowable maximum overshoot; **b** without overshoot

$$\bar{\boldsymbol{T}} = \begin{bmatrix} \frac{1}{\frac{e_{ss}}{M_p b_{c1}} + \mathrm{e}^{-l_c t}} & \cdots & 0 \\ \vdots & \ddots & \vdots \\ 0 & \cdots & \frac{1}{\frac{e_{ss}}{M_p b_{cm}} + \mathrm{e}^{-l_c t}} \end{bmatrix}, \tag{4.80}$$

where l_c represents the convergence speed of the bounds, and e_{ss} represents the steady-state error. From Eqs. (4.79) and (4.80), the initial upper bounds are obtained as $b_{\max} = b_{ci} + \frac{e_{ss}}{M_p}$, and the initial lower bounds are obtained as $b_{\min} = -M_p b_{ci} - e_{ss}$.

The initial position of the end effector can be measured using the task-space sensor. After the initial position is determined, the initial upper and lower bounds can be set to enclose the initial position with respect to the reference position, in such a way that the end effector is initially inside the performance bound, as illustrated in Fig. 4.17.

When a transient response without overshoot as illustrated in Fig. 4.17b is required, $\boldsymbol{f}_i(\Delta \boldsymbol{x}_T)$ is specified as

$$\boldsymbol{f}_i(\Delta \boldsymbol{x}_T) = \begin{bmatrix} \frac{\Delta x_{Ti}^2}{b_{ui}^2} - 1 \\ 1 - \frac{\Delta x_{Ti}^2}{b_{li}^2} \end{bmatrix} \leq \boldsymbol{0}, \tag{4.81}$$

where b_{ui} and b_{li} are constants. The transformation matrix $\bar{\boldsymbol{T}}$ is specified as

$$\bar{\boldsymbol{T}} = \begin{bmatrix} \frac{1}{\frac{e_{ss}}{b_{l1}} + \mathrm{e}^{-l_c t}} & \cdots & 0 \\ \vdots & \ddots & \vdots \\ 0 & \cdots & \frac{1}{\frac{e_{ss}}{b_{lm}} + \mathrm{e}^{-l_c t}} \end{bmatrix}. \tag{4.82}$$

From Eqs. (4.81) and (4.82), the initial upper bounds are obtained as $b_{\max} = b_{ui} + e_{ss}\frac{b_{ui}}{b_{li}}$, and the initial lower bounds are obtained as $b_{\min} = b_{li} + e_{ss}$. Similarly, the upper and lower bounds are set such that the end effector is inside the performance bound from the beginning and stays within it. The overshoot is eliminated when the end effector is restricted such that $\boldsymbol{f}_i(\Delta \boldsymbol{x}_T) \leq \boldsymbol{0}$. As the performance bound shrinks, the end effector is driven towards the reference position.

The unified bound can also be specified as sensory and nonsingularity bounds such that the robot is limited inside the regions where the sensory feedback is always available and the Jacobian matrix is nonsingular.

4.4.3 Potential Energy

As seen from the aforementioned, the unified objective bound is a general objective bound that can be used to specify a desired position, a desired trajectory, a performance bound, a sensory and nonsingularity bound, or a general dynamic region. The specifications of the bound according to various objectives are summarized as follows:

(i) When the reference position, the size of the bound, or the orientation of the bound is time-varying, the unified objective bound represents a general dynamic region.
(ii) When the bound is specified arbitrarily small and the reference position is the desired constant point ($\dot{\boldsymbol{x}}_o = \mathbf{0}$), it reduces to a setpoint.
(iii) When the bound is specified arbitrarily small and the reference position is time-varying ($\dot{\boldsymbol{x}}_o \neq \mathbf{0}$), it reduces to a trajectory.
(iv) When the initial bound encloses the initial error and the variation of the bound corresponds to the required transient response, the desired transient performance can be guaranteed.

The specifications of the unified bound for various objectives or applications are given in Table 4.1.

Based on the unified objective functions defined in (4.72), the potential energy function is designed as

$$P(\Delta \boldsymbol{x}_T) = \sum_{i=1}^{M} P_i(\Delta \boldsymbol{x}_T), \tag{4.83}$$

where

$$P_i(\Delta \boldsymbol{x}_T) = \frac{k_{xi}}{N}[\max(0, f_i(\Delta \boldsymbol{x}_T))]^N. \tag{4.84}$$

Table 4.1 Variation of unified bound

Type	Size of bound	Initial bound	$\boldsymbol{x}_o$	$\bar{\boldsymbol{T}}$
Setpoint	Arbitrarily small	Not required	Desired point	Identity matrix
Trajectory	Arbitrarily small	Not required	Desired trajectory	Identity matrix
Static region	Nonzero and constant	Not required	Constant position inside region	Constant
Dynamic region	Time-varying	Not required	Time-varying position inside region	Scaling or rotation matrix
Performance bound	From initial bound to small bound	Enclosing initial position error	Constant or time-varying	Based on transient requirement

That is,

$$P_i(\Delta \boldsymbol{x}_T) = \begin{cases} 0, & f_i(\Delta \boldsymbol{x}_T) \leq 0, \\ \frac{k_{xi}}{N}[f_i(\Delta \boldsymbol{x}_T)]^N, & f_i(\Delta \boldsymbol{x}_T) > 0 \end{cases} \tag{4.85}$$

where k_{xi} are positive constants. The above energy function is smooth and bounded below by zero.

Differentiating the potential energy function described by (4.84) with respect to $\Delta \boldsymbol{x}_T$ yields

$$\left(\frac{\partial P_i(\Delta \boldsymbol{x}_T)}{\partial \Delta \boldsymbol{x}_T}\right)^T = \begin{cases} 0, & f_i(\Delta \boldsymbol{x}_T) \leq 0, \\ k_{xi}[f_i(\Delta \boldsymbol{x}_T)]^{N-1}\left(\frac{\partial f_i(\Delta \boldsymbol{x}_T)}{\partial \Delta \boldsymbol{x}_T}\right)^T, & f_i(\Delta \boldsymbol{x}_T) > 0. \end{cases} \tag{4.86}$$

From Eq. (4.83), we have

$$\begin{aligned} \left(\frac{\partial P(\Delta \boldsymbol{x}_T)}{\partial \Delta \boldsymbol{x}_T}\right)^T &= \sum_{i=1}^{M}\left(\frac{\partial P_i(\Delta \boldsymbol{x}_T)}{\partial \Delta \boldsymbol{x}_T}\right)^T \\ &= \sum_{i=1}^{M} k_{xi}[\max(0, f_i(\Delta \boldsymbol{x}_T))]^{N-1}\left(\frac{\partial f_i(\Delta \boldsymbol{x}_T)}{\partial \Delta \boldsymbol{x}_T}\right)^T, \end{aligned} \tag{4.87}$$

which denotes the gradient of the potential energy. When the end effector enters the bound $f_i(\Delta \boldsymbol{x}_T) < 0$, the gradient reduces to zero, as illustrated in Fig. 4.18.

It is intuitive that reaching for bounds instead of a setpoint or trajectory should allow more flexibility in robot tasks. In fact, the potential energy $P_i(\Delta \boldsymbol{x}_T)$ remains zero if the end effector starts from an initial position within the bound and stays inside the bound where $f_i(\Delta \boldsymbol{x}_T) < 0$. In contrast, the potential energy using the conventional position error $\Delta \boldsymbol{x}$ is nonzero until the end effector reaches the reference position $\boldsymbol{x}_o$. Figure 4.19 illustrates the variation of potential energy $P_i(\Delta \boldsymbol{x}_T)$ as the

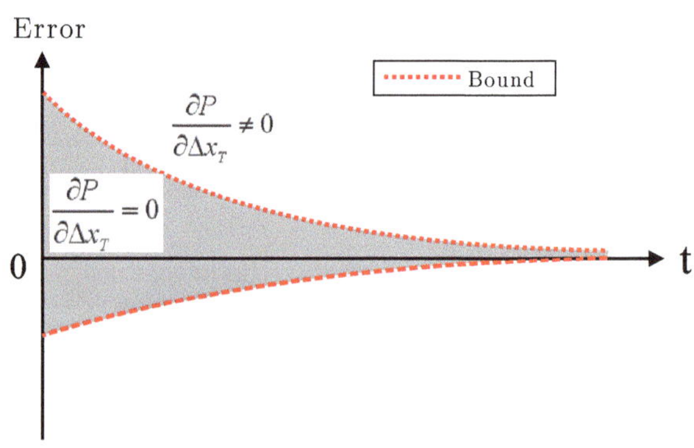

Fig. 4.18 The gradient is nonzero when the end effector is outside the bound, and it reduces to zero after the end effector enters the performance bound

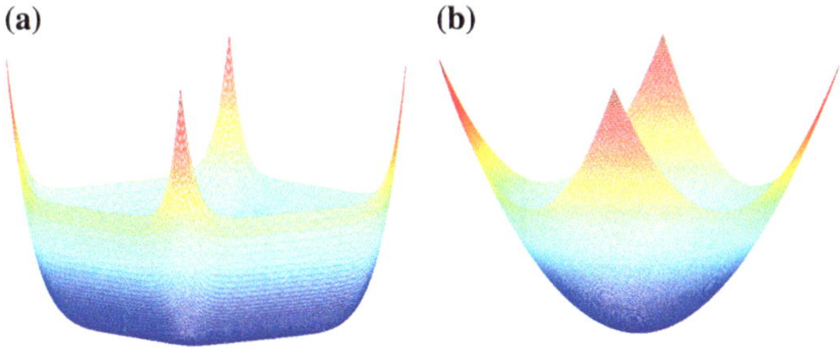

Fig. 4.19 The *bottom* of potential energy corresponds to the bound. As the bound shrinks, the *bottom* of potential energy converges to the desired position, **a** $t = 0$, **b** $t \to \infty$

desired bound shrinks. In Fig. 4.19, $P_i(\Delta \boldsymbol{x}_T)$ is zero when the end effector is inside the desired bound, and the bottom of $P_i(\Delta \boldsymbol{x}_T)$ shrinks to a point as the bound is scaled down.

When the unified bound is specified as a performance bound, the potential energy should be specified with an arbitrarily large gradient to keep the position error within the performance bound. To fulfill the requirement, a high-order potential energy $P_h(\Delta \boldsymbol{x}_T)$ is introduced as

$$P_h(\Delta \boldsymbol{x}_T) = \sum_{i=1}^{M} \frac{k_{hi}}{N_h} [\max(0, f_i(\Delta \boldsymbol{x}_T))]^{N_h}, \tag{4.88}$$

where k_{hi} are positive constants, and $N_h \geq 10$ is the order of the function, which is an even integer. The potential energy is then the combination of $P_h(\Delta \boldsymbol{x}_T)$ and the original potential energy $P(\Delta \boldsymbol{x}_T)$:

$$P_c(\Delta \boldsymbol{x}_T) = P_h(\Delta \boldsymbol{x}_T) + P(\Delta \boldsymbol{x}_T), \tag{4.89}$$

where $P_c(\Delta \boldsymbol{x}_T)$ represents the new potential energy. An illustration of $P_c(\Delta \boldsymbol{x}_T)$ is shown in Fig. 4.20a. Note that the parameters k_{xi} and k_{hi} in Eq. (4.89) are set large to create the steep gradient.

The steep gradient of $P_c(\Delta \boldsymbol{x}_T)$ may cause the oscillation of robot movement when the end effector is very near the boundary of the performance bound. To alleviate the problem, another potential energy function with a smaller reference bound can be introduced:

$$P_r(\Delta \boldsymbol{x}_T) = \sum_{i=1}^{M} \frac{k_{ri}}{N} [\max(0, f_{ri}(\Delta \boldsymbol{x}_T))]^{N}, \tag{4.90}$$

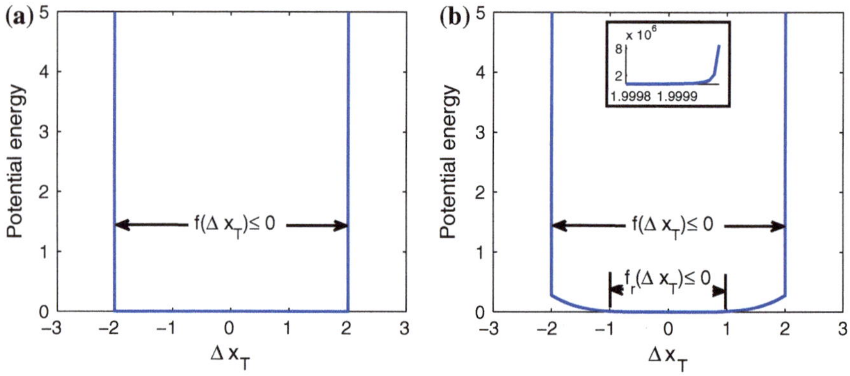

Fig. 4.20 Examples of potential energy functions with or without the reference region. **a** $P_c(\Delta x_T)$ (without $f_{ri}(\Delta x_T)$). **b** $P_o(\Delta x_T)$ (with $f_{ri}(\Delta x_T)$)

where k_{ri} are positive constants, and $f_{ri}(\Delta \boldsymbol{x}_T) \leq 0$ is the reference bound, which is inside $f_i(\Delta \boldsymbol{x}_T) \leq 0$. The parameters k_{ri} are not large, and thus the gradient of $P_r(\Delta \boldsymbol{x}_T)$ is smaller. Therefore, the overall potential energy can be specified as the summation of $P_r(\Delta \boldsymbol{x}_T)$ and $P_c(\Delta \boldsymbol{x}_T)$ as

$$P_o(\Delta \boldsymbol{x}_T) = P_c(\Delta \boldsymbol{x}_T) + P_r(\Delta \boldsymbol{x}_T), \tag{4.91}$$

where $P_o(\Delta \boldsymbol{x}_T)$ represents the overall potential energy. An illustration of $P_o(\Delta \boldsymbol{x}_T)$ is shown in Fig. 4.20b. The potential energy is smooth and bounded below by zero. If the end effector moves beyond the reference region where $f_{ri}(\Delta \boldsymbol{x}_T) > 0$, the gradient of $P_r(\Delta \boldsymbol{x}_T)$ becomes nonzero and drives the end effector back from the steep gradient. The reference region is hence introduced to reduce the possibility of oscillation in actual implementations. Note that the use of a reference region is optional.

Since the concept of a unified bound is general, it is also possible to formulate other potential energy functions with a steep gradient to keep the tracking errors inside the performance bound. For example, another potential energy function for the performance bound can be specified as

$$P_o(\Delta \boldsymbol{x}_T) = \sum_{i=1}^{M} \frac{k_{xi}}{2} \frac{[\max(0, f_{ri}(\Delta \boldsymbol{x}_T))]^2}{[f_i(\Delta \boldsymbol{x}_T)]^2}, \tag{4.92}$$

where an illustration of the potential energy $P_o(\Delta \boldsymbol{x}_T)$ in Eq. (4.92) is shown in Fig. 4.21. Since the initial bound encloses the initial error such that $f_{ri}(\Delta \boldsymbol{x}_T) < 0$, the initial potential energy is zero. When the end effector is very near the boundary where $f_i(\Delta \boldsymbol{x}_T) = 0$, the gradient of the potential energy becomes very large.

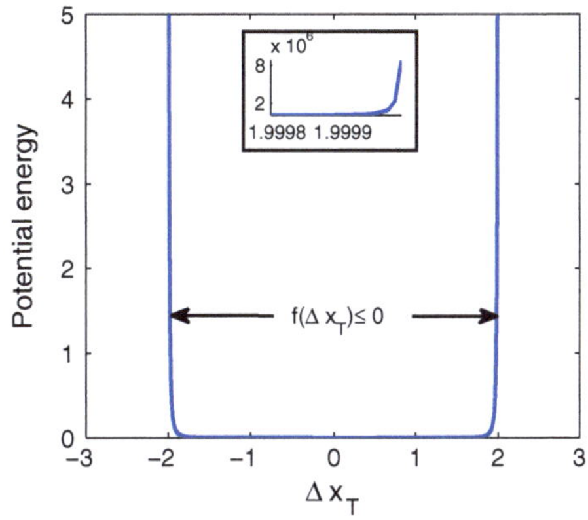

Fig. 4.21 An illustration of $P_o(\Delta \boldsymbol{x}_T)$

Similarly, differentiating $P_o(\Delta \boldsymbol{x}_T)$ with respect to $\Delta \boldsymbol{x}_T$ yields

$$\left(\frac{\partial P_o(\Delta \boldsymbol{x}_T)}{\partial \Delta \boldsymbol{x}_T}\right)^T \triangleq \Delta \varepsilon, \tag{4.93}$$

where $\Delta\varepsilon$ denotes a unified region error that is also the gradient of $P_o(\Delta \boldsymbol{x}_T)$. When the unified bound is not specified as the performance bound, k_{hi} and k_{ri} are set to zero, and $\Delta\varepsilon$ smoothly transits from nonzero to zero after the end effector enters the bound. When the unified bound is specified as the performance bound, k_{xi} and k_{hi} are set to be large, while k_{ri} can be small, and $\Delta\varepsilon$ becomes arbitrarily large if the end effector is outside the performance bound.

The potential energy function with the steep gradient $P_o(\Delta \boldsymbol{x}_T)$ can be used to keep the robot within the nonsingular regions and/or sensing zone (see Fig. 4.12). In general, by adding the corresponding region error $\Delta\varepsilon$ defined in (4.93) to a sensory-space feedback controller, the movement of the robot is always limited inside the sensory feedback regions and the nonsingular regions.

4.4.4 Adaptive Region-Tracking Control

We have seen that the desired objective for the robot end effector can be specified as a unified objective bound to provide flexibility in robot tasks. For example, when the bound is specified to be arbitrarily small, the control objective reduces to the conventional trajectory-tracking control. When the unified bound is specified as a performance bound, the transient response of the robot system is guaranteed. Based on the unified objective bound, we consider the following adaptive region-tracking controller:

$$\boldsymbol{\tau} = -\boldsymbol{K}_s \boldsymbol{s} - \hat{\boldsymbol{J}}^T\left(\boldsymbol{q}, \hat{\boldsymbol{\theta}}_k\right) \bar{\boldsymbol{T}}^T \Delta\varepsilon + \boldsymbol{Y}_d\left(\boldsymbol{q}, \dot{\boldsymbol{q}}, \dot{\boldsymbol{q}}_r, \ddot{\boldsymbol{q}}_r\right) \hat{\boldsymbol{\theta}}_d, \tag{4.94}$$

where s is a sliding vector, which is introduced as

$$s = \dot{q} - \dot{q}_r = \dot{q} - \hat{J}^+\left(q, \hat{\theta}_k\right)\left(\dot{x}_o - \bar{T}^{-1}\dot{\bar{T}}\Delta x\right) + \alpha\hat{J}^+\left(q, \hat{\theta}_k\right)\bar{T}^{-1}\Delta\varepsilon, \tag{4.95}$$

and

$$\dot{q}_r = \hat{J}^+\left(q, \hat{\theta}_k\right)\left(\dot{x}_o - \bar{T}^{-1}\dot{\bar{T}}\Delta x\right) - \alpha\hat{J}^+\left(q, \hat{\theta}_k\right)\bar{T}^{-1}\Delta\varepsilon \tag{4.96}$$

is a reference vector, $\hat{J}(q, \hat{\theta}_k) \in \Re^{m\times n}$ is the adaptive Jacobian matrix, $\hat{J}^+(q, \hat{\theta}_k) \in \Re^{n\times m}$ denotes the pseudoinverse of $\hat{J}(q, \hat{\theta}_k)$, α is a positive constant, $K_s \in \Re^{n\times n}$ is a diagonal positive definite matrix.

The estimated dynamic parameter $\hat{\theta}_d$ is updated by the following update law:

$$\dot{\hat{\theta}}_d = -L_d Y_d^T\left(q, \dot{q}, \dot{q}_r, \ddot{q}_r\right) s, \tag{4.97}$$

where $L_d \in \Re^{n_d\times n_d}$ is a diagonal positive definite matrix, and $Y_d(q, \dot{q}, \dot{q}_r, \ddot{q}_r) \in \Re^{n\times n_d}$ denotes a regressor matrix.

The estimated kinematic parameter $\hat{\theta}_k$ is updated by the following update law:

$$\dot{\hat{\theta}}_k = k_p L_k Y_k^T\left(q, \dot{q}\right)\bar{T}^T\Delta\varepsilon, \tag{4.98}$$

where $L_k \in \Re^{n_k\times n_k}$ is a diagonal positive definite matrix, and $Y_k(q, \dot{q}) \in \Re^{m\times n_k}$ is another regressor matrix.

Using the sliding vector s defined by Eq. (4.95), the robot dynamics described by Eq. (4.21) is written as

$$M(q)\dot{s} + \left(\tfrac{1}{2}\dot{M}(q) + S(q, \dot{q})\right)s + M(q)\ddot{q}_r + \left(\tfrac{1}{2}\dot{M}(q) + S(q, \dot{q})\right)\dot{q}_r + g(q) = \tau, \tag{4.99}$$

and the term $M(q)\ddot{q}_r + (\frac{1}{2}\dot{M}(q) + S(q, \dot{q}))\dot{q}_r + g(q)$ can be expressed in terms of the regressor and dynamic parameter according to *Property* 1.3 as

$$M(q)\ddot{q}_r + \left(\tfrac{1}{2}\dot{M}(q) + S(q, \dot{q})\right)\dot{q}_r + g(q) = Y_d\left(q, \dot{q}, \dot{q}_r, \ddot{q}_r\right)\theta_d. \tag{4.100}$$

Substituting Eqs. (4.94) and (4.100) into Eq. (4.99), we obtain the following closed-loop equation:

$$\begin{aligned} M(q)\dot{s} + \left(\tfrac{1}{2}\dot{M}(q) + S(q, \dot{q})\right)s + K_s s + \hat{J}^T\left(q, \hat{\theta}_k\right)\bar{T}^T\Delta\varepsilon \\ + Y_d\left(q, \dot{q}, \dot{q}_r, \ddot{q}_r\right)\Delta\theta_d = \mathbf{0}, \end{aligned} \tag{4.101}$$

where $\Delta\theta_d = \theta_d - \hat{\theta}_d$.

To prove the stability of the robot system, a Lyapunov-like candidate is defined as

$$V = \frac{1}{2}s^T M(q)s + P_o(\Delta x_T) + \frac{1}{2}\Delta\theta_d^T L_d^{-1}\Delta\theta_d + \frac{1}{2}\Delta\theta_k^T L_k^{-1}\Delta\theta_k. \quad (4.102)$$

Next, differentiating V with respect to time, we have

$$\dot{V} = s^T M(q)\dot{s} + \frac{1}{2}s^T \dot{M}(q)s + (\dot{\bar{T}}\Delta x + \bar{T}\Delta\dot{x})^T\Delta\varepsilon - \dot{\hat{\theta}}_d^T L_d^{-1}\Delta\theta_d - \dot{\hat{\theta}}_k^T L_k^{-1}\Delta\theta_k. \quad (4.103)$$

Substituting the closed-loop Eq. (4.101), the sliding vector in Eq. (4.95), and the update laws in Eqs. (4.97) and (4.98) into Eq. (4.103) and using the properties of robot dynamics, we have

$$\begin{aligned} \dot{V} &= -s^T K_s s - s^T\left\{\hat{J}^T(q,\hat{\theta}_k)\bar{T}^T\Delta\varepsilon + Y_d(q,\dot{q},\dot{q}_r,\ddot{q}_r)\Delta\theta_d\right\} \\ &\quad +(\dot{\bar{T}}\Delta x + \bar{T}\Delta\dot{x})^T\Delta\varepsilon - \dot{\hat{\theta}}_d^T L_d^{-1}\Delta\theta_d - \dot{\hat{\theta}}_k^T L_k^{-1}\Delta\theta_k \\ &= -s^T K_s s - \alpha\Delta\varepsilon^T\Delta\varepsilon. \end{aligned} \quad (4.104)$$

We define a compact set Ω as

$$\Omega = \{(q,\dot{q}) : V \leq \gamma\}, \quad (4.105)$$

where the potential energy for the region function $P_o(\Delta x_T)$ has a unique minimum at the desired region with a positive γ. If $(\dot{q}(0), q(0)) \in \Omega$, then $(q,\dot{q}) \in \Omega$, since $\dot{V} \leq 0$.

We now state the following theorem.

Theorem 4.5 *When the adaptive region-tracking control scheme and the update law in Eqs.* (4.94), (4.97), *and* (4.98) *are applied to a robotic manipulator, the closed-loop system gives rise to the convergence of the task-space position of the end effector to the unified bound.*

Proof Since $V > 0$ and $\dot{V} \leq 0$, V is bounded. Since V is bounded, s, $P_o(\Delta x_T)$, $\Delta\theta_d$, and $\Delta\theta_k$ are all bounded. The boundedness of $P_o(\Delta x_T)$ ensures the boundedness of the functions $f_i(\Delta x_T)$, as seen from the definition of $P_o(\Delta x_T)$ in Eq. (4.91). From the properties of the region function, Δx is also bounded. Since s, Δx, $\Delta\theta_d$, and $\Delta\theta_k$ are all bounded, the closed-loop system is stable.

In addition, the boundedness of Δx ensures the boundedness of x, since x_o is bounded. From the properties of the unified bound, the boundedness of $P_o(\Delta x_T)$ also ensures the boundedness of its partial derivative $\frac{\partial P_o(\Delta x_T)}{\partial \Delta x_T}$. Since both the functions of bound and its partial derivatives are bounded, the unified region error $\Delta\varepsilon$ is bounded, as seen from Eq. (4.93).

The boundedness of $\Delta\varepsilon$ ensures the boundedness of $\dot{q}_r$, as seen from Eq. (4.96). Since both s and $\dot{q}_r$ are bounded and $s = \dot{q} - \dot{q}_r$, $\dot{q}$ is bounded. The boundedness of $\dot{q}$ also ensures the boundedness of $\dot{x}$, since $\dot{x} = J(q)\dot{q}$ and $J(q)$ are trigonometric functions of $\boldsymbol{q}$. The boundedness of $\dot{\boldsymbol{x}}$ ensures the boundedness of the time derivative of the region error $\Delta\dot{\varepsilon}$, which implies that $\Delta\varepsilon$ is uniformly continuous. From Eq. (4.104), we obtain $\boldsymbol{s}$, $\Delta\varepsilon \in L_2(0, +\infty)$. Then it follows from Barbalat's lemma that

$$\Delta\varepsilon \rightarrow \mathbf{0}. \tag{4.106}$$

The convergence of $\Delta\varepsilon \rightarrow \mathbf{0}$ implies that $\boldsymbol{f}(\Delta\boldsymbol{x}_T) \leq \mathbf{0}$, which also indicates that the end effector stays inside the unified bound at steady state.
(i) When the unified bound is specified as the performance bound or the sensory and nonsingularity bound, the initial position of the end effector is inside the bound so that the initial region error satisfies $\Delta\varepsilon(0) = \mathbf{0}$. Suppose that the end effector exceeds the bound at a time instant when $f_i(\Delta\boldsymbol{x}_T) > 0$, $\Delta\varepsilon$ becomes arbitrarily large (see Figs. 4.20 and 4.21), which is contradicted with the conclusion that the region error $\Delta\varepsilon$ is bounded throughout the robot movement. Therefore, the end effector always stays inside the bound.
(ii) When the unified bound is specified as other types of regions, the end effector can start outside the bound during the transient stage, and it has been shown that the position of the end effector converges to the desired bound at steady state. $\triangle\triangle\triangle$

4.5 Summary and Notes

It is interesting to observe from human reaching tasks that the desired targets are regions with arbitrary shapes rather than points. Inspired by human reaching tasks, the concept of region control was first brought forward in Cheah and Wang (2005); Cheah et al. (2007). Specifying the objective as a region is a more general task, because when the desired region is very small, it reduces to a bound or a point. Region control provides more flexibility for robotic task specification, and it also consumes less energy than convectional setpoint control. In Cheah (2008b), it was shown that the concept of region control can be used to impose physical constraints on robot motion by adding additional regions. The desired region was extended to a performance bound in Li and Cheah (2014), where the transient response of the robot system is regulated when the end effector starts inside the performance bound and remains within it.

The results of region control are useful for multiple robot systems when the objective for the multiple robots is specified as a region. In Cheah et al. (2009), a shape-formation control scheme was proposed for a multiple-robot system in which the robots move as a group inside a desired region while maintaining a minimum separation distance among themselves. The results were also extended to coordination of of multigroup of robots in Haghighi and Cheah (2012). Decomposition of robot swarms to multiple groups increases versatility of the whole community in operating in intricate environments. The application of region control can also be found in cell manipulation using optical tweezers (Li and Cheah 2012a; Chen et al. 2013), where multiple microscopic particles are trapped by optical tweezers and manipulated toward a redefined region without collision.

Chapter 5
Regional Feedback Control of a Robot

Task-space sensory feedback control methods are valid only in a finite task space within a limited sensing zone where singularities of the Jacobian matrix are avoided. In this chapter, a regional feedback method is designed for task-space robot control. Based on the concept of regional feedback, a global task-space controller is developed for a robotic manipulator, which enables the end effector to start from any initial position outside the sensing zone and in the vicinity of singularities, and reach for a desired trajectory. The global task-space controller also allows the robot to enter vision-occluded areas or singular regions during the course of movement.

5.1 Regional Feedback

Cartesian-space feedback control fails when the robot is at or near a singular configuration and vision feedback is unavailable outside the field of view or inside the occluded areas. Therefore, task-space feedback information is inaccessible in the entire robot workspace. The stability analysis of standard task-space feedback control system is thus valid only in a finite workspace.

Given a desired trajectory $\boldsymbol{x}_d$, the objective of this chapter is to introduce a regional feedback control design method such that the global stability of the closed-loop system can be guaranteed. The task-space feedback $\boldsymbol{x}_T$ is considered a desired task variable in which the desired trajectory is specified. To achieve global stability, a reaching-task variable $\boldsymbol{r}_R$ that specifies the position of the robot in a different coordinate space with available feedback information is introduced. The variable $\boldsymbol{r}_R$ is used to drive the robot into the working range of $\boldsymbol{x}_T$. That is, in the regions where $\boldsymbol{x}_T$ is infeasible, the reaching-task variable $\boldsymbol{r}_R$ is adopted to drive the robot to leave those regions, as illustrated in Fig. 5.1.

C.C. Cheah and X. Li, *Task-Space Sensory Feedback Control of Robot Manipulators*, Intelligent Systems, Control and Automation: Science and Engineering 73,
DOI 10.1007/978-981-287-062-9_5

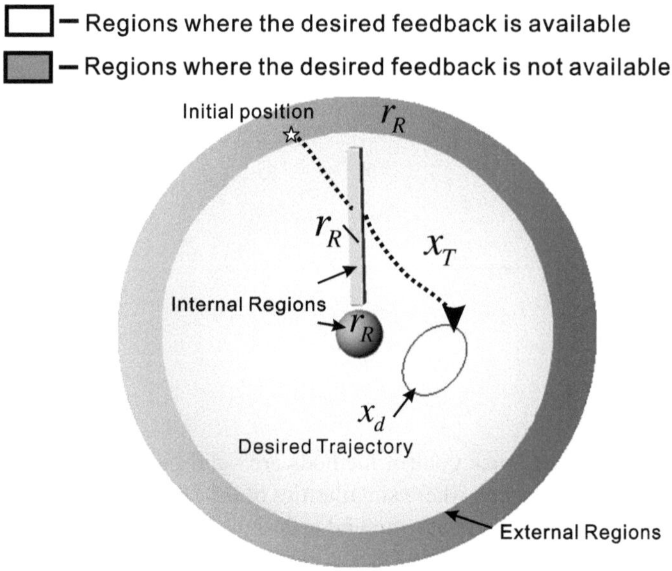

Fig. 5.1 Illustration of external and internal task-space regions

Using these two task variables, a regional feedback method is designed in this chapter for task-space robot control that integrates different feedback information smoothly into a single controller. Each piece of feedback information is adopted in a corresponding local region, and the combination of regional information ensures global convergence of the robot motion. The regional feedback information consists of the reaching-task variable $\boldsymbol{r}_R$, which drives the end effector to approach the desired task region, and the desired task variable $\boldsymbol{x}_T$, which converges to the desired trajectory.

Note that the purpose of the regional feedback method is not to avoid regions in which singularities or occlusions may occur, but to enable the robot to pass through such regions.

The entire robot workspace is divided into external regions and internal regions, as illustrated in Fig. 5.1. The desired task variable $\boldsymbol{x}_T$ is unavailable in these regions due to the limited field of view, occlusion, or singularities. Therefore, the reaching-task variable $\boldsymbol{r}_R$ is employed. By specifying the reaching-task variable $\boldsymbol{r}_R$ and the desired task variable $\boldsymbol{x}_T$ in different task coordinates according to specific robot tasks, it is shown that the regional feedback method is a unified formulation that is able to address various issues in robot task-space control. In the following, we first review some problems in robot task-space control and then introduce a unified formulation to address these issues.

5.1.1 Singularity Problem

In Cartesian-space control, the desired task variable $\boldsymbol{x}_T$ is specified in Cartesian space as

$$\boldsymbol{x}_T = \boldsymbol{r}, \tag{5.1}$$

where $\boldsymbol{r}$ represents the position of the end effector in Cartesian space.

The desired task feedback $\boldsymbol{x}_T$ is infeasible near the singular positions where the inverse Jacobian matrix approaches a singularity. Figure 5.2 shows the regions for a 2-DOF planar manipulator where the Cartesian-space feedback is feasible. To solve the problem, the joint position is specified as the reaching-task variable $\boldsymbol{r}_R$ and is employed near or at the singular positions. The variable $\boldsymbol{r}_R$ is thus specified in joint space as

$$\boldsymbol{r}_R = \boldsymbol{q}, \tag{5.2}$$

where $\boldsymbol{q}$ denotes the joint configuration.

For the 2-DOF planar manipulator in Example 1.3, the Jacobian matrix of the manipulator is described as

$$\boldsymbol{J}_m(\boldsymbol{q}) = \begin{bmatrix} -l_1\sin(q_1) - l_2\sin(q_1+q_2) & -l_2\sin(q_1+q_2) \\ l_1\cos(q_1) + l_2\cos(q_1+q_2) & l_2\cos(q_1+q_2) \end{bmatrix}, \tag{5.3}$$

and the determinant of the Jacobian matrix is obtained as

$$\det[\boldsymbol{J}_m(\boldsymbol{q})] = l_1 l_2 \sin(q_2) = 0. \tag{5.4}$$

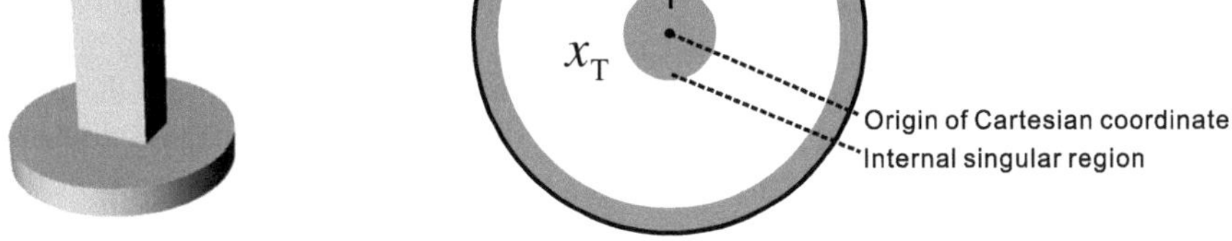

Fig. 5.2 Example of a singularity problem

Therefore, singularities occur where $q_2 = 0$ (external boundary singularity) and $q_2 = \pi$ (internal boundary singularity). Based on the singular configurations, two regions can be specified,

$$h_{E_1}(\boldsymbol{q}) = b_{E_1}^2 - (q_2)^2 \geq 0, \tag{5.5}$$

$$h_{I_{11}}(\boldsymbol{q}) = (q_2 - \pi)^2 - b_{I_{11}}^2 \leq 0, \tag{5.6}$$

where $h_{E_1}(\boldsymbol{q}) \geq 0$ is the external region, which covers the external boundary singularity, and $h_{I_{11}}(\boldsymbol{q}) \leq 0$ is the internal region, which covers the internal boundary singularity, and b_{E_1} and $b_{I_{11}}$ are positive constants that represent the sizes of the singular regions, which should be set to ensure that the manipulator is sufficiently far away from the singular configurations when the end effector leaves the singular regions. The joint information is used where $h_{E_1}(\boldsymbol{q}) \geq 0$ or $h_{I_{11}}(\boldsymbol{q}) \leq 0$ to drive the end effector away from the singular configurations. After the end effector has left the singular regions, the Cartesian-space feedback is activated to move the end effector toward the desired position.

For the 3-DOF manipulator illustrated in Example 1.4, the Jacobian matrix is obtained as

$$\boldsymbol{J}_m(\boldsymbol{q}) = \begin{bmatrix} -\mathrm{s}_1(l_2\mathrm{c}_2 + l_3\mathrm{c}_{23}) & -\mathrm{s}_2\mathrm{c}_1 l_2 - \mathrm{s}_{23} l_3 \mathrm{c}_1 & -\mathrm{s}_{23} l_3 \mathrm{c}_1 \\ -\mathrm{c}_1(l_2\mathrm{c}_2 + l_3\mathrm{c}_{23}) & -\mathrm{s}_2\mathrm{s}_1 l_2 - \mathrm{s}_{23} l_3 \mathrm{s}_1 & -\mathrm{s}_{23} l_3 \mathrm{s}_1 \\ 0 & -\mathrm{c}_2 l_2 - \mathrm{c}_{23} l_3 & -\mathrm{c}_{23} l_3 \end{bmatrix}, \tag{5.7}$$

and the determinant of the Jacobian matrix is obtained as

$$\det[\boldsymbol{J}_m(\boldsymbol{q})] = [l_2\cos(q_2) + l_3\cos(q_2 + q_3)]l_3 l_2 \sin(q_3) = 0. \tag{5.8}$$

Therefore, singularities occur when $q_3 = 0$, $q_3 = \pi$, and $[l_2\cos(q_2) + l_3\cos(q_2 + q_3)] = 0$. Similarly, three regions can be specified in joint space as

$$\begin{aligned} h_{E_1}(\boldsymbol{q}) &= b_{E_1}^2 - (q_3)^2 \geq 0, \\ h_{I_{11}}(\boldsymbol{q}) &= (q_3 - \pi)^2 - b_{I_{11}}^2 \leq 0, \\ h_{I_{12}}(\boldsymbol{q}) &= [l_2\cos(q_2) + l_3\cos(q_2 + q_3)]^2 - b_{I_{12}}^2 \leq 0, \end{aligned} \tag{5.9}$$

where $h_{E_1}(\boldsymbol{q}) \geq 0$ is the external region, which covers the external boundary singularity, and $h_{I_{11}}(\boldsymbol{q}) \leq 0$ is the internal region, which covers the internal boundary singularity, and $h_{I_{12}}(\boldsymbol{q}) \leq 0$ is another internal region, which covers the interior singularity, and $b_{I_{12}}$ is a positive constant. The joint information is used where $h_{E_1}(\boldsymbol{q}) \geq 0$, $h_{I_{11}}(\boldsymbol{q}) \leq 0$, or $h_{I_{12}}(\boldsymbol{q}) \leq 0$ to drive the end effector away from the singular configurations.

5.1.2 Limited Field of View

In vision-based control, the desired task variable $\boldsymbol{x}_T$ is specified in image space as

$$\boldsymbol{x}_T = \boldsymbol{x}, \tag{5.10}$$

where $\boldsymbol{x}$ denotes the position of the end effector in image space.

Due to the limited field of view of the camera or image occlusion, the visual feedback $\boldsymbol{x}$ cannot cover the entire robot workspace, as illustrated in Fig. 5.3. To solve this problem, the reaching-task variable $\boldsymbol{r}_R$ is specified in Cartesian space as

$$\boldsymbol{r}_R = \boldsymbol{r}. \tag{5.11}$$

Using the reaching task feedback $\boldsymbol{r}_R$, the end effector can start from an initial position outside the field of view or leave the field of view during the course of movement, and it can also move through the vision-occluded areas in the presence of occlusion.

Consider a standard fixed camera with a rectangular field of view, an external region can be specified in Cartesian space to cover those regions outside the field of view as

$$\boldsymbol{h}_{E_1}(\boldsymbol{r}) = \begin{bmatrix} \frac{(r_1 - r_{o_1})^2}{(r_{b_1} - r_{o_1})^2} - 1 \\ \frac{(r_2 - r_{o_2})^2}{(r_{b_2} - r_{o_2})^2} - 1 \\ \frac{(r_3 - r_{o_3})^2}{(r_{b_3} - r_{o_3})^2} - 1 \end{bmatrix} \geq \boldsymbol{0}, \tag{5.12}$$

where the vector $\boldsymbol{r}_b = [r_{b_1}, r_{b_2}, r_{b_3}]^T$ represents a set of boundary positions, and the vector $\boldsymbol{r}_o = [r_{o_1}, r_{o_2}, r_{o_3}]^T$ gives static reference positions. The function $\boldsymbol{h}_{E_1}(\boldsymbol{r})$ describes a rectangular block that is parallel to the field of view. When the block is mapped into the 2D plane of the field of view, it corresponds to a rectangle inside the field of view that thus matches the field of view. When the Cartesian-space feedback

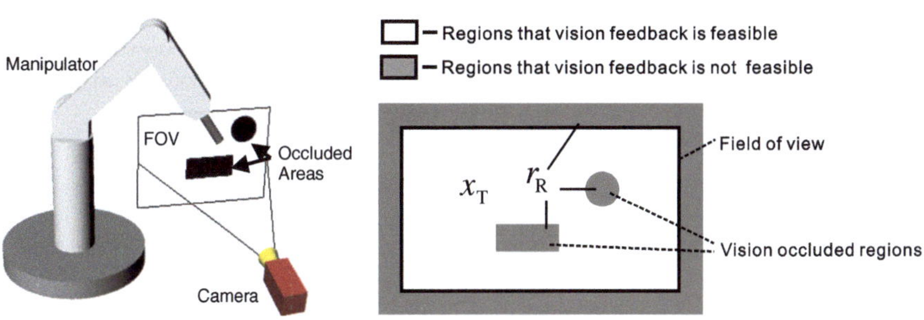

Fig. 5.3 Example of limited field of view

is employed such that the end effector is inside the Cartesian-space region where $\boldsymbol{h}_{E_1}(\boldsymbol{r}) \leq \boldsymbol{0}$, the end effector is also within the field of view. Since the objective is to bring the end effector into the field of view, the position information of the end effector is sufficient.

5.1.3 Micromanipulation and Mixed Cameras

In a micromanipulation system, the desired task variable $\boldsymbol{x}_T$ represents the position of the end effector measured by a microscope for the realization of the manipulation task. Therefore, the variable $\boldsymbol{x}_T$ is specified in the image space of the microscope as

$$\boldsymbol{x}_T = \boldsymbol{x}_z, \tag{5.13}$$

where $\boldsymbol{x}_z$ denotes the position of the end effector in the image space of the microscope.

Another camera with a wide angle can be used to control the movement of the end effector at the beginning stage, so as to drive the end effector toward the field of view of the microscope. Therefore, the reaching-task variable $\boldsymbol{r}_R$ is defined as the vision information obtained from the wide-angle camera as

$$\boldsymbol{r}_R = \boldsymbol{x}_w, \tag{5.14}$$

where $\boldsymbol{x}_w$ denotes the position of end effector in image space of the wide-angle camera.

Using regional feedback, the transition from the feedback $\boldsymbol{r}_R$ to the feedback $\boldsymbol{x}_T$ is smooth, which prevents vibration during the course of micromanipulation. The regional feedback control method can also be used in other types of mixed-camera control with a wide-angle lens such as a fish-eye camera, as illustrated in Fig. 5.4. Similarly, a wide-angle lens provides a large field of view for the robot's movement in the initial stage, while the zoom camera ensures the accuracy of robot movement by providing greater detail in the end stage.

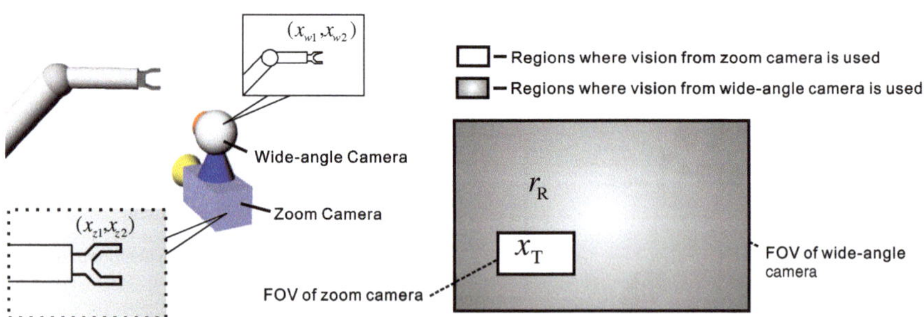

Fig. 5.4 Example of mixed cameras

An external region function in the image space of the wide-angle lens is specified to match the shape of the field of view of the zoom camera as

$$h_{E_1}(\boldsymbol{x}_w) = \frac{(x_{w1} - x_{E_1})^{20}}{a_1^{20}} + \frac{(x_{w2} - x_{E_2})^{20}}{a_2^{20}} - 1 \geq 0, \tag{5.15}$$

where $\boldsymbol{x}_w = [x_{w1}, x_{w2}]^T$ is the position of the end effector in the image space of the wide-angle lens, the order is set at 20 to form a rectangular region with rounded corners, and the constants a_1, a_2 determine the size of the rectangle, which are set to ensure that $h_{E_1}(\boldsymbol{x}_w) \leq 0$ is inside the field of view, and the reference position $\boldsymbol{x}_E = [x_{E_1}, x_{E_2}]^T$ is set as the center of the field of view. △△△

5.2 Region Function

5.2.1 Region Function for Reaching Tasks

As discussed in the previous section, there may exist some regions in the robot workspace where the desired task feedback $\boldsymbol{x}_T$ is infeasible due to singularities, a limited sensing zone, or image occlusion. Therefore, several functions are formulated in the coordinates of $\boldsymbol{r}_R$ to cover those regions, and the reaching task feedback $\boldsymbol{r}_R$ is used to drive the end effector out of such regions.

The regions formulated in the coordinates of $\boldsymbol{r}_R$ are classified into two categories: external regions and internal regions.

The external regions are specified to cover the positions that are beyond the external boundaries of the desired task variable $\boldsymbol{x}_T$, such as the external boundary singularity and the field of view. In general, the external region functions are specified in the coordinates of $\boldsymbol{r}_R$ as

$$\boldsymbol{h}_E(\boldsymbol{r}_R) = [h_{E_1}(\boldsymbol{r}_R), h_{E_2}(\boldsymbol{r}_R), \ldots, h_{E_M}(\boldsymbol{r}_R)]^T \geq \boldsymbol{0}, \tag{5.16}$$

where M is the total number of external regions. The robot employs the feedback $\boldsymbol{r}_R$ inside the external regions where $\boldsymbol{h}_E(\boldsymbol{r}_R) \geq \boldsymbol{0}$. The vector inequality $\boldsymbol{h}_E(\boldsymbol{r}_R) \geq \boldsymbol{0}$ implies that $h_{E_i}(\boldsymbol{r}_R) \geq 0$ for all $i = 1, \ldots, M$.

The specific forms of $\boldsymbol{h}_{E_i}(\boldsymbol{r}_R)$ depend on the robot's tasks. For example, consider the singularity problem of the 2-DOF manipulator in Sect. 5.1.1, in which the reaching-task variable $\boldsymbol{r}_R$ is specified in joint space (i.e., $\boldsymbol{r}_R = \boldsymbol{q}$), and the external boundary singularity occurs when the end effector is near the position $q_2 = 0$. Therefore, the external region is specified as in Eq. (5.5) as $h_{E_1}(\boldsymbol{q}) = b_{E_1}^2 - (q_2)^2 \geq 0$ to cover the external singular positions. The constant b_{E_1} represents the size of the external region, and it can be set to ensure that the manipulator is sufficiently far away from the singular configuration when the end effector leaves the external region.

Similarly, the internal regions are specified to cover positions that cannot be reached by the desired task variable $\boldsymbol{x}_T$ within the external boundaries, such as internal boundary singularities, interior singularities, and image-occluded areas. Each external region $h_{E_i}(\boldsymbol{r}_R)$ may be accompanied by several internal regions $h_{I_{i1}}(\boldsymbol{r}_R), h_{I_{i2}}(\boldsymbol{r}_R), \ldots, h_{I_{iN_i}}(\boldsymbol{r}_R)$. For example, in Fig. 5.2, there are two internal singular regions inside a single robot workspace. In Fig. 5.3, there are two occluded regions within a field of view. Therefore, the internal region functions are specified in general as

$$\begin{aligned}\boldsymbol{h}_I(\boldsymbol{r}_R) = \ & [\overbrace{h_{I_{11}}(\boldsymbol{r}_R), \ldots, h_{I_{1N_1}}(\boldsymbol{r}_R)}^{N_1}, \ldots, \overbrace{h_{I_{i1}}(\boldsymbol{r}_R), \ldots, h_{I_{iN_i}}(\boldsymbol{r}_R)}^{N_i}, \\ & \ldots, \overbrace{h_{I_{M1}}(\boldsymbol{r}_R), \ldots, h_{I_{MN_M}}(\boldsymbol{r}_R)}^{N_M}]^T \leq \boldsymbol{0}, \end{aligned} \tag{5.17}$$

where N_i is the number of internal regions corresponding to the ith external region $h_{E_i}(\boldsymbol{r}_R)$. The robot employs the feedback $\boldsymbol{r}_R$, where $\boldsymbol{h}_I(\boldsymbol{r}_R) \leq \boldsymbol{0}$. The vector inequality $\boldsymbol{h}_I(\boldsymbol{r}_R) \leq \boldsymbol{0}$ implies that $h_{I_{ij}}(\boldsymbol{r}_R) \leq 0$ for all $i = 1, \ldots, M$.

The specific forms of $h_{I_{ij}}(\boldsymbol{r}_R)$ also vary according to different robot tasks. Consider again the singularity problem of a 2-DOF manipulator. Since an internal boundary singularity occurs when the robot is near the position $q_2 = \pi$, the internal region is specified as in Eq. (5.6) as $h_{I_{11}}(\boldsymbol{q}) = (q_2 - \pi)^2 - b_{I_{11}}^2 \leq 0$ to cover the internal singular positions. The constant $b_{I_{11}}$ represents the size of the internal region, and it is also set to ensure that the manipulator is sufficiently far away from the singular configuration when the end effector leaves the internal region.

The feedback information $\boldsymbol{r}_R$ can be specified in different coordinates such as Cartesian space, image space, or joint space, and it is employed inside the external regions $\boldsymbol{h}_E(\boldsymbol{r}_R) \geq \boldsymbol{0}$ or the internal regions $\boldsymbol{h}_I(\boldsymbol{r}_R) \leq \boldsymbol{0}$.

5.2.2 Region Function for Desired Task

The end effector is driven by the desired task variable $\boldsymbol{x}_T$ after it leaves the external and internal regions where $\boldsymbol{h}_E(\boldsymbol{r}_R) < \boldsymbol{0}$ and $\boldsymbol{h}_I(\boldsymbol{r}_R) > \boldsymbol{0}$ in the coordinates of $\boldsymbol{r}_R$. To ensure a smooth transition from the feedback $\boldsymbol{r}_R$ to the feedback $\boldsymbol{x}_T$, another set of external and internal regions is formulated in the coordinates of $\boldsymbol{x}_T$ to cover the remaining workspace where the desired task feedback $\boldsymbol{x}_T$ is feasible.

The desired task variable $\boldsymbol{x}_T$ is specified as

$$\boldsymbol{x}_T = [\boldsymbol{x}_{T1}^T, \ldots, \boldsymbol{x}_{Ti}^T, \ldots, \boldsymbol{x}_{TM}^T]^T, \tag{5.18}$$

where $\boldsymbol{x}_{Ti} = [x_{Ti_1}, \ldots, x_{Ti_{n_i}}]^T \in \Re^{n_i}$ represents the ith feature point in the coordinates of the desired task feedback, $i = 1, 2, \ldots, M$, and M is the total number of features.

The external regions in the coordinates of $\boldsymbol{x}_T$ are formulated to include the regions where the desired task feedback is feasible, and the region functions are specified as

$$f_E(\boldsymbol{x}_T) = [f_{E_1}(\boldsymbol{x}_{T1}), f_{E_2}(\boldsymbol{x}_{T2}), \ldots, f_{E_M}(\boldsymbol{x}_{TM})]^T \leq \boldsymbol{0}. \quad (5.19)$$

The robot employs the feedback $\boldsymbol{x}_T$ where $f_E(\boldsymbol{x}_T) \leq \boldsymbol{0}$. The vector inequality $f_E(\boldsymbol{x}_T) \leq \boldsymbol{0}$ implies that $f_{E_i}(\boldsymbol{x}_{Ti}) \leq 0$ for all $i = 1, \ldots, M$.

Similarly, the internal regions in the coordinates of $\boldsymbol{x}_T$ are formulated to exclude the regions where the desired task feedback is infeasible, and the region functions are specified as

$$f_I(\boldsymbol{x}_T) = [\overbrace{f_{I_{11}}(\boldsymbol{x}_{T1}), \ldots, f_{I_{1N_1}}(\boldsymbol{x}_{T1})}^{N_1}, \ldots, \overbrace{f_{I_{M1}}(\boldsymbol{x}_{TM}), \ldots, f_{I_{MN_M}}(\boldsymbol{x}_{TM})}^{N_M}]^T \geq \boldsymbol{0}. \quad (5.20)$$

The robot employs the feedback $\boldsymbol{x}_T$ where $f_I(\boldsymbol{x}_T) \geq \boldsymbol{0}$. The vector inequality $f_I(\boldsymbol{x}_T) \geq \boldsymbol{0}$ implies that $f_{I_{ij}}(\boldsymbol{x}_{Ti}) \geq 0$ for all $i = 1, \ldots, M$.

The functions of $f_{E_i}(\boldsymbol{x}_{Ti})$ in Eq. (5.19) can be specified as follows:

$$f_{E_i}(\boldsymbol{x}_T) = \frac{(x_{Ti_1} - x_{Ei_1})^{n_{Ei}}}{a_{i_1}^{n_{Ei}}} + \cdots + \frac{(x_{Ti_{n_i}} - x_{Ei_{n_i}})^{n_{Ei}}}{a_{i_{n_i}}^{n_{Ei}}} - 1 \leq 0, \quad (5.21)$$

where $\boldsymbol{x}_{Ei} = [x_{Ei_1}, \ldots, x_{Ei_{n_i}}]^T \in \Re^{n_i}$ represent a set of reference positions, and $a_{i_1}, \ldots, a_{i_{n_i}}$ are positive constants and n_{Ei} are the orders of the region functions, which are even integers.

The functions of $f_{I_{ij}}(\boldsymbol{x}_{Ti})$ in Eq. (5.20) can be specified as follows:

$$f_{I_{ij}}(\boldsymbol{x}_{Ti}) = \frac{(x_{Ti_1} - x_{Iij_1})^{n_{Iij}}}{b_{ij_1}^{n_{Iij}}} + \cdots + \frac{(x_{Ti_{n_i}} - x_{Iij_{n_i}})^{n_{Iij}}}{b_{ij_{n_i}}^{n_{Iij}}} - 1 \geq 0, \quad (5.22)$$

where $\boldsymbol{x}_{Iij} = [x_{Iij_1}, \ldots, x_{Iij_{n_i}}]^T \in \Re^{n_i}$ is a set of reference positions and $b_{ij_1}, \ldots, b_{ij_{n_i}}$ are positive constants, and n_{Iij} are the orders of the region functions, which are also even integers.

Both the functions $f_{E_i}(\boldsymbol{x}_{Ti})$ and $f_{I_{ij}}(\boldsymbol{x}_{Ti})$ represent superellipses, and the position and shape of a superellipse can be varied by adjusting the reference positions $\boldsymbol{x}_{Ei}, \boldsymbol{x}_{Iij}$ and the orders n_{Ei}, n_{Iij} respectively, as illustrated in Fig. 5.5. The reference positions and orders of $f_{E_i}(\boldsymbol{x}_{Ti})$ and $f_{I_{ij}}(\boldsymbol{x}_{Ti})$ should be chosen such that the external regions $f_{E_i}(\boldsymbol{x}_{Ti}) \leq 0$ include the task space where $\boldsymbol{x}_T$ is available and the internal regions $f_{I_{ij}}(\boldsymbol{x}_{Ti}) \geq 0$ exclude the task space where $\boldsymbol{x}_T$ is infeasible.

Therefore, the region functions $f_{E_i}(\boldsymbol{x}_{Ti})$ and $f_{I_{ij}}(\boldsymbol{x}_{Ti})$ denote respectively the external boundary and internal boundary for the feedback $\boldsymbol{x}_T$. It is also possible that there is no internal boundary for $\boldsymbol{x}_T$ in some cases, such as no vision occlusion within the field of view. In that case, it is not required to specify the internal region functions $f_{I_{ij}}(\boldsymbol{x}_{Ti})$.

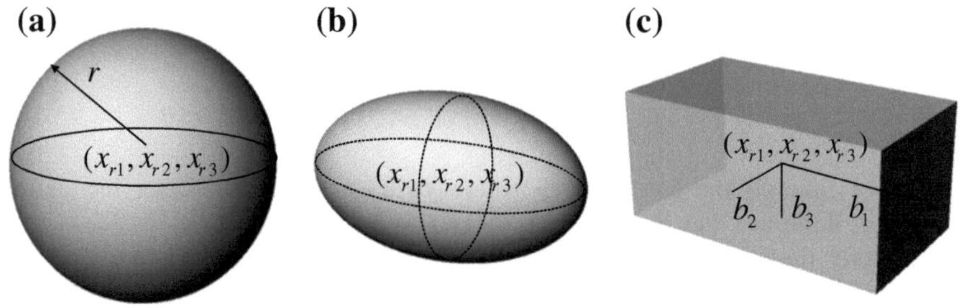

Fig. 5.5 Illustration of a superellipse. **a** Sphere (order = 2); **b** Ellipsoid (order = 2); **c** Rectangular block with rounded corners (order = 10)

The external regions $\boldsymbol{f}_E(\boldsymbol{x}_T) \leq \boldsymbol{0}$ and $\boldsymbol{h}_E(\boldsymbol{r}_R) \geq \boldsymbol{0}$ should match each other, to allow the robot to employ the feedback $\boldsymbol{x}_T$ within the external boundaries of $\boldsymbol{x}_T$ and use the feedback $\boldsymbol{r}_R$ outside the external boundaries. Similarly, the two internal regions $\boldsymbol{f}_I(\boldsymbol{x}_T) \geq \boldsymbol{0}$ and $\boldsymbol{h}_I(\boldsymbol{r}_R) \leq \boldsymbol{0}$ should also match each other, to allow the robot to use the feedback $\boldsymbol{r}_R$ in the regions where the feedback $\boldsymbol{x}_T$ is infeasible and use the feedback $\boldsymbol{x}_T$ after it leaves the regions. The combination of the regions in different coordinates covers the entire workspace and allows the global movement of the robot.

Example 5.1 For the planar 2-DOF manipulator in Sect. 5.1.1, the desired task feedback $\boldsymbol{x}_T$ is specified in Cartesian space (i.e., $\boldsymbol{x}_T = \boldsymbol{r}$), and it is activated after the end effector leaves the singular regions. The singular joint configuration $q_2 = 0$ corresponds to a circle in Cartesian space where the manipulator is fully stretched out, and the boundaries of the external region $h_{E_1}(\boldsymbol{q}) = b_{E_1}^2 - (q_2)^2 \geq 0$, which covers $q_2 = 0$, also corresponds to a circle in Cartesian space. To match $h_{E_1}(\boldsymbol{q}) \geq 0$, the external region function $f_{E_1}(\boldsymbol{r}_1)$ in Cartesian space is specified as

$$f_{E_1}(\boldsymbol{r}_1) = \frac{(r_{1_1} - r_{E1_1})^2}{a^2} + \frac{(r_{1_2} - r_{E1_2})^2}{a^2} - 1 \leq 0, \tag{5.23}$$

where $a_{1_1} = a_{1_2} = a$, and the order n_{E1} is set as $n_{E1} = 2$, so that the superellipse denoted by $f_{E_1}(\boldsymbol{r}_1)$ is a circle, and the reference position $\boldsymbol{r}_{E1} = [r_{E1_1}, r_{E1_2}]^T$ is the center of the circle, which is set as the origin of the Cartesian coordinates in this case. The constant a denotes the radius, and it is set to ensure a slight overlap between $h_{E_1}(\boldsymbol{q}) \geq 0$ and $f_{E_1}(\boldsymbol{r}_1) \leq 0$.

Similarly, the joint configuration $q_2 = \pi$ also corresponds to an inner circle in Cartesian space where the manipulator is fully folded back, and the boundaries of the internal region $h_{I_{11}}(\boldsymbol{q}) = (q_2 - \pi)^2 - b_{I_{11}}^2 \leq 0$, which covers $q_2 = \pi$, corresponds to a circle in Cartesian space as well. To match $h_{I_{11}}(\boldsymbol{q}) \leq 0$, the internal region function $f_{I_{11}}(\boldsymbol{r}_1)$ in Cartesian space is specified as

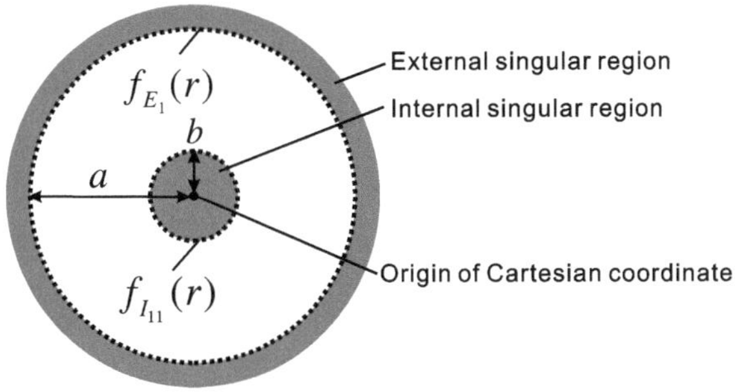

Fig. 5.6 External region $f_{E_1}(\boldsymbol{r}_1)$ and internal region $f_{I_{11}}(\boldsymbol{r}_1)$ for a 2-DOF planar manipulator

$$f_{I_{11}}(\boldsymbol{r}_1) = \frac{(r_{1_1} - r_{I11_1})^2}{b^2} + \frac{(r_{1_2} - r_{I11_2})^2}{b^2} - 1 \geq 0, \tag{5.24}$$

where the order n_{I11}, reference position $\boldsymbol{r}_{I11}$, and radius b are set in a similar argument. The region functions $f_{E_1}(\boldsymbol{r}_1)$ and $f_{I_{11}}(\boldsymbol{r}_1)$ are illustrated in Fig. 5.6. □

Example 5.2 For the problem of limited field of view illustrated in Sect. 5.1.2, the image-space regions are also specified to match the Cartesian-space regions, so that the end effector can transit smoothly from one region to another. For a rectangular field of view of a standard camera, the external region function in Eq. (5.21) is specified as

$$f_{E_1}(\boldsymbol{x}_1) = \frac{(x_{1_h} - x_{E1_h})^{20}}{a_1^{20}} + \frac{(x_{1_v} - x_{E1_v})^{20}}{a_2^{20}} - 1 \leq 0, \tag{5.25}$$

where the order n_{E1} is set high, $n_{E1} = 20$, to form a rectangular region with rounded corners, and the constants a_1, a_2 determine the size of the rectangle, which are set to ensure that $f_{E_1}(\boldsymbol{x}_1) \leq 0$ is inside the field of view, and the reference position $\boldsymbol{x}_{E1} = [x_{E1_h}, x_{E1_v}]^T$ is set as the center of the field of view.

Suppose that an oval occluded area exists within the field of view. Then the internal region function in Eq. (5.22) is specified as

$$f_{I_{11}}(\boldsymbol{x}_1) = \frac{(x_{1_1} - x_{I11_h})^2}{b_1^2} + \frac{(x_{1_2} - x_{I11_v})^2}{b_2^2} - 1 \geq 0, \tag{5.26}$$

where the order n_{I11}, reference position $\boldsymbol{x}_{I11}$, and constants b_1, b_2 are set so that the internal region excludes the occluded area. The region functions $f_{E_1}(\boldsymbol{x}_1)$ and $f_{I_{11}}(\boldsymbol{x}_1)$ are illustrated in Fig. 5.7. □

All the regions specified in the coordinates of $\boldsymbol{r}_R$ are assumed to be static. The singular regions of a robotic manipulator are known and can be obtained by analyzing

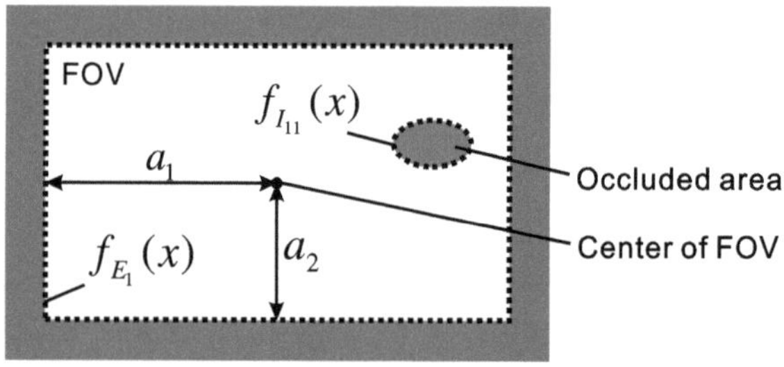

Fig. 5.7 External region $f_{E_1}(\boldsymbol{x}_1)$ and internal region $f_{I_{11}}(\boldsymbol{x}_1)$ for the rectangular field of view of a standard camera

the Jacobian matrix, while the size of the field of view can be determined by placing the markers that represent the boundaries, since the camera is fixed. Therefore, it is unnecessary to specify task-space regions online, and the construction of region functions is a one-time setup that does not vary with the desired motion.

The regions specified in the coordinates of $\boldsymbol{r}_R$ are slightly overlapped with the regions in the coordinates of $\boldsymbol{x}_T$, so that the end effector does not get stuck when it transits from the feedback $\boldsymbol{r}_R$ to $\boldsymbol{x}_T$. The singular regions are overlapped with the Cartesian-space regions introduced in Sect. 5.1.1, and the overlap can be ensured by setting the joint configurations that correspond to the boundaries of singular regions to be within the regions in Cartesian space. The Cartesian-space region is overlapped with the image-space region introduced in Sect. 5.1.2, and the overlap can be ensured by placing a set of land markers that represent the boundaries of the regions in Cartesian space to be within the regions in image space. Similarly, the wide-angle lens and the zoom camera in Sect. 5.1.3 are set to ensure that the robot is already inside the field of view of the zoom camera after it leaves the external region in the image space of the wide-angle lens.

Next, a set of task-oriented regions is introduced within the external regions so as to move the end effector toward the desired trajectory:

$$\boldsymbol{f}_T(\boldsymbol{x}_T) = [f_{T_1}(\boldsymbol{x}_{T1}), f_{T_2}(\boldsymbol{x}_{T2}), \ldots, f_{T_M}(\boldsymbol{x}_{TM})]^T \leq \boldsymbol{0}. \tag{5.27}$$

Let $\boldsymbol{x}_d = [\boldsymbol{x}_{d1}^T, \ldots, \boldsymbol{x}_{di}^T, \ldots \boldsymbol{x}_{dM}^T]^T$ denotes the desired trajectory, and $\boldsymbol{x}_{di} = [x_{di_1}, \ldots, x_{di_{n_i}}]^T \in \Re^{n_i}$ is the desired trajectory for the ith feature. The task-oriented regions are defined to enclose the desired trajectory as:

$$f_{T_i}(\boldsymbol{x}_{Ti}) = \frac{(x_{Ti_1} - x_{di_1})^2}{(x_{bi_1} - x_{di_1})^2} + \cdots + \frac{(x_{Ti_{n_i}} - x_{di_{n_i}})^2}{(x_{bi_{n_i}} - x_{di_{n_i}})^2} - 1 \leq 0, \tag{5.28}$$

where $\boldsymbol{x}_{bi} = [x_{bi_1}, \ldots, x_{bi_{n_i}}]^T \in \Re^{n_i}$ denote the boundary positions. The boundary positions are made up of several parts, and the desired position $\boldsymbol{x}_d(t)$ is time-varying and not necessarily the geometric center.

5.3 Potential Energy

In this section, the potential energy functions in coordinates of the desired task variable $\boldsymbol{x}_T$ and the reaching-task variable $\boldsymbol{r}_R$ are constructed. The potential energy in coordinates of $\boldsymbol{r}_R$ is used to construct a gradient force that drives the end effector to leave the regions where the feedback $\boldsymbol{x}_T$ is infeasible, and the potential energy in coordinates of $\boldsymbol{x}_T$ is used to construct another gradient force that drives the end effector to move toward the desired position.

5.3.1 Potential Energy for Reaching Task

Using the external regions in Eq. (5.16), potential energy functions are introduced as

$$P_{E_i}(\boldsymbol{r}_R) = \frac{k_{rE_i}}{N}[\max(0, h_{E_i}(\boldsymbol{r}_R))]^N, \tag{5.29}$$

where k_{rE_i} are positive constants, N is an even integer such that the potential energy is $\in C^2$. Similarly, the potential energy functions $P_{I_{ij}}(\boldsymbol{r}_R)$ for the internal regions in Eq. (5.17) are introduced as

$$P_{I_{ij}}(\boldsymbol{r}_R) = \frac{k_{rI_{ij}}}{N}[\min(0, h_{I_{ij}}(\boldsymbol{r}_R))]^N, \tag{5.30}$$

where $k_{rI_{ij}}$ are positive constants.

Note that $P_{E_i}(\boldsymbol{r}_R)$ and $P_{I_{ij}}(\boldsymbol{r}_R)$ are given as

$$P_{E_i}(\boldsymbol{r}_R) = \begin{cases} 0, & h_{E_i}(\boldsymbol{r}_R) \leq 0, \\ \frac{k_{rE_i}}{N}[h_{E_i}(\boldsymbol{r}_R)]^N, & h_{E_i}(\boldsymbol{r}_R) > 0, \end{cases} \tag{5.31}$$

$$P_{I_{ij}}(\boldsymbol{r}_R) = \begin{cases} 0, & h_{I_{ij}}(\boldsymbol{r}_R) \geq 0, \\ \frac{k_{rI_{ij}}}{N}[h_{I_{ij}}(\boldsymbol{r}_R)]^N, & h_{I_{ij}}(\boldsymbol{r}_R) < 0. \end{cases} \tag{5.32}$$

Therefore, the potential energy is smooth and bounded below by zero. It is zero when the position of the end effector is inside the internal and external regions.

The overall potential energy with the reaching-task variable $\boldsymbol{r}_R$ is the summation of $P_{E_i}(\boldsymbol{r}_R)$ and $P_{I_{ij}}(\boldsymbol{r}_R)$:

$$P_R(\boldsymbol{r}_R) = k_p\alpha_r \sum_{i=1}^{M} \left[P_{E_i}(\boldsymbol{r}_R) + \sum_{j=1}^{N_i} P_{I_{ij}}(\boldsymbol{r}_R) \right], \tag{5.33}$$

where k_p and α_r are positive constants. Differentiating $P_R(\boldsymbol{r}_R)$ with respect to $\boldsymbol{r}_R$ yields

$$\left(\frac{\partial P_R(\boldsymbol{r}_R)}{\partial \boldsymbol{r}_R}\right)^T = k_p\alpha_r \sum_{i=1}^{M}\left\{k_{rE_i}[\max(0, h_{E_i}(\boldsymbol{r}_R))]^{N-1}\left(\frac{\partial h_{E_i}(\boldsymbol{r}_R)}{\partial \boldsymbol{r}_R}\right)^T\right.$$
$$\left. + \sum_{j=1}^{N_i} k_{rI_{ij}}[\min(0, h_{I_{ij}}(\boldsymbol{r}_R))]^{N-1}\left(\frac{\partial h_{I_{ij}}(\boldsymbol{r}_R)}{\partial \boldsymbol{r}_R}\right)^T\right\} \triangleq k_p\alpha_r\Delta\boldsymbol{\varepsilon}_r, \tag{5.34}$$

where $\Delta\boldsymbol{\varepsilon}_r$ denotes the reaching regional feedback error, which is used to drive the end effector to leave the external and internal regions. From Eq. (5.34), note that $\Delta\boldsymbol{\varepsilon}_r = \mathbf{0}$ when $h_{E_i}(\boldsymbol{r}_R) \leq 0$ and $h_{I_{ij}}(\boldsymbol{r}_R) \geq 0$, which indicates that the regional feedback naturally reduces to zero after the end effector leaves those regions.

5.3.2 Potential Energy for Desired Task

For the construction of the potential energy in the coordinates of $\boldsymbol{x}_T$, two reference regions are formulated for the external region functions $f_{E_i}(\boldsymbol{x}_{Ti})$ and the internal region functions $f_{I_{ij}}(\boldsymbol{x}_{Ti})$ respectively.

First, a set of reference regions inside $f_{E_i}(\boldsymbol{x}_{Ti}) \leq 0$ is introduced as

$$f_{Er_i}(\boldsymbol{x}_{Ti}) = \frac{(x_{Ti_1} - x_{Ei_1})^{n_{Ei}}}{(\kappa_{E_i}a_{i_1})^{n_{Ei}}} + \cdots + \frac{(x_{Ti_{n_i}} - x_{Ei_{n_i}})^{n_{Ei}}}{(\kappa_{E_i}a_{i_{n_i}})^{n_{Ei}}} - 1 \leq 0, \tag{5.35}$$

where κ_{E_i} are positive constants and $\kappa_{E_i} < 1$.

Using $f_{E_i}(\boldsymbol{x}_{Ti})$ in Eq. (5.21) and $f_{Er_i}(\boldsymbol{x}_{Ti})$ in Eq. (5.35), the potential energy functions $P_{E_i}(\boldsymbol{x}_{Ti})$ are introduced as

$$P_{E_i}(\boldsymbol{x}_{Ti}) = \frac{k_{xE_i}}{N^2}\{\min[0, [\min(0, f_{E_i}(\boldsymbol{x}_{Ti}))]^N - (\kappa_{E_i}^{n_{Ei}} - 1)^N]\}^N, \tag{5.36}$$

which can also be written as:

$$P_{E_i}(\boldsymbol{x}_{Ti}) = \begin{cases} 0, & f_{Er_i}(\boldsymbol{x}_{Ti}) \leq 0, \\ \frac{k_{xE_i}}{N^2}\{[f_{E_i}(\boldsymbol{x}_{Ti})]^N - (\kappa_{E_i}^{n_{Ei}} - 1)^N\}^N, & f_{E_i}(\boldsymbol{x}_{Ti}) < 0, f_{Er_i}(\boldsymbol{x}_{Ti}) > 0 \\ \frac{k_{xE_i}}{N^2}(\kappa_{E_i}^{n_{Ei}} - 1)^{N^2}, & f_{E_i}(\boldsymbol{x}_{Ti}) \geq 0, \end{cases} \tag{5.37}$$

where k_{xE_i} are positive constants. An illustration of $P_{E_i}(\boldsymbol{x}_{Ti})$ in a 2-D space is shown in Fig. 5.8. From Eq. (5.37), it is seen that the $P_{E_i}(\boldsymbol{x}_{Ti})$ are smooth and bounded below

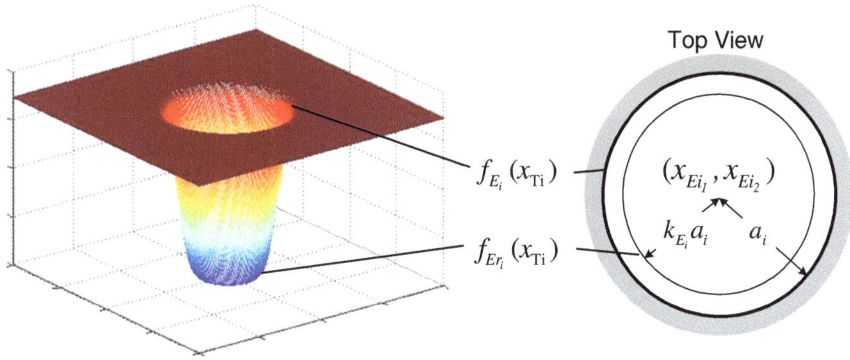

Fig. 5.8 Example of potential energy of $P_{E_i}(\boldsymbol{x}_{Ti})$ with $n_{Ei} = 2$ in 2-D space

by zero. The potential energy $P_{E_i}(\boldsymbol{x}_{Ti})$ is used to construct a gradient force to drive the end effector from $f_{E_i}(\boldsymbol{x}_{Ti}) \leq 0$ toward $f_{Er_i}(\boldsymbol{x}_{Ti}) \leq 0$ and hence $f_{T_i}(\boldsymbol{x}_{Ti}) \leq 0$.

Similarly, to introduce the potential energy functions for $f_{I_{ij}}(\boldsymbol{x}_{Ti})$, another set of reference regions enclosing $f_{I_{ij}}(\boldsymbol{x}_{Ti}) \geq 0$ is designed as

$$f_{Ir_{ij}}(\boldsymbol{x}_{Ti}) = \frac{(x_{Ti_1} - x_{Iij_1})^{n_{Iij}}}{(\kappa_{I_{ij}} b_{ij_1})^{n_{Iij}}} + \cdots + \frac{(x_{Ti_{n_i}} - x_{Iij_{n_i}})^{n_{Iij}}}{(\kappa_{I_{ij}} b_{ij_{n_i}})^{n_{Iij}}} - 1 \geq 0, \quad (5.38)$$

where $\kappa_{I_{ij}} > 1$ are constants.

Using the internal region functions and the reference region functions in Eqs. (5.22) and (5.38), the potential energy functions $P_{I_{ij}}(\boldsymbol{x}_{Ti})$ are designated as

$$P_{I_{ij}}(\boldsymbol{x}_{Ti}) = \frac{k_{xI_{ij}}}{N^2} \left\{ \min \left[0, [\min(0, f_{Ir_{ij}}(\boldsymbol{x}_{Ti}))]^N - \left(\frac{1}{\kappa_{I_{ij}}^{n_{Iij}}} - 1 \right)^N \right] \right\}^N . \quad (5.39)$$

The above equation can be written as

$$P_{I_{ij}}(\boldsymbol{x}_{Ti}) = \begin{cases} 0, & f_{I_{ij}}(\boldsymbol{x}_{Ti}) \leq 0, \\ \dfrac{k_{xI_{ij}}}{N^2} \left\{ [f_{Ir_{ij}}(\boldsymbol{x}_{Ti})]^N - \left(\dfrac{1}{\kappa_{I_{ij}}^{n_{Iij}}} - 1 \right)^N \right\}^N, & f_{Ir_{ij}}(\boldsymbol{x}_{Ti}) < 0, f_{I_{ij}}(\boldsymbol{x}_{Ti}) > 0 \\ \dfrac{k_{xI_{ij}}}{N^2} \left(\dfrac{1}{\kappa_{I_{ij}}^{n_{Iij}}} - 1 \right)^{N^2}, & f_{Ir_{ij}}(\boldsymbol{x}_{Ti}) \geq 0, \end{cases} \quad (5.40)$$

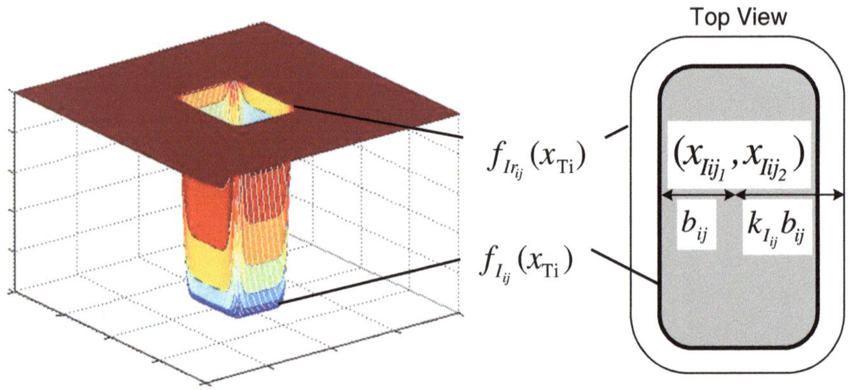

Fig. 5.9 Example of potential energy $P_{I_{ij}}(\boldsymbol{x}_{Ti})$ with $n_{Iij} = 8$ in 2-D space

where $k_{xI_{ij}}$ are positive constants. An illustration of the potential energy in a 2-D space is shown in Fig. 5.9. From Eq. (5.40), it is seen that the $P_{I_{ij}}(\boldsymbol{x}_{Ti})$ are also smooth and bounded below by zero. The potential energy $P_{I_{ij}}(\boldsymbol{x}_{Ti})$ is used to construct an energy function to avoid the use of the feedback $\boldsymbol{x}_T$ when it is infeasible.

The reference regions of the external and internal regions are used to introduce differences in the energy levels, and note that the internal regions are within each external region. The potential energy $P_{E_i}(\boldsymbol{x}_{Ti})$ is expressed in terms of the external region $f_{E_i}(\boldsymbol{x}_{Ti})$, so that $f_{E_i}(\boldsymbol{x}_{Ti})$ corresponds to the higher energy level of $P_{E_i}(\boldsymbol{x}_{Ti})$. The potential energy $P_{I_{ij}}(\boldsymbol{x}_{Ti})$ is expressed in terms of the reference region $f_{Ir_{ij}}(\boldsymbol{x}_{Ti})$, so that $f_{Ir_{ij}}(\boldsymbol{x}_{Ti})$ corresponds to the lower energy level of the offset energy term $\sum_{j=1}^{N_i} k_{c_{ij}}[\frac{k_{xI_{ij}}}{N^2}(\frac{1}{\frac{n_{Iij}}{\kappa_{I_{ij}}}} - 1)^{N^2} - P_{I_{ij}}(\boldsymbol{x}_{Ti})]$. In both cases, the regions and their reference regions correspond to the higher levels and the lower levels of the energy functions respectively.

Using the task-oriented regions in Eq. (5.28), the corresponding potential energy functions $P_{T_i}(\boldsymbol{x}_{Ti})$ are specified as follows:

$$P_{T_i}(\boldsymbol{x}_{Ti}) = \frac{k_{T_i}}{N}\{1 - [\min(0, f_{T_i}(\boldsymbol{x}_{Ti}))]^N\}, \tag{5.41}$$

where k_{T_i} are positive constants. An illustration of $P_{T_i}(\boldsymbol{x}_{Ti})$ in 2-D space is shown in Fig. 5.10, where the bottom of $P_{T_i}(\boldsymbol{x}_{Ti})$ corresponds to the desired position, and the potential field ensures the convergence after the end effector enters the task-oriented regions.

The overall potential energy $P_D(\boldsymbol{x}_T)$ with the desired task variable $\boldsymbol{x}_T$ is specified as the combination of $P_{E_i}(\boldsymbol{x}_{Ti})$, $P_{I_{ij}}(\boldsymbol{x}_{Ti})$, and $P_{T_i}(\boldsymbol{x}_{Ti})$ as follows:

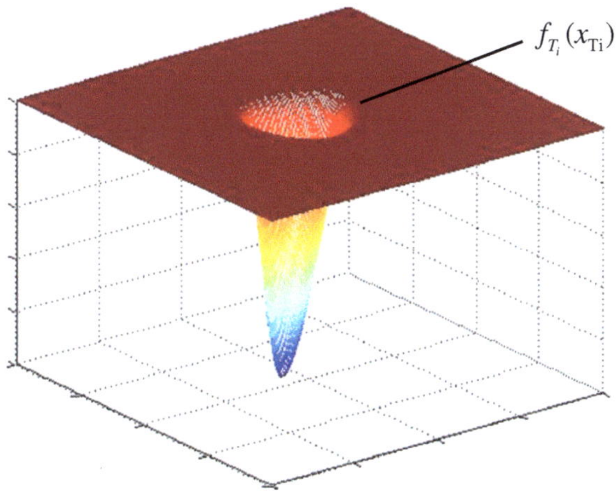

Fig. 5.10 Example of potential energy $P_{T_i}(\boldsymbol{x}_{Ti})$ in 2-D space

$$P_D(\boldsymbol{x}_T) = k_p \alpha_x \sum_{i=1}^{M} \left\{ P_i(\boldsymbol{x}_{Ti}) \prod_{j=1}^{N_i} P_{I_{ij}}(\boldsymbol{x}_{Ti}) + \sum_{j=1}^{N_i} k_{c_{ij}} \left[\frac{k_{xI_{ij}}}{N^2} \left(\frac{1}{\kappa_{I_{ij}}^{n_{Iij}}} - 1 \right)^{N^2} - P_{I_{ij}}(\boldsymbol{x}_{Ti}) \right] \right\}, \quad (5.42)$$

where α_x is a positive constant and $k_{c_{ij}}$ are positive constants, and $P_i(\boldsymbol{x}_{Ti}) = P_{T_i}(\boldsymbol{x}_{Ti}) + P_{E_i}(\boldsymbol{x}_{Ti})$. An illustration of the combination is shown in Fig. 5.11, and the reasons for the combination of potential energy functions in Eq. (5.42) are summarized as follows:

(i) The top contour of potential energy $P_{E_i}(\boldsymbol{x}_{Ti})$ (see Fig. 5.11a) is fixed to match $h_{E_i}(\boldsymbol{r}_R)$, but its bottom is flat, while the potential energy $P_{T_i}(\boldsymbol{x}_{Ti})$ (see Fig. 5.11b) drives the end effector towards the desired position but its top contour cannot match $h_{E_i}(\boldsymbol{r}_R)$. To combine the advantages, $P_i(\boldsymbol{x}_{Ti})$ (see Fig. 5.11c) is defined as the summation of $P_{E_i}(\boldsymbol{x}_{Ti})$ and $P_{T_i}(\boldsymbol{x}_{Ti})$. Therefore, as $\boldsymbol{x}_{di}$ varies, the bottom of the potential energy $P_i(\boldsymbol{x}_{Ti})$ changes, while the top contour remains the same. Since the top contour of $P_i(\boldsymbol{x}_{Ti})$ corresponds to the external region functions $f_{E_i}(\boldsymbol{x}_{Ti})$ and its bottom is the desired position, the potential field of $P_i(\boldsymbol{x}_{Ti})$ enables the end effector to move toward the desired position after it enters the external regions.

(ii) The potential energy $P_i(\boldsymbol{x}_{Ti})$ is further multiplied by the potential energy $P_{I_{ij}}(\boldsymbol{x}_{Ti})$ (see Fig. 5.11d). The multiplied potential energy is flattened at $f_{I_{ij}}(\boldsymbol{x}_{Ti}) < 0$, where the gradient reduces to zero. It avoids the use of feedback $\boldsymbol{x}_T$, where $f_{I_{ij}}(\boldsymbol{x}_{Ti}) < 0$, and the reaching-task feedback $\boldsymbol{r}_R$ is employed in those regions.

(iii) After that, the potential energy obtained in step (ii) is offset inside the regions $f_{I_{ij}}(\boldsymbol{x}_{Ti}) < 0$ by adding the term $\sum_{j=1}^{N_i} k_{c_{ij}}[\frac{k_{xI_{ij}}}{N^2}(\frac{1}{\kappa_{I_{ij}}^{n_{Iij}}} - 1)^{N^2} - P_{I_{ij}}(\boldsymbol{x}_{Ti})]$ (see Fig. 5.11e). Note that the potential energy functions $P_{I_{ij}}(\boldsymbol{x}_{Ti})$ are zero when $f_{I_{ij}}(\boldsymbol{x}_{Ti}) \leq 0$

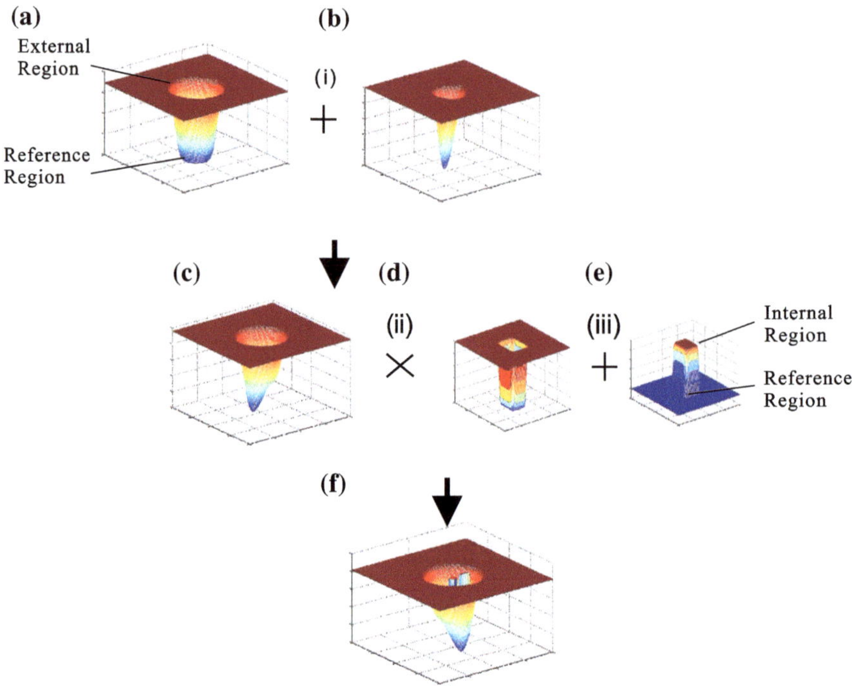

Fig. 5.11 An illustration of the combination of potential energy functions in 2-D space. **a** Potential energy for external regions $P_{E_i}(x_{Ti})$. **b** Potential energy for task-oriented regions $P_{T_i}(x_{Ti})$. **c** $P_i(x_{Ti})$. **d** Potential energy for internal regions $P_{I_{ij}}(x_{Ti})$. **e** Offset. **f** Overall potential energy $P_D(x_T)$

(see Fig. 5.11d). The offset term is therefore introduced to raise the potential energy to enable the end effector to move away from the internal regions after leaving them. By adjusting the value of $k_{c_{ij}}$ in the offset term, the energy level can be varied to allow the end effector to pass through the internal regions or avoid them.

The aim of this combination is to move the end effector toward the desired position and keep it away from the regions where the desired feedback $\boldsymbol{x}_T$ is infeasible. An example of the overall potential energy $P_D(\boldsymbol{x}_T)$ is shown in Fig. 5.12.

The external regions $f_{E_i}(\boldsymbol{x}_{Ti}) \leq 0$ and the internal regions $f_{I_{ij}}(\boldsymbol{x}_{Ti}) \geq 0$ in the coordinates of $\boldsymbol{x}_T$ are defined to cover the task space where the desired task feedback $\boldsymbol{x}_T$ is feasible. The external region functions $f_{E_i}(\boldsymbol{x}_{Ti})$ denote the external boundary of the feedback $\boldsymbol{x}_T$. Within the external boundary, there may be several regions excluded by $f_{I_{ij}}(\boldsymbol{x}_{Ti}) \geq 0$ where $\boldsymbol{x}_T$ is infeasible. This is a general formulation, and

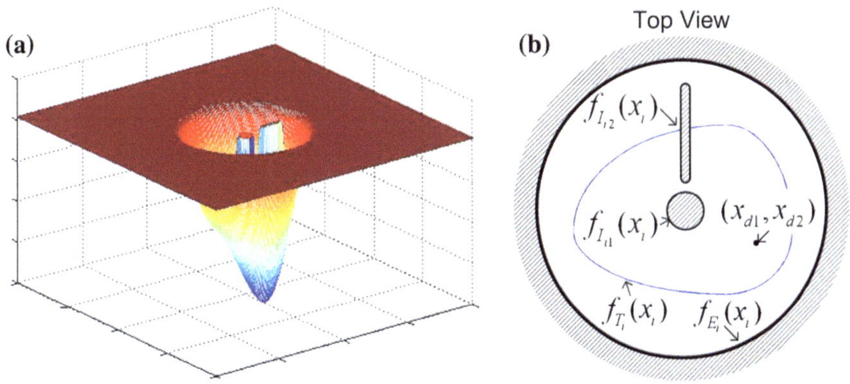

Fig. 5.12 Example of overall potential energy in 2-D space. **a** Potential energy $P_D(\boldsymbol{x}_T)$. **b** *Top view*

it is also possible that there is no internal region where $\boldsymbol{x}_T$ is infeasible inside the external task-space boundary. For example, there may be no occlusion within the field of view or no internal singularity inside the external task-space boundary. In those cases, the potential energy $P_D(\boldsymbol{x}_T)$ in Eq. (5.42) can be simplified to

$$P_D(\boldsymbol{x}_T) = k_p\alpha_x \sum_{i=1}^{M} P_i(\boldsymbol{x}_{Ti}) = k_p\alpha_x \sum_{i=1}^{M} [P_{E_i}(\boldsymbol{x}_{Ti}) + P_{T_i}(\boldsymbol{x}_{Ti})]. \quad (5.43)$$

Differentiating $P_D(\boldsymbol{x}_T)$ in Eq. (5.42) with respect to $\boldsymbol{x}_T$ yields

$$\begin{aligned}\left(\frac{\partial P_D(\boldsymbol{x}_T)}{\partial \boldsymbol{x}_T}\right)^T &= k_p\alpha_x \sum_{i=1}^{M} \left\{ \left(\frac{\partial P_i(\boldsymbol{x}_{Ti})}{\partial \boldsymbol{x}_T}\right)^T \prod_{j=1}^{N_i} P_{I_{ij}}(\boldsymbol{x}_{Ti}) \right.\\ &\quad + P_i(\boldsymbol{x}_{Ti}) \sum_{j=1}^{N_i} \left[\left(\frac{\partial P_{I_{ij}}(\boldsymbol{x}_{Ti})}{\partial \boldsymbol{x}_T}\right)^T \prod_{k \neq j}^{N_i} P_{I_{ik}}(\boldsymbol{x}_{Ti}) \right] \\ &\quad \left. - \sum_{j=1}^{N_i} k_{c_{ij}} \left(\frac{\partial P_{I_{ij}}(\boldsymbol{x}_{Ti})}{\partial \boldsymbol{x}_T}\right)^T \right\} \triangleq k_p\alpha_x \Delta\varepsilon_x, \quad (5.44)\end{aligned}$$

where $\Delta\varepsilon_x$ is the desired regional feedback error, which can be used to drive the end effector toward the desired trajectory, and it is activated automatically after the end effector is inside the external and internal regions in the coordinates of $\boldsymbol{x}_T$.

From Eq. (5.40), $(\frac{\partial P_{I_{ij}}(\boldsymbol{x}_{Ti})}{\partial \boldsymbol{x}_T})^T$ in Eq. (5.44) is given as

$$\left(\frac{\partial P_{I_{ij}}(\boldsymbol{x}_{Ti})}{\partial \boldsymbol{x}_T}\right)^T = \begin{cases} 0, & f_{I_{ij}}(\boldsymbol{x}_{Ti}) \leq 0, \\ k_{xI_{ij}} \left\{ [f_{Ir_{ij}}(\boldsymbol{x}_{Ti})]^N - \left(\frac{1}{\kappa_{I_{ij}}^{n_{Iij}}} - 1 \right)^N \right\}^{N-1} f_{Ir_{ij}}(\boldsymbol{x}_{Ti})^{N-1} \left(\frac{\partial f_{Ir_{ij}}(\boldsymbol{x}_{Ti})}{\partial \boldsymbol{x}_T} \right)^T, & f_{I_{ij}}(\boldsymbol{x}_{Ti}) > 0,\, f_{Ir_{ij}}(\boldsymbol{x}_{Ti}) < 0, \\ 0, & f_{Ir_{ij}}(\boldsymbol{x}_{Ti}) \geq 0. \end{cases} \tag{5.45}$$

which can also be written as

$$\left(\frac{\partial P_{I_{ij}}(\boldsymbol{x}_{Ti})}{\partial \boldsymbol{x}_T}\right)^T = k_{xI_{ij}} \min \left\{ 0, [\min(0, f_{Ir_{ij}}(\boldsymbol{x}_{Ti}))]^N - \left(\frac{1}{\kappa_{I_{ij}}^{n_{Iij}}} - 1 \right)^N \right\}^{N-1} [\min(0, f_{Ir_{ij}}(\boldsymbol{x}_{Ti}))]^{N-1} \times \left(\frac{\partial f_{Ir_{ij}}(\boldsymbol{x}_{Ti})}{\partial \boldsymbol{x}_T} \right)^T. \tag{5.46}$$

In addition, $(\frac{\partial P_i(\boldsymbol{x}_{Ti})}{\partial \boldsymbol{x}_T})^T$ in Eq. (5.44) is given as

$$\left(\frac{\partial P_i(\boldsymbol{x}_{Ti})}{\partial \boldsymbol{x}_T}\right)^T = \left(\frac{\partial P_{E_i}(\boldsymbol{x}_{Ti})}{\partial \boldsymbol{x}_T}\right)^T + \left(\frac{\partial P_{T_i}(\boldsymbol{x}_{Ti})}{\partial \boldsymbol{x}_T}\right)^T, \tag{5.47}$$

where

$$\left(\frac{\partial P_{E_i}(\boldsymbol{x}_{Ti})}{\partial \boldsymbol{x}_T}\right)^T = \begin{cases} 0, & f_{Er_i}(\boldsymbol{x}_{Ti}) \leq 0, \\ k_{xE_i} \left\{ [f_{E_i}(\boldsymbol{x}_{Ti})]^N - \left(\kappa_{E_i}^{n_{Ei}} - 1 \right)^N \right\}^{N-1} f_{E_i}(\boldsymbol{x}_{Ti})^{N-1} \left(\frac{\partial f_{E_i}(\boldsymbol{x}_{Ti})}{\partial \boldsymbol{x}_T} \right)^T, & f_{Er_i}(\boldsymbol{x}_{Ti}) > 0,\, f_{E_i}(\boldsymbol{x}_{Ti}) < 0, \\ 0, & f_{E_i}(\boldsymbol{x}_{Ti}) \geq 0. \end{cases} \tag{5.48}$$

which can also be written as

$$\left(\frac{\partial P_{E_i}(\boldsymbol{x}_{Ti})}{\partial \boldsymbol{x}_T}\right)^T = k_{xE_i} \{ \min\{0, [\min(0, f_{E_i}(\boldsymbol{x}_{Ti}))]^N - (\kappa_{E_i}^{n_{Ei}} - 1)^N \}\}^{N-1} [\min(0, f_{E_i}(\boldsymbol{x}_{Ti}))]^{N-1} \times \left(\frac{\partial f_{E_i}(\boldsymbol{x}_{Ti})}{\partial \boldsymbol{x}_T} \right)^T \tag{5.49}$$

and

$$\left(\frac{\partial P_{T_i}(\boldsymbol{x}_{Ti})}{\partial \boldsymbol{x}_T}\right)^T = -k_{T_i}[\min(0, f_{T_i}(\boldsymbol{x}_{Ti}))]^{N-1}\left(\frac{\partial f_{T_i}(\boldsymbol{x}_{Ti})}{\partial \boldsymbol{x}_T}\right)^T, \tag{5.50}$$

where the partial derivative $(\frac{\partial P_{T_i}(\boldsymbol{x}_{Ti})}{\partial \boldsymbol{x}_T})^T$ is the gradient of potential energy $P_{T_i}(\boldsymbol{x}_{Ti})$ for the task-oriented regions. From Eq. (5.50), if the end effector is outside the task-oriented regions where $f_{T_i}(\boldsymbol{x}_{Ti}) > 0$, the gradient of $P_{T_i}(\boldsymbol{x}_{Ti})$ reduces to zero. After the end effector enters the task-oriented regions, $f_{T_i}(\boldsymbol{x}_{Ti}) \leq 0$, and the gradient of $P_{T_i}(\boldsymbol{x}_{Ti})$ becomes

$$\left(\frac{\partial P_{T_i}(\boldsymbol{x}_{Ti})}{\partial \boldsymbol{x}_T}\right)^T = -k_{T_i}[f_{T_i}(\boldsymbol{x}_{Ti})]^{N-1}\left(\frac{\partial f_{T_i}(\boldsymbol{x}_{Ti})}{\partial \boldsymbol{x}_T}\right)^T,$$

where $(\frac{\partial f_{T_i}(\boldsymbol{x}_{Ti})}{\partial \boldsymbol{x}_T})^T = [0, \ldots, \frac{2(x_{Ti_1} - x_{di_1})}{(x_{bi_1} - x_{di_1})^2}, \ldots, \frac{2(x_{Tin_i} - x_{din_i})}{(x_{bin_i} - x_{din_i})^2}, \ldots, 0]^T \in \Re^{\sum_{i=1}^{M} n_i}$, so it is nonzero until the end effector reaches the desired position.

5.4 Task-Space Controller with Regional Feedback

The velocity of the end effector in the task-space coordinate of $\boldsymbol{x}_T$ is related to the joint velocity by

$$\dot{\boldsymbol{x}}_T = \boldsymbol{J}_x(\boldsymbol{q})\dot{\boldsymbol{q}}, \tag{5.51}$$

where $\boldsymbol{J}_x(\boldsymbol{q})$ is the Jacobian matrix from joint space to task space. When the desired task variable $\boldsymbol{x}_T$ is specified in Cartesian space, $\boldsymbol{J}_x(\boldsymbol{q}) = \boldsymbol{J}_m(\boldsymbol{q})$ represents the mapping from joint space to Cartesian space as described by Eq. (1.24). When $\boldsymbol{x}_T$ is specified in image space, $\boldsymbol{J}_x(\boldsymbol{q}) = \boldsymbol{J}_s(\boldsymbol{r})\boldsymbol{J}_m(\boldsymbol{q})$ represents the mapping from joint space to image space described by Eq. (1.38). Similarly, the velocity of the end effector in the different coordinate space of $\boldsymbol{r}_R$ is related to the joint velocity by

$$\dot{\boldsymbol{r}}_R = \boldsymbol{J}_r(\boldsymbol{q})\dot{\boldsymbol{q}}, \tag{5.52}$$

where $\boldsymbol{J}_r(\boldsymbol{q})$ is the Jacobian matrix from joint space to the corresponding task space.

After the feedback variables $\boldsymbol{r}_R$ and $\boldsymbol{x}_T$ are specified in particular coordinates, the reaching region error $\Delta\varepsilon_r$ and the desired region error $\Delta\varepsilon_x$ in Eqs. (5.34) and (5.44) are obtained accordingly. Using the region errors, the regional feedback controller for the robotic manipulator is designed.

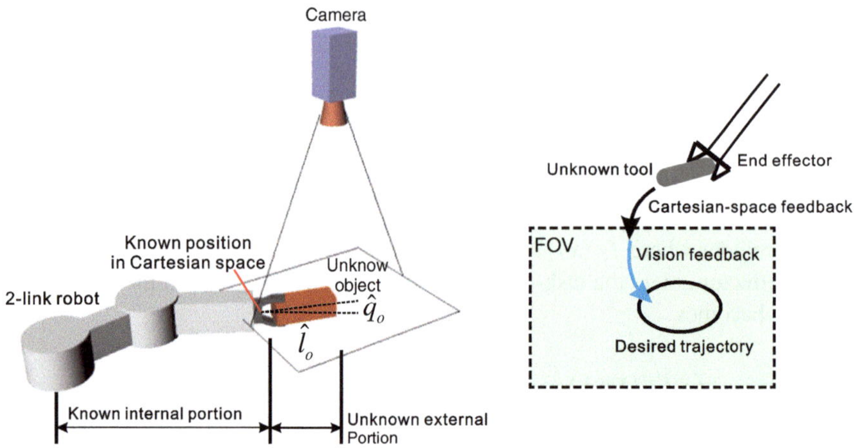

Fig. 5.13 A manipulator with an unknown object

We now consider regional feedback control for a robotic manipulator with uncertain kinematics. It is assumed that the robot kinematics can be partitioned into a known internal portion (manipulator) and an unknown external portion (object). The known portion is specified in the coordinates of the reaching-task variable, and the unknown portion is specified in the coordinates of the desired task variable. The reaching region error $\Delta\varepsilon_r$ is used for the region reaching control, of known internal portion, and the desired region error $\Delta\varepsilon_x$ is used for the adaptive tracking control, of unknown external portion.

Example 5.3 Consider a two-link robot holding an unknown object, as shown in Fig. 5.13, where a fixed camera is used to measure the position of the object.

For the robot system in Fig. 5.13, if the position of the end effector is known in Cartesian space, it can be specified as the reaching-task variable. The manipulator kinematics is thus considered the known portion. Since the robot is holding an unknown object, the position of the endpoint of the object is unavailable in Cartesian space. Thus, the position of the object is obtained in image space and is used as the desired task variable. The object is the unknown portion of the overall kinematics.

The Jacobian matrix for the internal portion represents the mapping from joint space to Cartesian space as

$$J_m(q) = \begin{bmatrix} -l_1\sin(q_1) - l_2\sin(q_1+q_2) & -l_2\sin(q_1+q_2) \\ l_1\cos(q_1) + l_2\cos(q_1+q_2) & l_2\cos(q_1+q_2) \end{bmatrix}, \tag{5.53}$$

where l_1, and l_2 denote the link lengths, and q_1 and q_2 denote the joint angles for the first and second links.

The Jacobian matrix for the external portion represents the mapping from joint space to image space as

$$\hat{\boldsymbol{J}}(\boldsymbol{q}) = \hat{\boldsymbol{J}}_s(\boldsymbol{r})\hat{\boldsymbol{J}}_m(\boldsymbol{q}) = \begin{bmatrix} \frac{f}{\hat{z}} & 0 \\ 0 & \frac{f}{\hat{z}} \end{bmatrix}$$
$$\times \begin{bmatrix} -l_1\sin(q_1) - l_2\sin(q_1+q_2) - \hat{l}_o\sin(q_1+q_2+\hat{q}_o) & -l_2\sin(q_1+q_2) - \hat{l}_o\sin(q_1+q_2+\hat{q}_o) \\ l_1\cos(q_1) + l_2\cos(q_1+q_2) + \hat{l}_o\cos(q_1+q_2+\hat{q}_o) & l_2\cos(q_1+q_2) + \hat{l}_o\cos(q_1+q_2+\hat{q}_o) \end{bmatrix}, \tag{5.54}$$

where f is the focal length, $\hat{z}$ is the unknown constant depth, $\hat{l}_o$ is the length of the unknown object, $\hat{q}_o$ is the angle of the unknown object, $\hat{\boldsymbol{J}}_m(\boldsymbol{q})$ denotes the Jacobian matrix from joint space to Cartesian space, $\hat{\boldsymbol{J}}_s(\boldsymbol{r})$ denotes the Jacobian matrix from Cartesian space to image space. Since exact knowledge of the depth and the kinematic parameters of the object is unknown, both $\hat{\boldsymbol{J}}_m(\boldsymbol{q})$ and $\hat{\boldsymbol{J}}_s(\boldsymbol{r})$ are uncertain, and hence the Jacobian matrix from joint space to image space $\hat{\boldsymbol{J}}(\boldsymbol{q})$ is also uncertain. □

Note that the desired position is specified only in the region functions $f_{T_i}(\boldsymbol{x}_{Ti})$. The desired velocity is specified as

$$\dot{\boldsymbol{x}}_{di} = \boldsymbol{0}, \quad \text{if} \quad f_{Er_i}(\boldsymbol{x}_{Ti}) \geq 0 \ \text{ or } \ f_{Ir_{ij}}(\boldsymbol{x}_{Ti}) \leq 0. \tag{5.55}$$

That is, the desired velocity is zero if the end effector is outside the external and internal reference regions in the coordinates of $\boldsymbol{x}_T$.

Equation (5.55) can be satisfied using the weight factors in the desired trajectory. In general, the desired trajectory $\boldsymbol{x}_{di}$ consists of a time-varying part and a constant part, which is denoted by

$$\boldsymbol{x}_{di}(t) = \boldsymbol{x}_{ci} + \boldsymbol{x}_{vi}(t), \tag{5.56}$$

where $\boldsymbol{x}_{ci} = [x_{ci_1}, \ldots, x_{ci_{n_i}}]^T$ is the constant part, and $\boldsymbol{x}_{vi}(t) = [x_{vi_1}, \ldots, x_{vi_{n_i}}]^T$ is the time-varying part. Using the weight factors, $\boldsymbol{x}_{di}$ is revised as

$$\boldsymbol{x}_{di}(t) = \boldsymbol{x}_{ci} + w_i(\boldsymbol{x}_{Ti})\boldsymbol{x}_{vi}(t), \tag{5.57}$$

where $w_i(\boldsymbol{x}_{Ti})$ are weight factors, which are defined so that $w_i(\boldsymbol{x}_{Ti}) = 0$ when $f_{Er_i}(\boldsymbol{x}_{Ti}) \geq 0$ or $f_{Ir_{ij}}(\boldsymbol{x}_{Ti}) \leq 0$, and $w_i(\boldsymbol{x}_{Ti})$ smoothly transit to 1 when $f_{Er_i}(\boldsymbol{x}_{Ti}) < 0$ and $f_{Ir_{ij}}(\boldsymbol{x}_{Ti}) > 0$.

To satisfy Eq. (5.55), weight factors are introduced as

$$w_i(\boldsymbol{x}_{Ti}) = w_{E_i}(\boldsymbol{x}_{Ti}) \prod_{j=1}^{N_i} w_{I_{ij}}(\boldsymbol{x}_{Ti}), \tag{5.58}$$

where $w_{E_i}(\boldsymbol{x}_{Ti})$ and $w_{I_{ij}}(\boldsymbol{x}_{Ti})$ are the weight factors corresponding to the ith external region and the ijth internal region respectively. The weight factor $w_{E_i}(\boldsymbol{x}_{Ti})$

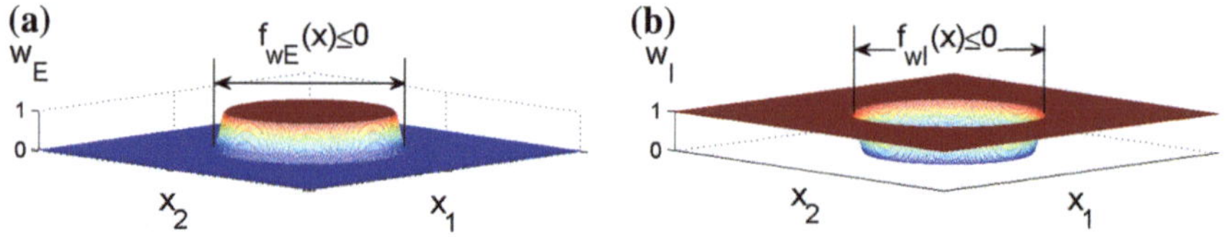

Fig. 5.14 Weight factor. **a** $w_{E_i}(\boldsymbol{x}_{Ti})$. **b** $w_{I_{ij}}(\boldsymbol{x}_{Ti})$

smoothly change from 0 to 1 when the end effector transits from $f_{Er_i}(\boldsymbol{x}_{Ti}) \geq 0$ to $f_{Er_i}(\boldsymbol{x}_{Ti}) < 0$. The weight factor $w_{I_{ij}}(\boldsymbol{x}_{Ti})$ smoothly changes from 1 to 0 when the end effector transits from from $f_{Ir_{ij}}(\boldsymbol{x}_{Ti}) \geq 0$ to $f_{Ir_{ij}}(\boldsymbol{x}_{Ti}) < 0$. Both $w_{E_i}(\boldsymbol{x}_{Ti})$ and $w_{I_{ij}}(\boldsymbol{x}_{Ti})$ must be continuous and smooth between 0 and 1. An illustration is given in Fig. 5.14.

To construct $w_{E_i}(\boldsymbol{x}_{Ti})$, two regions inside $f_{Er_i}(\boldsymbol{x}_{Ti})$ are defined as

$$
\begin{aligned}
f_{wE_i}(\boldsymbol{x}_{Ti}) &= \frac{(x_{Ti_1} - x_{Ei_1})^{nEi}}{c_{i_1}{}^{nEi}} + \cdots + \frac{(x_{Ti_{n_i}} - x_{Ei_{n_i}})^{nEi}}{c_{i_{n_i}}{}^{nEi}} - 1 \leq 0, \\
f_{wEr_i}(\boldsymbol{x}_{Ti}) &= \frac{(x_{Ti_1} - x_{Ei_1})^{nEi}}{(\kappa_{w_i} c_{i_1})^{nEi}} + \cdots + \frac{(x_{Ti_{n_i}} - x_{Ei_{n_i}})^{nEi}}{(\kappa_{w_i} c_{i_{n_i}})^{nEi}} - 1 \leq 0,
\end{aligned}
\tag{5.59}
$$

where $c_{i_1}, \ldots, c_{i_{n_i}}$ are positive constants, and $\kappa_{w_i} < 1$ are also positive constants. Hence the size of $f_{wEr_i}(\boldsymbol{x}_{Ti}) \leq 0$ is smaller than the size of $f_{wE_i}(\boldsymbol{x}_{Ti}) \leq 0$, and the size of $f_{wE_i}(\boldsymbol{x}_{Ti}) \leq 0$ is smaller than the size of $f_{Er_i}(\boldsymbol{x}_{Ti}) \leq 0$. Based on $f_{wE_i}(\boldsymbol{x}_{Ti})$ and $f_{wEr_i}(\boldsymbol{x}_{Ti})$, the weight factors are given as

$$
w_{E_i}(\boldsymbol{x}_{Ti}) = \begin{cases} 0, & f_{wE_i}(\boldsymbol{x}_{Ti}) \geq 0, \\ 1 - \frac{\{[f_{wE_i}(\boldsymbol{x}_{Ti})]^4 - (\kappa_{w_i}^{nEi} - 1)^4\}^4}{(\kappa_{w_i}^{nEi} - 1)^{16}}, & f_{wEr_i}(\boldsymbol{x}_{Ti}) > 0,\ f_{wE_i}(\boldsymbol{x}_{Ti}) < 0, \\ 1, & f_{wEr_i}(\boldsymbol{x}_{Ti}) \leq 0. \end{cases}
\tag{5.60}
$$

Similarly, to construct $w_{I_{ij}}(\boldsymbol{x}_{Ti})$, two regions are defined as

$$
\begin{aligned}
f_{wI_{ij}}(\boldsymbol{x}_{Ti}) &= \frac{(x_{Ti_1} - x_{Iij_1})^{nIij}}{d_{ij_1}{}^{nIij}} + \cdots + \frac{(x_{Ti_{n_i}} - x_{Iij_{n_i}})^{nIij}}{d_{ij_{n_i}}{}^{nIij}} - 1 \leq 0, \\
f_{wIr_{ij}}(\boldsymbol{x}_{Ti}) &= \frac{(x_{Ti_1} - x_{Iij_1})^{nIij}}{(\kappa_{w_{ij}} d_{ij_1})^{nIij}} + \cdots + \frac{(x_{Ti_{n_i}} - x_{Iij_{n_i}})^{nIij}}{(\kappa_{w_{ij}} d_{ij_{n_i}})^{nIij}} - 1 \leq 0,
\end{aligned}
\tag{5.61}
$$

where $d_{ij_1}, \ldots, d_{ij_{n_i}}$ are positive constants, and $\kappa_{w_{ij}} < 1$ are also positive constants. Based on $f_{wI_{ij}}(\boldsymbol{x}_{Ti})$ and $f_{wIr_{ij}}(\boldsymbol{x}_{Ti})$, the weight factors are specified as

$$w_{I_{ij}}(\boldsymbol{x}_{Ti}) = \begin{cases} 0, & f_{w_{Ir_{ij}}}(\boldsymbol{x}_{Ti}) \leq 0, \\ \frac{\{[f_{w_{I_{ij}}}(\boldsymbol{x}_{Ti})]^4 - (\kappa_{w_{ij}}^{n_{Iij}} - 1)^4\}^4}{(\kappa_{w_{ij}}^{n_{Iij}} - 1)^{16}}, & f_{w_{Ir_{ij}}}(\boldsymbol{x}_{Ti}) > 0,\ f_{w_{I_{ij}}}(\boldsymbol{x}_{Ti}) < 0, \\ 1, & f_{w_{I_{ij}}}(\boldsymbol{x}_{Ti}) \geq 0. \end{cases} \tag{5.62}$$

Therefore, when the end effector is outside the external and internal reference regions, $w_i(\boldsymbol{x}_{Ti}) = 0$, and thus $\boldsymbol{x}_{di}(t) = \boldsymbol{x}_{ci}$ and $\dot{\boldsymbol{x}}_{di} = \boldsymbol{0}$, then Eq. (5.55) is satisfied. When the end effector enters the external and internal reference regions, $w_i(\boldsymbol{x}_{Ti})$ smoothly increase to 1, and hence $\boldsymbol{x}_{di}(t) = \boldsymbol{x}_{ci} + \boldsymbol{x}_{vi}(t)$, which is the actual trajectory.

Next, a reference vector $\dot{\boldsymbol{x}}_a = [\dot{\boldsymbol{x}}_{a1}^T, \ldots, \dot{\boldsymbol{x}}_{aM}^T]^T$ is introduced, where $\dot{\boldsymbol{x}}_{ai}$ are specified as

$$\dot{\boldsymbol{x}}_{ai} = \left[\dot{x}_{di_1} - \dot{x}_{di_1} \frac{x_{Ti_1} - x_{di_1}}{x_{bi_1} - x_{di_1}}, \ldots, \dot{x}_{di_{n_i}} - \dot{x}_{di_{n_i}} \frac{x_{Ti_{n_i}} - x_{di_{n_i}}}{x_{bi_{n_i}} - x_{di_{n_i}}} \right]^T. \tag{5.63}$$

Using $\dot{\boldsymbol{x}}_a$ in Eq. (5.63), a sliding vector is designated as

$$s_q = \dot{\boldsymbol{q}} - \hat{\boldsymbol{J}}_x^+(\boldsymbol{q}, \hat{\boldsymbol{\theta}}_k)\dot{\boldsymbol{x}}_a + \alpha_x \hat{\boldsymbol{J}}_x^T(\boldsymbol{q}, \hat{\boldsymbol{\theta}}_k)\Delta\varepsilon_x + \alpha_r \boldsymbol{J}_r^T(\boldsymbol{q})\Delta\varepsilon_r, \tag{5.64}$$

where $\hat{\boldsymbol{J}}_x^+(\boldsymbol{q}, \hat{\boldsymbol{\theta}}_k)$ is the approximate pseudoinverse matrix, $\hat{\boldsymbol{\theta}}_k$ represents the uncertain kinematic parameters, and $\Delta\varepsilon_x$ and $\Delta\varepsilon_r$ denote the desired region error and the reaching region error respectively.

From Eq. (5.57), note that the reference vectors $\dot{\boldsymbol{x}}_{ai}$ described by Eq. (5.63) can be expressed as

$$\dot{\boldsymbol{x}}_{ai} = \boldsymbol{A}_i \dot{\boldsymbol{x}}_i + \dot{\boldsymbol{x}}_{ri}, \tag{5.65}$$

where

$$\boldsymbol{A}_i = \begin{bmatrix} \frac{\partial w_i(\boldsymbol{x}_{Ti})}{\partial x_{Ti_1}} \left(x_{vi_1} - x_{vi_1} \frac{x_{Ti_1} - x_{di_1}}{x_{bi_1} - x_{di_1}} \right) & \cdots & \frac{\partial w_i(\boldsymbol{x}_{Ti})}{\partial x_{Ti_{n_i}}} \left(x_{vi_1} - x_{vi_1} \frac{x_{Ti_1} - x_{di_1}}{x_{bi_1} - x_{di_1}} \right) \\ \vdots & \ddots & \vdots \\ \frac{\partial w_i(\boldsymbol{x}_{Ti})}{\partial x_{Ti_1}} \left(x_{vi_{n_i}} - x_{vi_{n_i}} \frac{x_{Ti_{n_i}} - x_{di_{n_i}}}{x_{bi_{n_i}} - x_{di_{n_i}}} \right) & \cdots & \frac{\partial w_i(\boldsymbol{x}_{Ti})}{\partial x_{Ti_{n_i}}} \left(x_{vi_{n_i}} - x_{vi_{n_i}} \frac{x_{Ti_{n_i}} - x_{di_{n_i}}}{x_{bi_{n_i}} - x_{di_{n_i}}} \right) \end{bmatrix}, \tag{5.66}$$

and

$$\dot{\boldsymbol{x}}_{ri} = \left[w_i(\boldsymbol{x}_{Ti})\dot{x}_{vi_1} - w_i(\boldsymbol{x}_{Ti})\dot{x}_{vi_1} \frac{x_{Ti_1} - x_{di_1}}{x_{bi_1} - x_{di_1}}, \ldots, w_i(\boldsymbol{x}_{Ti})\dot{x}_{vi_{n_i}} \right.$$
$$\left. - w_i(\boldsymbol{x}_{Ti})\dot{x}_{vi_{n_i}} \frac{x_{Ti_{n_i}} - x_{di_{n_i}}}{x_{bi_{n_i}} - x_{di_{n_i}}} \right]^T. \tag{5.67}$$

Next, a compound matrix is introduced as

$$A = \begin{bmatrix} A_1 & \mathbf{0} & \cdots & \mathbf{0} \\ \mathbf{0} & A_2 & \cdots & \mathbf{0} \\ \vdots & \vdots & \ddots & \vdots \\ \mathbf{0} & \mathbf{0} & \cdots & A_M \end{bmatrix}, \tag{5.68}$$

where the matrix A is introduced to avoid the use of joint acceleration in the controller.

Using Eq. (5.68), the sliding vector s_q in Eq. (5.64) can be rewritten as

$$\begin{aligned} s_q = & [I_n - \hat{J}_x^+(q, \hat{\theta}_k) A \hat{J}_x(q, \hat{\theta}_k)]\dot{q} - \hat{J}_x^+(q, \hat{\theta}_k)\dot{x}_r \\ & + \alpha_x \hat{J}_x^T(q, \hat{\theta}_k)\Delta\varepsilon_x + \alpha_r J_r^T(q)\Delta\varepsilon_r, \end{aligned} \tag{5.69}$$

where $I_n \in \Re^{n \times n}$ is the identity matrix, and $\dot{x}_r = [\dot{x}_{r1}^T, \ldots, \dot{x}_{rM}^T]^T$.

Next, a new sliding vector is defined as

$$\begin{aligned} s = \dot{q} - \dot{q}_r = & \dot{q} - [I_n - \hat{J}_x^+(q, \hat{\theta}_k) A \hat{J}_x(q, \hat{\theta}_k)]^{-1} \\ & \times [\hat{J}_x^+(q, \hat{\theta}_k)\dot{x}_r - \alpha_x \hat{J}_x^T(q, \hat{\theta}_k)\Delta\varepsilon_x - \alpha_r J_r^T(q)\Delta\varepsilon_r], \end{aligned} \tag{5.70}$$

where $\dot{q}_r = [I_n - \hat{J}_x^+(q, \hat{\theta}_k) A \hat{J}_x(q, \hat{\theta}_k)]^{-1}[\hat{J}_x^+(q, \hat{\theta}_k)\dot{x}_r - \alpha_x \hat{J}_x^T(q, \hat{\theta}_k)\Delta\varepsilon_x - \alpha_r J_r^T(q)\Delta\varepsilon_r]$ and $s_q = [I_n - \hat{J}_x^+(q, \hat{\theta}_k) A \hat{J}_x(q, \hat{\theta}_k)]s$.

The regional feedback controller for the robotic manipulator is achieved using the region errors and the sliding vector as

$$\begin{aligned} \tau = & -[I_n - \hat{J}_x^+(q, \hat{\theta}_k) A \hat{J}_x(q, \hat{\theta}_k)]^T [k_p \alpha_x \hat{J}_x^T(q, \hat{\theta}_k)\Delta\varepsilon_x \\ & + k_p \alpha_r J_r^T(q)\Delta\varepsilon_r] - K_s s + Y_d(q, \dot{q}, \dot{q}_r, \ddot{q}_r)\hat{\theta}_d, \end{aligned} \tag{5.71}$$

where K_s is a positive definite matrix. When the end effector reaches the regions where the feedback x_T is infeasible due to singularities, occlusion, or limited sensing zone, $\dot{x}_a$ and $\Delta\varepsilon_x$ reduce to zero, and the reaching regional feedback $\Delta\varepsilon_r$ enables the end effector to pass through those regions. After the end effector leaves the regions, $\Delta\varepsilon_r$ reduces to zero, and the desired regional feedback $\Delta\varepsilon_x$ is activated to drive the end effector to the desired trajectory. Therefore, the region errors $\Delta\varepsilon_x$ and $\Delta\varepsilon_r$ work within the corresponding regions, and the controller is continuous, since the region errors are continuous.

The estimated parameters $\hat{\theta}_d$ are updated by the following update law:

$$\dot{\hat{\theta}}_d = -L_d Y_d^T(q, \dot{q}, \dot{q}_r, \ddot{q}_r)s, \tag{5.72}$$

where $L_d \in \Re^{n_d \times n_d}$ is a positive definite matrix. The uncertain kinematic parameters $\hat{\theta}_k$ are updated by the following update law:

$$\dot{\hat{\boldsymbol{\theta}}}_k = k_p \alpha_x \boldsymbol{L}_k \boldsymbol{Y}_k^T(\boldsymbol{q}, \dot{\boldsymbol{q}}) \Delta \boldsymbol{\varepsilon}_x, \tag{5.73}$$

where $\boldsymbol{L}_k \in \Re^{n_k \times n_k}$ is a positive definite matrix.

In the control law described by Eq. (5.71), note that the pseudoinverse Jacobian $\hat{\boldsymbol{J}}_x^+(\boldsymbol{q}, \hat{\boldsymbol{\theta}}_k)$ can be replaced by $\hat{\boldsymbol{J}}_x^{-1}(\boldsymbol{q}, \hat{\boldsymbol{\theta}}_k)$ in the case of nonredundant robots. Hence the controller can be applied to both nonredundant and redundant robots.

From Eq. (1.5), the robot dynamic equation is stated as

$$\boldsymbol{M}(\boldsymbol{q})\ddot{\boldsymbol{q}} + \left(\tfrac{1}{2}\dot{\boldsymbol{M}}(\boldsymbol{q}) + \boldsymbol{S}(\boldsymbol{q}, \dot{\boldsymbol{q}})\right)\dot{\boldsymbol{q}} + \boldsymbol{g}(\boldsymbol{q}) = \boldsymbol{\tau}. \tag{5.74}$$

Using Eq. (5.70) and *Property* 1.3, Eq. (5.74) can be written as

$$\boldsymbol{M}(\boldsymbol{q})\dot{\boldsymbol{s}} + \left(\tfrac{1}{2}\dot{\boldsymbol{M}}(\boldsymbol{q}) + \boldsymbol{S}(\boldsymbol{q}, \dot{\boldsymbol{q}})\right)\boldsymbol{s} + \boldsymbol{Y}_d(\boldsymbol{q}, \dot{\boldsymbol{q}}, \dot{\boldsymbol{q}}_r, \ddot{\boldsymbol{q}}_r)\boldsymbol{\theta}_d = \boldsymbol{\tau}. \tag{5.75}$$

The closed-loop equation is obtained by substituting Eq. (5.71) into Eq. (5.75) to give

$$\begin{aligned}&\boldsymbol{M}(\boldsymbol{q})\dot{\boldsymbol{s}} + \left(\tfrac{1}{2}\dot{\boldsymbol{M}}(\boldsymbol{q}) + \boldsymbol{S}(\boldsymbol{q}, \dot{\boldsymbol{q}})\right)\boldsymbol{s} + [\boldsymbol{I}_n - \hat{\boldsymbol{J}}_x^+(\boldsymbol{q}, \hat{\boldsymbol{\theta}}_k)\boldsymbol{A}\hat{\boldsymbol{J}}_x(\boldsymbol{q}, \hat{\boldsymbol{\theta}}_k)]^T \\ &\quad \times [k_p \alpha_x \hat{\boldsymbol{J}}_x^T(\boldsymbol{q}, \hat{\boldsymbol{\theta}}_k)\Delta\boldsymbol{\varepsilon}_x + k_p \alpha_r \boldsymbol{J}_r^T(\boldsymbol{q})\Delta\boldsymbol{\varepsilon}_r] + \boldsymbol{K}_s \boldsymbol{s} + \boldsymbol{Y}_d(\boldsymbol{q}, \dot{\boldsymbol{q}}, \dot{\boldsymbol{q}}_r, \ddot{\boldsymbol{q}}_r)\Delta\boldsymbol{\theta}_d = \boldsymbol{0},\end{aligned} \tag{5.76}$$

where $\Delta\boldsymbol{\theta}_d = \boldsymbol{\theta}_d - \hat{\boldsymbol{\theta}}_d$.

To prove stability, consider the following Lyapunov-like candidate:

$$V = \tfrac{1}{2}\boldsymbol{s}^T \boldsymbol{M}(\boldsymbol{q})\boldsymbol{s} + P_D(\boldsymbol{x}_T) + P_R(\boldsymbol{r}_R) + \tfrac{1}{2}\Delta\boldsymbol{\theta}_d^T \boldsymbol{L}_d^{-1}\Delta\boldsymbol{\theta}_d + \tfrac{1}{2}\Delta\boldsymbol{\theta}_k^T \boldsymbol{L}_k^{-1}\Delta\boldsymbol{\theta}_k, \tag{5.77}$$

where $\Delta\boldsymbol{\theta}_k = \boldsymbol{\theta}_k - \hat{\boldsymbol{\theta}}_k$.

Next, note that the time derivative of $P_D(\boldsymbol{x}_T)$ in Eq. (5.42) is given as

$$\begin{aligned}\dot{P}_D(\boldsymbol{x}_T) = k_p \alpha_x \sum_{i=1}^{M} \Bigg\{ &\dot{P}_i(\boldsymbol{x}_{Ti}) \prod_{j=1}^{N_i} P_{I_{ij}}(\boldsymbol{x}_{Ti}) \\ &+ P_i(\boldsymbol{x}_{Ti}) \sum_{j=1}^{N_i} \left[\dot{P}_{I_{ij}}(\boldsymbol{x}_{Ti}) \prod_{k \neq j}^{N_i} P_{I_{ik}}(\boldsymbol{x}_{Ti}) \right] - \sum_{j=1}^{N_i} k_{C_{ij}} \dot{P}_{I_{ij}}(\boldsymbol{x}_{Ti}) \Bigg\},\end{aligned} \tag{5.78}$$

where $\dot{P}_i(\boldsymbol{x}_{Ti})$ in Eq. (5.78) is given as

$$\dot{P}_i(\boldsymbol{x}_{Ti}) = \dot{P}_E(\boldsymbol{x}_{Ti}) + \dot{P}_T(\boldsymbol{x}_{Ti}). \tag{5.79}$$

From Eq. (5.41), the time derivative of $P_{T_i}(\boldsymbol{x}_{Ti})$ is given as

$$\begin{aligned}
\dot{P}_{T_i}(\boldsymbol{x}_{Ti}) &= -k_{T_i}[\min(0, f_{T_i}(\boldsymbol{x}_{Ti}))]^{N-1}\dot{f}_{T_i}(\boldsymbol{x}_{Ti}) = -k_{T_i}[\min(0, f_{T_i}(\boldsymbol{x}_{Ti}))]^{N-1} \\
&\quad \times \left(\frac{2(x_{Ti_1}-x_{di_1})(\dot{x}_{Ti_1}-\dot{x}_{di_1})}{(x_{bi_1}-x_{di_1})^2} + \frac{2(x_{Ti_1}-x_{di_1})^2\dot{x}_{di_1}}{(x_{bi_1}-x_{di_1})^3} + \cdots \right. \\
&\quad \left. + \frac{2(x_{Ti_{n_i}}-x_{di_{n_i}})(\dot{x}_{Ti_{n_i}}-\dot{x}_{di_{n_i}})}{(x_{bi_{n_i}}-x_{di_{n_i}})^2} + \frac{2(x_{Ti_{n_i}}-x_{di_{n_i}})^2\dot{x}_{di_{n_i}}}{(x_{bi_{n_i}}-x_{di_{n_i}})^3}\right) \\
&= \left[\dot{x}_{Ti_1} - \dot{x}_{di_1} + \frac{\dot{x}_{di_1}(x_{Ti_1}-x_{di_1})}{(x_{bi_1}-x_{di_1})}, \ldots, \dot{x}_{Ti_{n_i}} - \dot{x}_{di_{n_i}} + \frac{\dot{x}_{di_{n_i}}(x_{Ti_{n_i}}-x_{di_{n_i}})}{(x_{bi_{n_i}}-x_{di_{n_i}})}\right] \\
&\quad \times \left[-\frac{2k_{T_i}[\min(0, f_{T_i}(\boldsymbol{x}_{Ti}))]^{N-1}(x_{Ti_1}-x_{di_1})}{(x_{bi_1}-x_{di_1})^2}, \ldots, \right. \\
&\quad \left. -\frac{2k_{T_i}[\min(0, f_{T_i}(\boldsymbol{x}_{Ti}))]^{N-1}(x_{Ti_{n_i}}-x_{di_{n_i}})}{(x_{bi_{n_i}}-x_{di_{n_i}})^2}\right]^T \\
&= (\dot{\boldsymbol{x}}_{Ti} - \dot{\boldsymbol{x}}_{a_i})^T\left(\frac{\partial P_{T_i}(\boldsymbol{x}_{Ti})}{\partial \boldsymbol{x}_{Ti}}\right)^T = (\dot{\boldsymbol{x}}_T - \dot{\boldsymbol{x}}_a)^T\left(\frac{\partial P_{T_i}(\boldsymbol{x}_{Ti})}{\partial \boldsymbol{x}_T}\right)^T,
\end{aligned} \tag{5.80}$$

where the corresponding terms in $\dot{\boldsymbol{x}}_T - \dot{\boldsymbol{x}}_a$ are multiplied by zero in $(\frac{\partial P_{T_i}(\boldsymbol{x}_{Ti})}{\partial \boldsymbol{x}_T})^T$.

Substituting Eqs. (5.79) and (5.80) into Eq. (5.78) yields

$$\begin{aligned}
\dot{P}_D(\boldsymbol{x}_T) &= k_p\alpha_x \sum_{i=1}^{M}\left\{\left[(\dot{\boldsymbol{x}}_T - \dot{\boldsymbol{x}}_a)^T\left(\frac{\partial P_{T_i}(\boldsymbol{x}_{Ti})}{\partial \boldsymbol{x}_T}\right)^T\right.\right. \\
&\quad \left. + \dot{\boldsymbol{x}}_T^T\left(\frac{\partial P_{E_i}(\boldsymbol{x}_{Ti})}{\partial \boldsymbol{x}_T}\right)^T\right]\prod_{j=1}^{N_i} P_{I_{ij}}(\boldsymbol{x}_{Ti}) - \sum_{j=1}^{N_i}\left[k_{c_{ij}}\dot{\boldsymbol{x}}_T^T\left(\frac{\partial P_{I_{ij}}(\boldsymbol{x}_{Ti})}{\partial \boldsymbol{x}_T}\right)^T\right] \\
&\quad \left. + [P_{T_i}(\boldsymbol{x}_{Ti}) + P_{E_i}(\boldsymbol{x}_{Ti})]\sum_{j=1}^{N_i}\left[\dot{\boldsymbol{x}}_T^T\left(\frac{\partial P_{I_{ij}}(\boldsymbol{x}_{Ti})}{\partial \boldsymbol{x}_T}\right)^T\prod_{k\neq j}^{N_i} P_{I_{ik}}(\boldsymbol{x}_{Ti})\right]\right\} \\
&= k_p\alpha_x\left\{\dot{\boldsymbol{x}}_T^T\Delta\varepsilon_x - \sum_{i=1}^{M}\left[\dot{\boldsymbol{x}}_a^T\left(\frac{\partial P_{T_i}(\boldsymbol{x}_{Ti})}{\partial \boldsymbol{x}_T}\right)^T\prod_{j=1}^{N_i} P_{I_{ij}}(\boldsymbol{x}_{Ti})\right]\right\}.
\end{aligned} \tag{5.81}$$

Differentiating Eq. (5.77) with respect to time and substituting Eq. (5.81) into it yields the following equation:

$$\begin{aligned}
\dot{V} &= \boldsymbol{s}^T\boldsymbol{M}(\boldsymbol{q})\dot{\boldsymbol{s}} + \tfrac{1}{2}\boldsymbol{s}^T\dot{\boldsymbol{M}}(\boldsymbol{q})\boldsymbol{s} + k_p\alpha_x\dot{\boldsymbol{x}}_T^T\Delta\varepsilon_x + k_p\alpha_r\dot{\boldsymbol{r}}_R^T\Delta\varepsilon_r \\
&\quad - k_p\alpha_x\sum_{i=1}^{M}\left[\dot{\boldsymbol{x}}_a^T\left(\frac{\partial P_{T_i}(\boldsymbol{x}_{Ti})}{\partial \boldsymbol{x}_T}\right)^T\prod_{j=1}^{N_i} P_{I_{ij}}(\boldsymbol{x}_i)\right] - \dot{\hat{\boldsymbol{\theta}}}_d^T\boldsymbol{L}_d^{-1}\Delta\boldsymbol{\theta}_d - \dot{\hat{\boldsymbol{\theta}}}_k^T\boldsymbol{L}_k^{-1}\Delta\boldsymbol{\theta}_k.
\end{aligned} \tag{5.82}$$

On substituting Eqs. (5.72) and (5.76) into Eq. (5.82) and using *Property* 1.2 and $s_q = [I_n - \hat{J}_x^+(q, \hat{\theta}_k)A\hat{J}_x(q, \hat{\theta}_k)]s$, one has

$$
\begin{aligned}
\dot{V} &= -s^T K_s s - s^T[I_n - \hat{J}_x^+(q, \hat{\theta}_k)A\hat{J}_x(q, \hat{\theta}_k)]^T \\
&\quad \times [k_p\alpha_x \hat{J}_x^T(q, \hat{\theta}_k)\Delta\varepsilon_x + k_p\alpha_r J_r^T(q)\Delta\varepsilon_r] \\
&\quad - s^T Y_d(q, \dot{q}, \dot{q}_r, \ddot{q}_r)\Delta\theta_d + k_p\alpha_r \dot{r}_R^T\Delta\varepsilon_r - \dot{\hat{\theta}}_d^T L_d^{-1}\Delta\theta_d - \dot{\hat{\theta}}_k^T L_k^{-1}\Delta\theta_k \\
&= -s^T K_s s - s_q^T[k_p\alpha_x \hat{J}_x^T(q, \hat{\theta}_k)\Delta\varepsilon_x + k_p\alpha_r J_r^T(q)\Delta\varepsilon_r] \\
&\quad - s^T Y_d(q, \dot{q}, \dot{q}_r, \ddot{q}_r)\Delta\theta_d + k_p\alpha_r \dot{r}_R^T\Delta\varepsilon_r - \dot{\hat{\theta}}_d^T L_d^{-1}\Delta\theta_d - \dot{\hat{\theta}}_k^T L_k^{-1}\Delta\theta_k \\
&\quad + k_p\alpha_x\left\{\dot{x}_T^T\Delta\varepsilon_x - \sum_{i=1}^{M}\left[\dot{x}_a^T\left(\frac{\partial P_{T_i}(x_{Ti})}{\partial x_T}\right)^T\prod_{j=1}^{N_i}P_{I_{ij}}(x_{Ti})\right]\right\} \\
&= -s^T K_s s + s_q^T[-k_p\alpha_x \hat{J}_x^T(q, \hat{\theta}_k)\Delta\varepsilon_x - k_p\alpha_r J_r^T(q)\Delta\varepsilon_r] + k_p\alpha_r \dot{r}_R^T\Delta\varepsilon_r \\
&\quad + k_p\alpha_x\left\{\dot{x}_T^T\Delta\varepsilon_x - \sum_{i=1}^{M}\left[\dot{x}_a^T\left(\frac{\partial P_{T_i}(x_{Ti})}{\partial x_T}\right)^T\prod_{j=1}^{N_i}P_{I_{ij}}(x_{Ti})\right]\right\} - \dot{\hat{\theta}}_k^T L_k^{-1}\Delta\theta_k.
\end{aligned}
\tag{5.83}
$$

Then substituting Eqs. (5.64) and (5.73) into Eq. (5.83) yields

$$
\begin{aligned}
\dot{V} &= -s^T K_s s - k_p[\alpha_x \hat{J}_x^T(q, \hat{\theta}_k)\Delta\varepsilon_x + \alpha_r J_r^T(q)\Delta\varepsilon_r]^T \\
&\quad \times [\alpha_x \hat{J}_x^T(q, \hat{\theta}_k)\Delta\varepsilon_x + \alpha_r J_r^T(q)\Delta\varepsilon_r] \\
&\quad + [\dot{q} - \hat{J}_x^+(q, \hat{\theta}_k)\dot{x}_a]^T[-k_p\alpha_x \hat{J}_x^T(q, \hat{\theta}_k)\Delta\varepsilon_x - k_p\alpha_r J_r^T(q)\Delta\varepsilon_r] \\
&\quad + k_p\alpha_x\left\{\dot{x}_T^T\Delta\varepsilon_x - \sum_{i=1}^{M}\left[\dot{x}_a^T\left(\frac{\partial P_{T_i}(x_{Ti})}{\partial x_T}\right)^T\prod_{j=1}^{N_i}P_{I_{ij}}(x_{Ti})\right]\right\} \\
&\quad + k_p\alpha_r \dot{r}_R^T\Delta\varepsilon_r - \dot{\hat{\theta}}_k^T L_k^{-1}\Delta\theta_k \\
&= -s^T K_s s - k_p[\alpha_x \hat{J}_x^T(q, \hat{\theta}_k)\Delta\varepsilon_x + \alpha_r J_r^T(q)\Delta\varepsilon_r]^T \\
&\quad \times [\alpha_x \hat{J}_x^T(q, \hat{\theta}_k)\Delta\varepsilon_x + \alpha_r J_r^T(q)\Delta\varepsilon_r] \\
&\quad + k_p\alpha_r \dot{x}_a^T \hat{J}_x^{+T}(q, \hat{\theta}_k)J_r^T(q)\Delta\varepsilon_r + k_p\alpha_x \dot{x}_a^T\Delta\varepsilon_x \\
&\quad - k_p\alpha_x\sum_{i=1}^{M}\left[\dot{x}_a^T\left(\frac{\partial P_{T_i}(x_{Ti})}{\partial x_T}\right)^T\prod_{j=1}^{N_i}P_{I_{ij}}(x_{Ti})\right],
\end{aligned}
\tag{5.84}
$$

where the term $\dot{x}_a^T\Delta\varepsilon_x$ is given as

$$\dot{\boldsymbol{x}}_a^T \Delta\varepsilon_x = \sum_{i=1}^{M} \left\{ \dot{\boldsymbol{x}}_a^T \left[\left(\frac{\partial P_{E_i}(\boldsymbol{x}_{Ti})}{\partial \boldsymbol{x}_T} \right)^T + \left(\frac{\partial P_{T_i}(\boldsymbol{x}_{Ti})}{\partial \boldsymbol{x}_T} \right)^T \right] \prod_{j=1}^{N_i} P_{I_{ij}}(\boldsymbol{x}_{Ti}) \right.$$
$$+ [P_{T_i}(\boldsymbol{x}_{Ti}) + P_{E_i}(\boldsymbol{x}_{Ti})] \sum_{j=1}^{N_i} \left[\dot{\boldsymbol{x}}_a^T \left(\frac{\partial P_{I_{ij}}(\boldsymbol{x}_{Ti})}{\partial \boldsymbol{x}_T} \right)^T \prod_{k \neq j}^{N_i} P_{I_{ik}}(\boldsymbol{x}_{Ti}) \right]$$
$$\left. - \sum_{j=1}^{N_i} \left[k_{c_{ij}} \dot{\boldsymbol{x}}_a^T \left(\frac{\partial P_{I_{ij}}(\boldsymbol{x}_{Ti})}{\partial \boldsymbol{x}_T} \right)^T \right] \right\}. \tag{5.85}$$

Substituting Eq. (5.85) into Eq. (5.84) leads to

$$\dot{V} = -\boldsymbol{s}^T \boldsymbol{K}_s \boldsymbol{s} - k_p[\alpha_x \hat{\boldsymbol{J}}_x^T(\boldsymbol{q}, \hat{\boldsymbol{\theta}}_k)\Delta\varepsilon_x + \alpha_r \boldsymbol{J}_r^T(\boldsymbol{q})\Delta\varepsilon_r]^T$$
$$\times [\alpha_x \hat{\boldsymbol{J}}_x^T(\boldsymbol{q}, \hat{\boldsymbol{\theta}}_k)\Delta\varepsilon_x + \alpha_r \boldsymbol{J}_r^T(\boldsymbol{q})\Delta\varepsilon_r] + k_p \alpha_r \dot{\boldsymbol{x}}_a^T \hat{\boldsymbol{J}}_x^{+T}(\boldsymbol{q}, \hat{\boldsymbol{\theta}}_k) \boldsymbol{J}_r^T(\boldsymbol{q})\Delta\varepsilon_r$$
$$+ k_p \alpha_x \sum_{i=1}^{M} \left\{ \dot{\boldsymbol{x}}_a^T \left(\frac{\partial P_{E_i}(\boldsymbol{x}_{Ti})}{\partial \boldsymbol{x}_T} \right)^T \prod_{j=1}^{N_i} P_{I_{ij}}(\boldsymbol{x}_{Ti}) \sum_{j=1}^{N_i} \left[k_{c_{ij}} \dot{\boldsymbol{x}}_a^T \left(\frac{\partial P_{I_{ij}}(\boldsymbol{x}_{Ti})}{\partial \boldsymbol{x}_T} \right)^T \right] \right.$$
$$\left. + [P_{T_i}(\boldsymbol{x}_{Ti}) + P_{E_i}(\boldsymbol{x}_{Ti})] \sum_{j=1}^{N_i} \left[\dot{\boldsymbol{x}}_a^T \left(\frac{\partial P_{I_{ij}}(\boldsymbol{x}_{Ti})}{\partial \boldsymbol{x}_T} \right)^T \prod_{k \neq j}^{N_i} P_{I_{ik}}(\boldsymbol{x}_{Ti}) \right] \right\}. \tag{5.86}$$

From Eqs. (5.55) and (5.63), note that

$$\dot{\boldsymbol{x}}_{ai} = \boldsymbol{0}, \text{ if } f_{Er_i}(\boldsymbol{x}_{Ti}) \geq 0 \text{ or } f_{Ir_{ij}}(\boldsymbol{x}_{Ti}) \leq 0. \tag{5.87}$$

This condition indicates that $f_{Er_i}(\boldsymbol{x}_{Ti}) \geq 0$ or $f_{Ir_{ij}}(\boldsymbol{x}_{Ti}) \leq 0$ includes that $h_{E_i}(\boldsymbol{r}_R) \geq 0$ or $h_{I_{ij}}(\boldsymbol{r}_R) \leq 0$, where the reaching regional feedback $\Delta\varepsilon_r$ is nonzero. Thus $\boldsymbol{x}_{ai}$ and $\Delta\varepsilon_r$ cannot be nonzero at the same time.

Equation (5.87) also indicates that $\dot{\boldsymbol{x}}_a$ is nonzero where $f_{Er_i}(\boldsymbol{x}_{Ti}) < 0$ and $f_{Ir_{ij}}(\boldsymbol{x}_{Ti}) > 0$. The gradient of potential energy $P_{E_i}(\boldsymbol{x}_{Ti})$ reduces to zero where $f_{Er_i}(\boldsymbol{x}_{Ti}) < 0$, while the gradient of potential energy $P_{I_{ij}}(\boldsymbol{x}_{Ti})$ reduces to zero where $f_{Ir_{ij}}(\boldsymbol{x}_{Ti}) > 0$. Thus, $\boldsymbol{x}_{ai}$ and $(\frac{\partial P_{E_i}(\boldsymbol{x}_{Ti})}{\partial \boldsymbol{x}_T})^T$, $\boldsymbol{x}_{ai}$ and $(\frac{\partial P_{I_{ij}}(\boldsymbol{x}_{Ti})}{\partial \boldsymbol{x}_T})^T$ cannot be nonzero at the same time.

Therefore, the last two terms in Eq. (5.86) are always zero, so that

$$\dot{V} = -\boldsymbol{s}^T \boldsymbol{K}_s \boldsymbol{s} - k_p[\alpha_x \hat{\boldsymbol{J}}_x^T(\boldsymbol{q}, \hat{\boldsymbol{\theta}}_k)\Delta\varepsilon_x + \alpha_r \boldsymbol{J}_r^T(\boldsymbol{q})\Delta\varepsilon_r]^T$$
$$\times [\alpha_x \hat{\boldsymbol{J}}_x^T(\boldsymbol{q}, \hat{\boldsymbol{\theta}}_k)\Delta\varepsilon_x + \alpha_r \boldsymbol{J}_r^T(\boldsymbol{q})\Delta\varepsilon_r]. \tag{5.88}$$

We may now state the following theorem:

Theorem 5.1 *The regional feedback control law* (5.71) *and the update law* (5.72) *for the robot system* (5.74) *guarantee the convergence of the tracking errors. That is,* $\boldsymbol{x} \to \boldsymbol{x}_d$, $\dot{\boldsymbol{x}} \to \dot{\boldsymbol{x}}_d$ *as* $t \to \infty$.

Proof Since $V > 0$ and $\dot{V} \leq 0$, V is bounded. Hence $\boldsymbol{s}$, $\Delta\boldsymbol{\theta}_d$, $\Delta\boldsymbol{\theta}_k$, $P_R(\boldsymbol{r}_R)$, and $P_D(\boldsymbol{x}_T)$ are bounded. Since $P_R(\boldsymbol{r}_R)$ is bounded, $P_{E_i}(\boldsymbol{r}_R)$ and $P_{I_{ij}}(\boldsymbol{r}_R)$ are bounded. The boundedness of $P_{E_i}(\boldsymbol{r}_R)$ and $P_{I_{ij}}(\boldsymbol{r}_R)$ ensures the boundedness of $h_{E_i}(\boldsymbol{r}_R)$ and $h_{I_{ij}}(\boldsymbol{r}_R)$. In addition, $P_{E_i}(\boldsymbol{x}_{Ti})$, $P_{I_{ij}}(\boldsymbol{x}_{Ti})$ and $P_{T_i}(\boldsymbol{x}_{Ti})$ are also bounded, since $P_D(\boldsymbol{x}_T)$ is bounded. The boundedness of $P_{E_i}(\boldsymbol{x}_{Ti})$, $P_{I_{ij}}(\boldsymbol{x}_{Ti})$, and $P_{T_i}(\boldsymbol{x}_{Ti})$ ensures the boundedness of $f_{E_i}(\boldsymbol{x}_{Ti})$, $f_{I_{ij}}(\boldsymbol{x}_{Ti})$, and $f_{T_i}(\boldsymbol{x}_{Ti})$. Since the region functions are bounded, the reaching-task variable $\boldsymbol{r}_R$ and the desired task variable $\boldsymbol{x}_T$ are bounded. Hence, $\frac{\partial h_{E_i}(\boldsymbol{r}_R)}{\partial \boldsymbol{r}_R}$, $\frac{\partial h_{I_{ij}}(\boldsymbol{r}_R)}{\partial \boldsymbol{r}_R}$, $\frac{\partial f_{E_i}(\boldsymbol{x}_{Ti})}{\partial \boldsymbol{x}_T}$, $\frac{\partial f_{I_{ij}}(\boldsymbol{x}_{Ti})}{\partial \boldsymbol{x}_T}$, and $\frac{\partial f_{T_i}(\boldsymbol{x}_{Ti})}{\partial \boldsymbol{x}_T}$ are also bounded. Therefore, $\Delta\boldsymbol{\varepsilon}_r$ and $\Delta\boldsymbol{\varepsilon}_x$ are bounded. Since $\boldsymbol{x}_T$ is bounded, $\dot{\boldsymbol{x}}_a$ is bounded if $\dot{\boldsymbol{x}}_d$ is bounded. Hence, $\dot{\boldsymbol{q}}_r$ is bounded. From Eq. (5.70), $\dot{\boldsymbol{q}}$ is bounded because $\boldsymbol{s}$ is bounded. The boundedness of $\dot{\boldsymbol{q}}$ guarantees the boundedness of $\dot{\boldsymbol{x}}_T$ and $\dot{\boldsymbol{r}}_R$, since both $\boldsymbol{J}_x(\boldsymbol{q})$ and $\boldsymbol{J}_r(\boldsymbol{q})$ are trigonometric functions of $\boldsymbol{q}$ or constant. Therefore, $\Delta\dot{\boldsymbol{\varepsilon}}_r$ and $\Delta\dot{\boldsymbol{\varepsilon}}_x$ are bounded. Then $\ddot{\boldsymbol{q}}_r$ is bounded if $\ddot{\boldsymbol{x}}_d$ is bounded. From the closed-loop equation (5.76), it is concluded that $\dot{\boldsymbol{s}}$ is bounded. Thus, $\ddot{V}$ is bounded, since $\dot{\boldsymbol{s}}, \boldsymbol{s}, \Delta\dot{\boldsymbol{\varepsilon}}_r, \Delta\boldsymbol{\varepsilon}_r, \Delta\dot{\boldsymbol{\varepsilon}}_x, \Delta\boldsymbol{\varepsilon}_x$ are bounded. Therefore, $\dot{V}$ is uniformly continuous. Applying Barbalat's Lemma, we obtain $\dot{V} \to 0$, which also indicates that

$$\begin{aligned} \alpha_x \hat{\boldsymbol{J}}_x^T(\boldsymbol{q}, \hat{\boldsymbol{\theta}}_k)\Delta\boldsymbol{\varepsilon}_x + \alpha_r \boldsymbol{J}_r^T(\boldsymbol{q})\Delta\boldsymbol{\varepsilon}_r &\to \boldsymbol{0}, \\ \boldsymbol{s} &\to \boldsymbol{0}. \end{aligned} \tag{5.89}$$

If the end effector is located where $h_{E_i}(\boldsymbol{r}_R) \geq 0$ or $h_{I_{ij}}(\boldsymbol{r}_R) \leq 0$, then $\Delta\boldsymbol{\varepsilon}_x = \boldsymbol{0}$, $\Delta\boldsymbol{\varepsilon}_r \neq \boldsymbol{0}$, which contradicts Eq. (5.89), since $\boldsymbol{J}_r(\boldsymbol{q})$ is nonsingular. Therefore, the end effector can settle down only where $f_{E_i}(\boldsymbol{x}_{Ti}) \leq 0$ and $f_{I_{ij}}(\boldsymbol{x}_{Ti}) \geq 0$, and hence $\Delta\boldsymbol{\varepsilon}_r = \boldsymbol{0}$. Since $\boldsymbol{J}_x(\boldsymbol{q})$ is also nonsingular, from Eq. (5.89), we have $\Delta\boldsymbol{\varepsilon}_x = \boldsymbol{0}$. From Eq. (5.44), we see that $\Delta\boldsymbol{\varepsilon}_x = \boldsymbol{0}$ can be satisfied only where $f_{T_i}(\boldsymbol{x}_{Ti}) \leq 0$, $f_{Er_i}(\boldsymbol{x}_{Ti}) \leq 0$, and $f_{Ir_{ij}}(\boldsymbol{x}_{Ti}) \geq 0$. Hence $\Delta\boldsymbol{\varepsilon}_x = 0$ means that $\frac{\partial f_{T_i}(\boldsymbol{x}_{Ti})}{\partial \boldsymbol{x}_T} = \boldsymbol{0}$. That is, $\boldsymbol{x}_T \to \boldsymbol{x}_d$ as $t \to \infty$. From the definition of $\dot{\boldsymbol{x}}_a$ in Eq. (5.63), $\boldsymbol{x}_T \to \boldsymbol{x}_d$ indicates that $\dot{\boldsymbol{x}}_a \to \dot{\boldsymbol{x}}_d$. Then from the definition of $\boldsymbol{s}$ in Eq. (5.70), we have that $\dot{\boldsymbol{x}}_a \to \dot{\boldsymbol{x}}_d$, $\Delta\boldsymbol{\varepsilon}_r \to \boldsymbol{0}$, $\Delta\boldsymbol{\varepsilon}_x \to \boldsymbol{0}$, and $\boldsymbol{s} \to \boldsymbol{0}$ implies $\dot{\boldsymbol{x}}_T \to \dot{\boldsymbol{x}}_d$ as $t \to \infty$. $\triangle\triangle\triangle$

The combination of potential energies does not result in a local minimum for the tracking control task. This can be shown by analyzing the partial derivatives of the potential energy in Eqs. (5.34) and (5.44) and the controller in Eq. (5.71).

(i) If the end effector is located where $f_{E_i}(\boldsymbol{x}_{Ti}) \leq 0$, $f_{Er_i}(\boldsymbol{x}_{Ti}) > 0$, and $f_{Ir_{ij}}(\boldsymbol{x}_{Ti}) \geq 0$, the gradient of $P_i(\boldsymbol{x}_{Ti})$ is nonzero, and the gradient of $P_{I_{ij}}(\boldsymbol{x}_{Ti})$ reduces to zero. Therefore, $\Delta\boldsymbol{\varepsilon}_x = \sum_{i=1}^{M}[(\frac{\partial P_i(\boldsymbol{x}_{Ti})}{\partial \boldsymbol{x}_T})^T \prod_{j=1}^{N_i} P_{I_{ij}}(\boldsymbol{x}_{Ti})]$, which is nonzero until the end effector is inside the external reference regions such that $f_{Er_i}(\boldsymbol{x}_{Ti}) \leq 0$.

(ii) After the end effector enters the external reference regions, it is inside the task-oriented regions such that $f_{T_i}(\boldsymbol{x}_{Ti}) \leq 0$. If the end effector is located where $f_{T_i}(\boldsymbol{x}_{Ti}) \leq 0$ and $f_{Ir_{ij}}(\boldsymbol{x}_{Ti}) \geq 0$, then both the gradient of $P_{E_i}(\boldsymbol{x}_{Ti})$ and $P_{I_{ij}}(\boldsymbol{x}_{Ti})$ are zero, and $P_i(\boldsymbol{x}_i) = P_{T_i}(\boldsymbol{x}_{Ti})$. Therefore, $\Delta\boldsymbol{\varepsilon}_x = \sum_{i=1}^{M}[(\frac{\partial P_{T_i}(\boldsymbol{x}_{Ti})}{\partial \boldsymbol{x}_T})^T \prod_{j=1}^{N_i} P_{I_{ij}}(\boldsymbol{x}_{Ti})]$, which is nonzero until the end effector reaches the desired position.

(iii) When the end effector starts outside the external or internal regions such that $f_{E_i}(\boldsymbol{x}_{Ti}) > 0$ or $f_{I_{ij}}(\boldsymbol{x}_{Ti}) < 0$, the reaching task feedback $\boldsymbol{r}_R$ is activated to enable the end effector to leave those regions.
(iv) However, there is one saddle point between the internal regions and the internal reference regions where $f_{I_{ij}}(\boldsymbol{x}_{Ti}) \geq 0$ and $f_{Ir_{ij}}(\boldsymbol{x}_{Ti}) < 0$, and the gradient of potential energy $(\frac{\partial P_D(\boldsymbol{x}_T)}{\partial \boldsymbol{x}_T})^T$ at the saddle point is zero. The saddle point exists in the presence of internal regions. This is because the potential energy $P_i(\boldsymbol{x}_{Ti})$ is first multiplied by $P_{I_{ij}}(\boldsymbol{x}_{Ti})$ and then added to the offset term. If the end effector reaches the saddle position, both $\Delta\varepsilon_x$ and $\Delta\varepsilon_r$ reduce to zero. However, both $\boldsymbol{s} = \dot{\boldsymbol{q}} - [\boldsymbol{I}_n - \hat{\boldsymbol{J}}_x^+(\boldsymbol{q}, \hat{\boldsymbol{\theta}}_k)\boldsymbol{A}\hat{\boldsymbol{J}}_x(\boldsymbol{q}, \hat{\boldsymbol{\theta}}_k)]^{-1}[\hat{\boldsymbol{J}}_x^+(\boldsymbol{q}, \hat{\boldsymbol{\theta}}_k)\dot{\boldsymbol{x}}_r - \alpha_x \hat{\boldsymbol{J}}_x^T(\boldsymbol{q}, \hat{\boldsymbol{\theta}}_k)\Delta\varepsilon_x - \alpha_r \boldsymbol{J}_r^T(\boldsymbol{q})\Delta\varepsilon_r]$ and $\boldsymbol{\tau}$ are nonzero, because the velocity control term $\dot{\boldsymbol{q}} - [\boldsymbol{I}_n - \hat{\boldsymbol{J}}_x^+(\boldsymbol{q}, \hat{\boldsymbol{\theta}}_k)\boldsymbol{A}\hat{\boldsymbol{J}}_x(\boldsymbol{q}, \hat{\boldsymbol{\theta}}_k)]^{-1}\hat{\boldsymbol{J}}_x^+(\boldsymbol{q}, \hat{\boldsymbol{\theta}}_k)\dot{\boldsymbol{x}}_r$ is nonzero. When the system starts from rest at the saddle point with $\dot{\boldsymbol{x}}_d = \boldsymbol{0}$, the initial estimates of the uncertain parameters $\hat{\boldsymbol{\theta}}_d$ can be set to nonzero values, so that $\boldsymbol{Y}_d(\boldsymbol{q}, \dot{\boldsymbol{q}}, \dot{\boldsymbol{q}}_r, \ddot{\boldsymbol{q}}_r)\hat{\boldsymbol{\theta}}_d$ is nonzero, resulting in a nonzero initial control input. This enables the end effector to leave the saddle point. If all elements in the regressor matrix reduce to zero, the initial control input may reduce to zero, and an overparameterization of the dynamic model is needed to move the end effector out of the saddle point. □

In the presence of uncertain depth information, the adaptive regional feedback controller is modified as

$$\begin{aligned}\boldsymbol{\tau}_a = &-[\boldsymbol{I}_n - \hat{\boldsymbol{J}}^+(\boldsymbol{q}, \hat{\boldsymbol{\theta}}_k)\hat{\boldsymbol{Z}}(\boldsymbol{q}, \hat{\boldsymbol{\theta}}_z)\boldsymbol{A}\hat{\boldsymbol{Z}}^{-1}(\boldsymbol{q}, \hat{\boldsymbol{\theta}}_z)\hat{\boldsymbol{J}}(\boldsymbol{q}, \hat{\boldsymbol{\theta}}_k)]^T[k_p\alpha_r \boldsymbol{J}_r^T(\boldsymbol{q})\Delta\varepsilon_r \\ &+ k_p\alpha_x \hat{\boldsymbol{J}}^T(\boldsymbol{q}, \hat{\boldsymbol{\theta}}_k)\hat{\boldsymbol{Z}}^{-1}(\boldsymbol{q}, \hat{\boldsymbol{\theta}}_z)\Delta\varepsilon_x] - \boldsymbol{K}_s\hat{\boldsymbol{s}}_a + \boldsymbol{Y}_d(\boldsymbol{q}, \dot{\boldsymbol{q}}, \dot{\hat{\boldsymbol{q}}}_{ar}, \ddot{\hat{\boldsymbol{q}}}_{ar})\hat{\boldsymbol{\theta}}_d, \quad (5.90)\end{aligned}$$

where the estimated parameters $\hat{\boldsymbol{\theta}}_d$, $\hat{\boldsymbol{\theta}}_k$, and $\hat{\boldsymbol{\theta}}_z$ are updated using the following update laws:

$$\begin{aligned}\dot{\hat{\boldsymbol{\theta}}}_d &= -\boldsymbol{L}_d\boldsymbol{Y}_d^T(\boldsymbol{q}, \dot{\boldsymbol{q}}, \dot{\hat{\boldsymbol{q}}}_{ar}, \ddot{\hat{\boldsymbol{q}}}_{ar})\hat{\boldsymbol{s}}_a, \\ \dot{\hat{\boldsymbol{\theta}}}_k &= -k_p\alpha_x\boldsymbol{L}_k\boldsymbol{Y}_k^T(\boldsymbol{q}, \dot{\boldsymbol{q}})\hat{\boldsymbol{Z}}^{-1}(\boldsymbol{q}, \hat{\boldsymbol{\theta}}_z)\Delta\varepsilon_x, \\ \dot{\hat{\boldsymbol{\theta}}}_z &= k_p\alpha_x\boldsymbol{L}_z\boldsymbol{Y}_z^T(\boldsymbol{q}, \dot{\boldsymbol{x}})\hat{\boldsymbol{Z}}^{-1}(\boldsymbol{q}, \hat{\boldsymbol{\theta}}_z)\Delta\varepsilon_x, \quad (5.91)\end{aligned}$$

where $\boldsymbol{L}_k$ and $\boldsymbol{L}_z$ are symmetric positive definite matrices, $\boldsymbol{Y}_k(\boldsymbol{q}, \dot{\boldsymbol{q}})$ is the kinematic regressor, $\boldsymbol{Y}_z(\boldsymbol{q}, \dot{\boldsymbol{x}})$ is the depth regressor, $\hat{\boldsymbol{Z}}(\boldsymbol{q}, \hat{\boldsymbol{\theta}}_z)$ is the approximate depth matrix, $\hat{\boldsymbol{J}}(\boldsymbol{q}, \hat{\boldsymbol{\theta}}_k)$ denotes an approximated Jacobian matrix, and $\hat{\boldsymbol{s}}_a$ is a sliding vector defined as

$$\begin{aligned}\hat{\boldsymbol{s}}_a = \dot{\boldsymbol{q}} - \dot{\hat{\boldsymbol{q}}}_{ar} = \dot{\boldsymbol{q}} - &[\boldsymbol{I}_n - \hat{\boldsymbol{J}}^+(\boldsymbol{q}, \hat{\boldsymbol{\theta}}_k)\hat{\boldsymbol{Z}}(\boldsymbol{q}, \hat{\boldsymbol{\theta}}_z)\boldsymbol{A}\hat{\boldsymbol{Z}}^{-1}(\boldsymbol{q}, \hat{\boldsymbol{\theta}}_z)\hat{\boldsymbol{J}}(\boldsymbol{q}, \hat{\boldsymbol{\theta}}_k)]^{-1} \\ &\times [\hat{\boldsymbol{J}}^+(\boldsymbol{q}, \hat{\boldsymbol{\theta}}_k)\hat{\boldsymbol{Z}}(\boldsymbol{q}, \hat{\boldsymbol{\theta}}_z)\dot{\boldsymbol{x}}_r - \alpha_x\hat{\boldsymbol{J}}^T(\boldsymbol{q}, \hat{\boldsymbol{\theta}}_k)\hat{\boldsymbol{Z}}^{-1}(\boldsymbol{q}, \hat{\boldsymbol{\theta}}_z)\Delta\varepsilon_x - \alpha_r\boldsymbol{J}_r^T(\boldsymbol{q})\Delta\varepsilon_r]. \quad (5.92)\end{aligned}$$

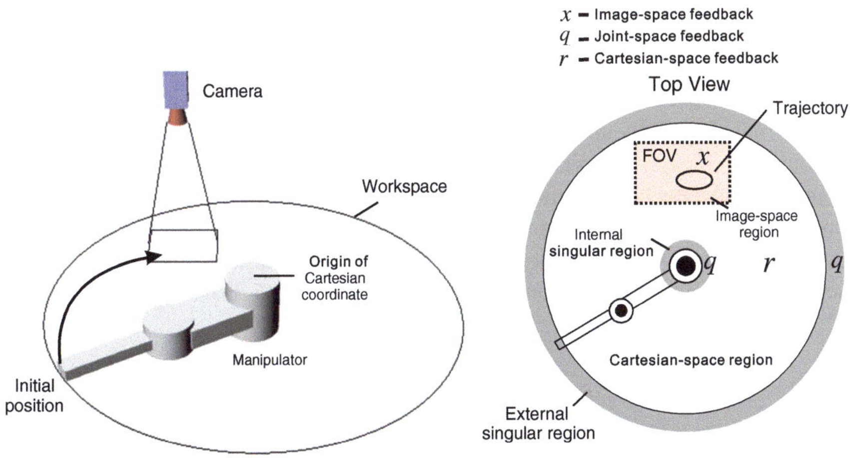

Fig. 5.15 Multiple regional feedback for a 2-DOF planar robot

5.5 Multiple Regional Feedback

In the previous section, the regional feedback method integrates dual task-space variables into one controller, which can be used to solve the problems of singularity and limited field of view separately. In this section, the results are extended to a multiple regional feedback controller, which allows the issues of singularity and limited field of view to be addressed concurrently.

The multiple regional feedback method consists of two reaching task variables and one desired task variable. The joint-space information and the Cartesian-space feedback serve together as the reaching task feedback, while the vision is specified as the desired task feedback. Note that the Cartesian-space feedback is necessary for the multiple regional feedback method, which enables the robot to transit from joint space to image space without solving the inverse kinematics.

The concept of multiple regional feedback can be illustrated by considering a 2-DOF manipulator system with a fixed camera configuration as shown in Fig. 5.15. In that figure, the entire robot workspace is divided into three categories. The joint-space regions are defined around the singular positions, and the robot employs joint information inside the joint-space regions to leave singular configurations. The image-space regions are defined in the vicinity of the desired trajectory within the fields of view of cameras, and the robot utilizes vision feedback only after the end effector enters the image-space region. Finally, the Cartesian-space regions cover the remaining workspace to ensure a smooth transition from joint-space regions to image-space regions. Each piece of feedback information is employed in a corresponding local region, and the combination of regional feedback guarantees the global stability of robot movement.

5.5.1 Joint-Space Feedback

As discussed in Sect. 5.1.1, the singularities of the robotic manipulator can be divided into external singularity and internal singularity, where the former is called the external boundary singularity, and the latter includes the internal boundary singularity and the interior singularity. For multiple task-space control, the joint-space region functions are all formulated to cover the singular positions.

Since the joint-space feedback serves as the reaching-task variable, the potential energy for the joint-space regions is introduced as in Eq. (5.33) as

$$P_J(\boldsymbol{q}) = k_p \alpha_q \sum_{i=1}^{M} \left[P_{JE_i}(\boldsymbol{q}) + \sum_{j=1}^{N_i} P_{JI_{ij}}(\boldsymbol{q}) \right] = k_p \alpha_q$$
$$\times \sum_{i=1}^{M} \frac{k_{qE_i}}{N} [\max(0, h_{JE_i}(\boldsymbol{q}))]^N + \sum_{j=1}^{N_i} \frac{k_{qI_{ij}}}{N} [\min(0, h_{JI_{ij}}(\boldsymbol{q}))]^N \}, \quad (5.93)$$

where k_{qE_i}, $k_{qI_{ij}}$, and α_q are positive constants, and $h_{JE_i}(\boldsymbol{q})$ and $h_{JI_{ij}}(\boldsymbol{q})$ represent the external and internal singular regions, which are described in Eqs. (5.16) and (5.17). The robot employs only the joint information within the singular regions, so that the inverse Jacobian matrix is not required. In Eq. (5.93), $P_{JE_i}(\boldsymbol{q})$ and $P_{JI_{ij}}(\boldsymbol{q})$ are the potential energy functions for the external and internal singular regions. Note that the potential energy in joint space is bounded below by zero, and it smoothly reduces to zero where $h_{JE_i}(\boldsymbol{q}) < 0$ and $h_{JI_{ij}}(\boldsymbol{q}) > 0$. Therefore, the potential energy $P_J(\boldsymbol{q})$ will be used to move the end effector out of the singular regions.

Example 5.4 Consider the manipulator system displayed in Fig. 5.15. Singularities occur when $q_2 = 0$, where the manipulator is fully stretched out, and $q_2 = \pi$, where the manipulator is fully folded back. Therefore, two joint-space regions are specified to cover the singular positions $q_2 = 0$ and $q_2 = \pi$ respectively as

$$\begin{aligned} h_{JE_1}(\boldsymbol{q}) &= b_{E_1}^2 - (q_2)^2 \geq 0, \\ h_{JI_{11}}(\boldsymbol{q}) &= (q_2 - \pi)^2 - b_{I_{11}}^2 \leq 0, \end{aligned} \quad (5.94)$$

where b_{E_1} and $b_{I_{11}}$ are positive constants that represent the sizes of the singular regions. The singular regions $h_{JE_1}(\boldsymbol{q})$ and $h_{JI_{11}}(\boldsymbol{q})$ correspond to circles in Cartesian space, as illustrated in Fig. 5.16. □

5.5.2 Image-Space Feedback

Since the vision feedback serves as the desired task variable, the potential energy for the image-space regions is introduced as in Eq. (5.42) as

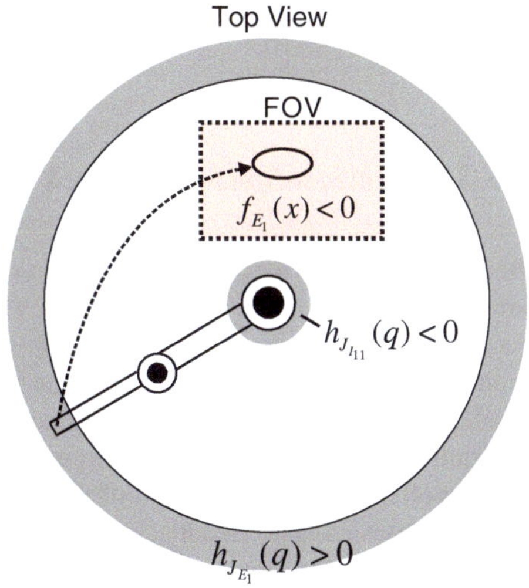

Fig. 5.16 The singular regions $h_{J_{E_1}}(\boldsymbol{q})$ and $h_{J_{I_{11}}}(\boldsymbol{q})$ correspond to circular paths in Cartesian space

$$P_D(\boldsymbol{x}) = k_p\alpha_x \sum_{i=1}^{M}\{P_i(\boldsymbol{x}_i)\prod_{j=1}^{N_i} P_{I_{ij}}(\boldsymbol{x}_i) + \sum_{j=1}^{N_i} k_{c_{ij}}[\frac{k_{xI_{ij}}}{N^2}(\frac{1}{\kappa_{I_{ij}}^{n_{Iij}}} - 1)^{N^2} - P_{I_{ij}}(\boldsymbol{x}_i)]\}, \tag{5.95}$$

where M is the number of feature points, $P_i(\boldsymbol{x}_i) = P_{E_i}(\boldsymbol{x}_i) + P_{T_i}(\boldsymbol{x}_i)$, $P_{E_i}(\boldsymbol{x}_i)$, $P_{I_{ij}}(\boldsymbol{x}_i)$, and $P_{T_i}(\boldsymbol{x}_i)$ are the respective potential energy functions for the external regions, internal regions, and task-oriented regions in image space. The image-space regions are specified based on the general expressions in Eqs. (5.21), (5.22), and (5.28). The potential energy in image space $P_D(\boldsymbol{x})$ is constant when the end effector is outside the image-space region, and the bottom of potential energy corresponds to the desired position, so that the convergence of tracking errors is guaranteed.

Example 5.5 Consider again the manipulator system in Fig. 5.15. The external region in image space $f_{E_1}(\boldsymbol{x}_1)$ described by Eq. (5.21) is specified to match the rectangular field of view as

$$f_{E_1}(\boldsymbol{x}_1) = \frac{(x_{1_h} - x_{E1_h})^{20}}{a_1^{20}} + \frac{(x_{1_v} - x_{E1_v})^{20}}{a_2^{20}} - 1 \leq 0, \tag{5.96}$$

where a_1, a_2 are positive constants that are set to ensure that $f_{E_1}(\boldsymbol{x}_1) \leq 0$ is inside the field of view, and the reference position $\boldsymbol{x}_{E1} = [x_{E1_h}, x_{E1_v}]^T$ is set as the center of the field of view. In addition, the task-oriented region in Eq. (5.28) is introduced to ensure the convergence of tracking errors as

$$f_{T_1}(\boldsymbol{x}_1) = \frac{(x_{1_h} - x_{d1_h})^2}{(x_{b1_h} - x_{d1_h})^2} + \frac{(x_{1_v} - x_{d1_v})^2}{(x_{b1_v} - x_{d1_v})^2} - 1 \leq 0, \tag{5.97}$$

where $\boldsymbol{x}_{b1} = [x_{b1_h}, x_{b1_v}]^T$ is the boundary position. The image-space region is illustrated in Fig. 5.16. □

5.5.3 Cartesian-Space Feedback

Besides the joint-space regions and image-space regions, the Cartesian-space regions are specified to cover the remaining robot workspace. Therefore, the Cartesian-space feedback serves as another reaching-task variable to guarantee a smooth transition from joint-space regions to image-space regions. The Cartesian-space feedback is used for the multiple regional feedback method, since solving the inverse kinematics is required if the joint information is employed as the reaching-task variable to drive the end effector to enter the field of view of each camera.

The Cartesian-space feedback variable $\boldsymbol{r}$ is specified as

$$\boldsymbol{r} = [\boldsymbol{r}_1^T, \ldots, \boldsymbol{r}_i^T, \ldots, \boldsymbol{r}_M^T]^T, \tag{5.98}$$

where $\boldsymbol{r}_i = [r_{i_1}, \ldots, r_{i_{n_i}}]^T \in \Re^{n_i}$ denotes the ith feature point in Cartesian space, $i = 1, 2, \ldots, M$, and M is the total number of features. Since the joint-space regions and image-space regions correspond to different shapes in Cartesian space, the Cartesian-space regions are specified as a set of subregions to match the joint-space regions and image-space regions respectively.

5.5.3.1 Subregions $h_{C_{E_i}}(\boldsymbol{r}_i)$ to Match the External Singular Regions

To match the external singular region $h_{J_{E_i}}(\boldsymbol{q})$, Cartesian-space subregions are specified as

$$h_{C_{E_i}}(\boldsymbol{r}_i) = \frac{(r_{i_1} - r_{Ei_1})^{n_{CEi}}}{a_{Ci_1}^{n_{CEi}}} + \cdots + \frac{(r_{in_i} - r_{Ein_i})^{n_{CEi}}}{a_{Cin_i}^{n_{CEi}}} - 1 \leq 0, \tag{5.99}$$

where the n_{CEi} denote the orders of region functions, which are even integers, $\boldsymbol{r}_{Ei} = [r_{Ei_1}, \ldots, r_{Ein_i}]^T$ are reference positions, and $a_{Ci_1}, \ldots, a_{Cin_i}$ is a set of positive constants. The functions $h_{C_{E_i}}(\boldsymbol{r}_i)$ are specified to exclude the external singularity in Cartesian space, and thus the $h_{C_{E_i}}(\boldsymbol{r}_i)$ represent the outer boundaries of Cartesian-space feedback. Therefore, the robot employs joint information where $h_{C_{E_i}}(\boldsymbol{r}_i) > 0$ and Cartesian-space feedback where $h_{C_{E_i}}(\boldsymbol{r}_i) \leq 0$.

To construct the potential energy, reference regions $h_{C_{Er_i}}(\boldsymbol{r}_i)$ within $h_{C_{E_i}}(\boldsymbol{r}_i)$ are introduced as

$$h_{C_{Er_i}}(\boldsymbol{r}_i) = \frac{(r_{i_1} - r_{Ei_1})^{n_{CEi}}}{(\kappa_{CE_i} a_{Ci_1})^{n_{CEi}}} + \cdots + \frac{(r_{in_i} - r_{Ein_i})^{n_{CEi}}}{(\kappa_{CE_i} a_{Cin_i})^{n_{CEi}}} - 1 \leq 0, \tag{5.100}$$

where κ_{CE_i} are positive constants and $\kappa_{CE_i} < 1$.

Using $h_{CE_i}(\boldsymbol{r}_i)$ and $h_{CEr_i}(\boldsymbol{r}_i)$, the potential energy functions are introduced as

$$P_{CE_i}(\boldsymbol{r}_i) = \frac{k_{CE_i}}{N^2}\{\min[0, [\min(0, h_{CE_i}(\boldsymbol{r}_i))]^N - (\kappa_{CE_i}^{n_{CEi}} - 1)^N]\}^N, \quad (5.101)$$

which can also be written as

$$P_{CE_i}(\boldsymbol{r}_i) = \begin{cases} 0, & h_{CEr_i}(\boldsymbol{r}_i) \leq 0, \\ \frac{k_{CE_i}}{N^2}\{[h_{CE_i}(\boldsymbol{r}_i)]^N - (\kappa_{CE_i}^{n_{CEi}} - 1)^N\}^N, & h_{CE_i}(\boldsymbol{r}_i) < 0, h_{CEr_i}(\boldsymbol{r}_i) > 0, \\ \frac{k_{CE_i}}{N^2}(\kappa_{CE_i}^{n_{CEi}} - 1)^{N^2}, & h_{CE_i}(\boldsymbol{r}_i) \geq 0, \end{cases} \quad (5.102)$$

where k_{CE_i} are positive constants. From Eq. (5.101), it can be seen that $P_{CE_i}(\boldsymbol{r}_i)$ are smooth and bounded below by zero.

5.5.3.2 Subregions $h_{CI_{ij}}(\boldsymbol{r}_i)$ to Match the Internal Singular Regions

Similarly, to match the internal singular regions $h_{JI_{ij}}(\boldsymbol{q})$, a set of sub-Cartesian-space regions is specified as

$$h_{CI_{ij}}(\boldsymbol{r}_i) = \frac{(r_{i1} - r_{Iij_1})^{n_{CIij}}}{b_{Cij_1}^{n_{CIij}}} + \cdots + \frac{(r_{in_i} - r_{Iijn_i})^{n_{CIij}}}{b_{Cijn_i}^{n_{CIij}}} - 1 \geq 0, \quad (5.103)$$

$\boldsymbol{r}_{Iij} = [r_{Iij_1}, \ldots, r_{Iijn_i}]^T$ represents reference points within $h_{CI_{ij}}(\boldsymbol{r}_i)$, $b_{Cij_1}, \ldots, b_{Cijn_i}$ are positive constants, and n_{CIij} denote the orders of the region functions, which are even integers. Unlike subregions, $h_{CE_i}(\boldsymbol{r}_i)$, $h_{CI_{ij}}(\boldsymbol{r}_i)$ enclose singular regions, so that $h_{CI_{ij}}(\boldsymbol{r}_i)$ represent the inner boundaries of Cartesian-space feedback, and the robot employs joint information where $h_{CI_{ij}}(\boldsymbol{r}_i) < 0$ and Cartesian-space feedback where $h_{CI_{ij}}(\boldsymbol{r}_i) \geq 0$. The orders n_{CIij} in Eq. (5.103) can be varied to match different types of internal singularities in Cartesian space.

Next, reference regions that enclose $h_{CI_{ij}}(\boldsymbol{r}_i)$ are specified as

$$h_{CIr_{ij}}(\boldsymbol{r}_i) = \frac{(r_{i1} - r_{Iij_1})^{n_{CIij}}}{(\kappa_{CI_{ij}} b_{Cij_1})^{n_{CIij}}} + \cdots + \frac{(r_{in_i} - r_{Iijn_i})^{n_{CIij}}}{(\kappa_{CI_{ij}} b_{Cijn_i})^{n_{CIij}}} - 1 \geq 0, \quad (5.104)$$

where $\kappa_{CI_{ij}} > 1$ are constants.

Using the subregions $h_{CI_{ij}}(\boldsymbol{r}_i)$ and the corresponding reference regions described by Eqs. (5.103) and (5.104), the potential energy functions $P_{CI_{ij}}(\boldsymbol{r}_i)$ are specified as

$$P_{CI_{ij}}(\boldsymbol{r}_i) = \frac{k_{CI_{ij}}}{N^2}\{\min[0, [\min(0, h_{CIr_{ij}}(\boldsymbol{r}_i))]^N - (\frac{1}{\kappa_{CI_{ij}}^{n_{CIij}}} - 1)^N]\}^N. \quad (5.105)$$

The above equation can be written as

$$P_{CI_{ij}}(\boldsymbol{r}_i) = \begin{cases} 0, & h_{CI_{ij}}(\boldsymbol{r}_i) \leq 0, \\ \frac{k_{CI_{ij}}}{N^2}\left\{\left[h_{CIr_{ij}}(\boldsymbol{r}_i)\right]^N - \left(\frac{1}{\kappa_{CI_{ij}}^{n_{CIij}}} - 1\right)^N\right\}^N, & h_{CI_{ij}}(\boldsymbol{r}_i) > 0, h_{CIr_{ij}}(\boldsymbol{r}_i) < 0, \\ \frac{k_{CI_{ij}}}{N^2}\left(\frac{1}{\kappa_{CI_{ij}}^{n_{CIij}}} - 1\right)^{N^2}, & h_{CIr_{ij}}(\boldsymbol{r}_i) \geq 0, \end{cases} \tag{5.106}$$

where $k_{CI_{ij}}$ are positive constants.

Example 5.6 For the robot system in Fig. 5.15, since the end effector evolves in a plane, the Cartesian-space regions are all specified in a 2-D space. The external singular region $h_{J_{E_1}}(\boldsymbol{q})$ in Eq. (5.94) corresponds to a circle in Cartesian space. To match the external singular region $h_{J_{E_1}}(\boldsymbol{q})$, the subregion in Eq. (5.99) is specified as a circle:

$$h_{C_{E_1}}(\boldsymbol{r}_1) = \frac{(r_{1_1} - r_{E1_1})^2}{a^2} + \frac{(r_{1_2} - r_{E1_2})^2}{a^2} - 1 \leq 0, \tag{5.107}$$

where $\boldsymbol{r}_{E1} = [r_{E1_1}, r_{E1_2}]^T$ is the reference position, which is set as the origin of Cartesian coordinates, and a denotes the radius. Similarly, to match the internal singular region $h_{J_{I_{11}}}(\boldsymbol{q})$ in Eq. (5.94), the subregion in Eq. (5.103) is also specified as a circle:

$$h_{C_{I_{11}}}(\boldsymbol{r}_1) = \frac{(r_{1_1} - r_{I11_1})^2}{b^2} + \frac{(r_{1_2} - r_{I11_2})^2}{b^2} - 1 \geq 0, \tag{5.108}$$

where $\boldsymbol{r}_{I11} = [r_{I11_1}, r_{I11_2}]^T$ is the reference position, which is also set as the origin of Cartesian coordinates, and b denotes the radius. □

5.5.3.3 Subregions $h_{C_{V_i}}(\boldsymbol{r}_i)$ to Match the Image-Space Regions

To match the image-space region, another set of subregions is formulated as

$$h_{C_{V_i}}(\boldsymbol{r}_i) = \frac{(r_{i_1} - r_{Vi_1})^{n_{CVi}}}{c_{Ci_1}^{n_{CVi}}} + \cdots + \frac{(r_{in_i} - r_{Vin_i})^{n_{CVi}}}{c_{Cin_i}^{n_{CVi}}} - 1 \geq 0, \tag{5.109}$$

where $\boldsymbol{r}_{Vi} = [r_{Vi_1}, \ldots, r_{Vin_i}]^T$ are reference points inside the subregions, $c_{Ci_1}, \ldots, c_{Cin_i}$ are positive constants, and n_{CVi} are the orders of the region functions, which are even integers. The robot employs vision feedback where $h_{C_{V_i}}(\boldsymbol{r}_i) < 0$ and Cartesian-space feedback where $h_{C_{V_i}}(\boldsymbol{r}_i) \geq 0$. In addition, the regions $h_{C_{V_i}}(\boldsymbol{r}_i) < 0$ are within the task-oriented image-space regions $f_{T_i}(\boldsymbol{x}_i) < 0$, so that the end effector does not get stuck during the transition from Cartesian space to image space.

A set of reference regions larger than $h_{CV_i}(\boldsymbol{r}_i)$ is specified as

$$h_{CVr_i}(\boldsymbol{r}_i) = \frac{(r_{i_1} - r_{Vi_1})^{n_{CVi}}}{(\kappa_{CV_i} c_{Ci_1})^{n_{CVi}}} + \cdots + \frac{(r_{in_i} - r_{Vin_i})^{n_{CVi}}}{(\kappa_{CV_i} c_{Cin_i})^{n_{CVi}}} - 1 \geq 0, \tag{5.110}$$

where $\kappa_{CV_i} > 1$ are constants.

Based on the region functions of $h_{CV_i}(\boldsymbol{r}_i)$ and the corresponding reference regions, the potential energies $P_{CVi}(\boldsymbol{r}_i)$ for the subregion are specified as

$$P_{CVi}(\boldsymbol{r}_i) = \frac{k_{CV_i}}{N^2}\{\min[0, [\min(0, h_{CVr_i}(\boldsymbol{r}_i))]^N - (\frac{1}{\kappa_{CV_i}^{n_{CVi}}} - 1)^N]\}^N. \tag{5.111}$$

The above equation can be written as

$$P_{CV_i}(\boldsymbol{r}_i) = \begin{cases} 0, & h_{CV_i}(\boldsymbol{r}_i) \leq 0, \\ \frac{k_{CV_i}}{N^2}\left\{\left[h_{CVr_i}(\boldsymbol{r}_i)\right]^N - \left(\frac{1}{\kappa_{CVi}^{n_{CVi}}} - 1\right)^N\right\}^N, & h_{CV_i}(\boldsymbol{r}_i) > 0, h_{CVr_i}(\boldsymbol{r}_i) < 0, \\ \frac{k_{CV_i}}{N^2}\left(\frac{1}{\kappa_{CV_i}^{n_{CVi}}} - 1\right)^{N^2}, & h_{CVr_i}(\boldsymbol{r}_i) \geq 0, \end{cases}$$

where k_{CV_i} are positive constants.

Example 5.7 Since the region function $f_{E_1}(\boldsymbol{x}_1)$ in Eq. (5.96) corresponds to a rectangle in image space, the subregion in Eq. (5.109) is specified to match $f_{E_1}(\boldsymbol{x}_1)$ as

$$h_{CV_1}(\boldsymbol{r}_1) = \frac{(r_{1_1} - r_{V1_1})^{20}}{c_1^{20}} + \frac{(r_{1_2} - r_{V1_2})^{20}}{c_2^{20}} - 1 \geq 0, \tag{5.112}$$

where $\boldsymbol{r}_{V1} = [r_{V1_1}, r_{V1_2}]^T$ is the reference position that is set as the center of the field of view, and c_1, c_2 are positive constants. □

5.5.3.4 Subregions $h_{CEV_i}(\boldsymbol{r}_i)$ to Drive End Effector from $h_{CE_i}(\boldsymbol{r}_i) \leq 0$ to $h_{CV_i}(\boldsymbol{r}_i) \leq 0$

Note that the $P_{CE_i}(\boldsymbol{r}_i)$ are zero when $h_{CEr_i}(\boldsymbol{r}_i) \leq 0$, and $P_{CV_i}(\boldsymbol{r}_i)$ are constant when $h_{CVr_i}(\boldsymbol{r}_i) \geq 0$. Therefore, another set of subregions $h_{CEV_i}(\boldsymbol{r}_i)$ is introduced so that the end effector can be driven toward the image-space regions after it enters the Cartesian-space regions. The subregions $h_{CEV_i}(\boldsymbol{r}_i)$ are introduced as

$$h_{CEV_i}(\boldsymbol{r}_i) = \frac{(r_{i_1} - r_{EVi_1})^2}{(r_{bi_1} - r_{EVi_1})^2} + \cdots + \frac{(r_{in_i} - r_{EVin_i})^2}{(r_{bin_i} - r_{EVin_i})^2} - 1 \leq 0, \tag{5.113}$$

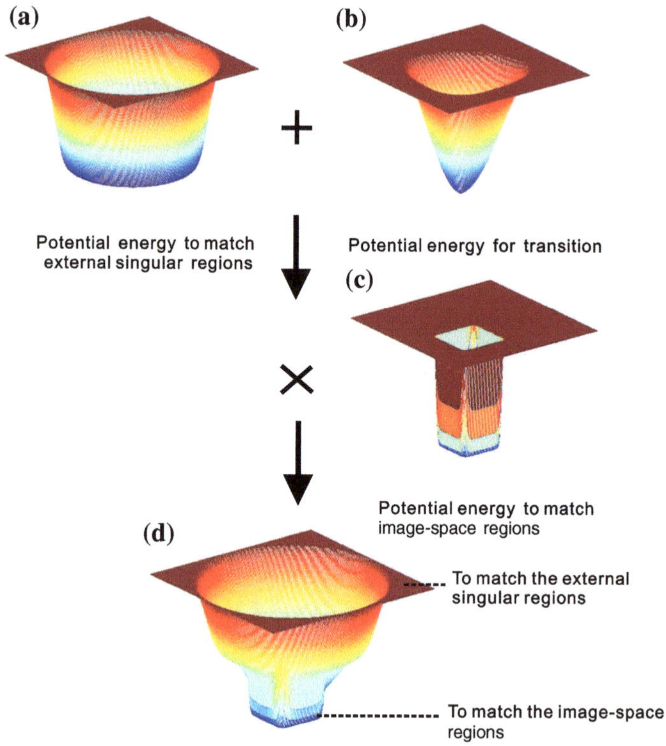

Fig. 5.17 Illustration of the potential energy $[P_{C_{E_i}}(\boldsymbol{r}_i) + P_{C_{EV_i}}(\boldsymbol{r}_i)]P_{C_{V_i}}(\boldsymbol{r}_i)$ in 2-D space. **a** $P_{C_{Ei}}(r)$. **b** $P_{C_{EVi}}(r)$. **c** $P_{C_{Vi}}(r)$. **d** $[P_{C_{Ei}}(r) + P_{C_{EVi}}(r)]P_{C_{Vi}}(r)$

where $\boldsymbol{r}_{bi} = [r_{bi_1}, \ldots, r_{bi_{n_i}}]^T$ denote the boundary positions and $\boldsymbol{r}_{EVi} = [r_{EVi_1}, \ldots, r_{EVi_{n_i}}]^T$ are reference positions. The regions described by Eq. (5.113) are divided into several parts, so that the geometrical centers $\boldsymbol{r}_{EVi}$ are unnecessary.

The potential energy functions for the subregions $h_{C_{EV_i}}(\boldsymbol{r}_i)$ are designed as

$$P_{C_{EV_i}}(\boldsymbol{r}_i) = \frac{k_{CEV_i}}{N}\{1 - [\min(0, h_{C_{EV_i}}(\boldsymbol{r}_i))]^N\}, \tag{5.114}$$

where k_{CEV_i} are positive constants. Note that the gradient of $P_{C_{EVi}}(\boldsymbol{r}_i)$ is nonzero until the end effector converges to the reference position $\boldsymbol{r}_{EVi}$.

5.5.3.5 Cartesian-Space Potential Energy

The overall potential energy $P_C(\boldsymbol{r})$ in Cartesian space is formulated as a combination of potential energies $P_{C_{E_i}}(\boldsymbol{r}_i)$, $P_{C_{I_{ij}}}(\boldsymbol{r}_i)$, $P_{C_{V_i}}(\boldsymbol{r}_i)$, and $P_{C_{EV_i}}(\boldsymbol{r}_i)$ as

$$P_C(\boldsymbol{r}) = k_p \alpha_r \sum_{i=1}^{M} \left\{ \left[P_{C_{E_i}}(\boldsymbol{r}_i) + P_{C_{EV_i}}(\boldsymbol{r}_i) \right] P_{C_{V_i}}(\boldsymbol{r}_i) \prod_{j=1}^{N_i} P_{C_{I_{ij}}}(\boldsymbol{r}_i) \right.$$
$$\left. + \sum_{j=1}^{N_i} k_{p_{ij}} \left[\frac{k_{CI_{ij}}}{N^2} \left(\frac{1}{\kappa_{CI_{ij}}^{n_{CIij}}} - 1 \right)^{N^2} - P_{C_{I_{ij}}}(\boldsymbol{r}_i) \right] \right\}, \qquad (5.115)$$

where $k_{p_{ij}}$ are positive constants. In Eq. (5.115), the term $P_{C_{E_i}}(\boldsymbol{r}_i)+P_{C_{EV_i}}(\boldsymbol{r}_i)$ enables the end effector to move toward the image-space region after it leaves the external singular positions. Then $[P_{C_{E_i}}(\boldsymbol{r}_i)+P_{C_{EV_i}}(\boldsymbol{r}_i)]P_{C_{V_i}}(\boldsymbol{r}_i)$ is flattened at $h_{C_{V_i}}(\boldsymbol{r}_i) < 0$, where the gradient reduces to zero. It avoids the use of Cartesian-space feedback inside the image-space region, and the vision feedback is employed there. The combination of $[P_{C_{E_i}}(\boldsymbol{r}_i) + P_{C_{EV_i}}(\boldsymbol{r}_i)]P_{C_{V_i}}(\boldsymbol{r}_i)$ is illustrated in Fig. 5.17.

After that, the potential energy is multiplied by $\prod_{j=1}^{N_i} P_{C_{I_{ij}}}(\boldsymbol{r}_i)$, so that it is flattened at $h_{C_{I_{ij}}}(\boldsymbol{r}_i) < 0$, where the gradient reduces to zero. It avoids the use of Cartesian-space

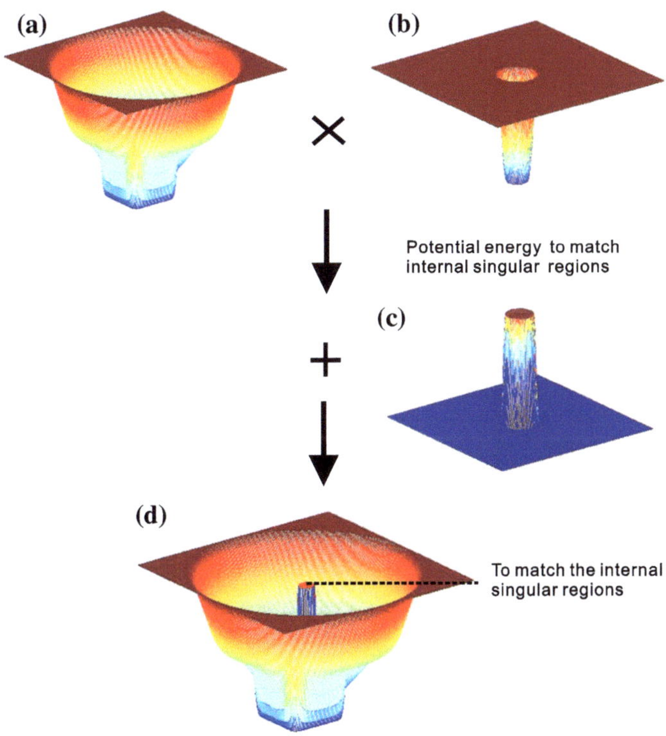

Fig. 5.18 Illustration of the potential energy $P_C(\boldsymbol{r})$ in 2-D space. **a** $[P_{C_{Ei}}(r) + P_{C_{EVi}}(r)]P_{C_{Vi}}(r)$. **b** $P_{C_{Iij}}(r)$. **c** Offset. **d** $P_C(r)$

feedback inside the internal singular regions, and the joint information is employed in those regions. Next, the potential energy is offset inside the regions $h_{C_{I_{ij}}}(\boldsymbol{r}_i) < 0$ by adding the term $\sum_{j=1}^{N_i} k_{p_{ij}}[\frac{k_{CI_{ij}}}{N^2}(\frac{1}{\frac{n_{CIij}}{\kappa_{CI_{ij}}}}-1)^{N^2} - P_{C_{I_{ij}}}(\boldsymbol{r}_i)]$. The offset term is introduced to raise the potential energy to enable the end effector to move away from the internal singular regions after leaving them. By adjusting the value of $k_{p_{ij}}$ in the offset term, the energy level can be varied to allow the end effector to pass through the internal singular regions or avoid them.

The aim of $P_C(\boldsymbol{r})$ is to move the end effector toward the image-space region while avoiding singularities during movement. An illustration of $P_C(\boldsymbol{r})$ is given in Fig. 5.18.

5.6 Multiple Regional Feedback Controller

From Eqs. (1.24) and (1.38), the velocities of the end effector in Cartesian space and image space are related to the joint velocity by

$$\begin{aligned} \dot{\boldsymbol{x}} &= \boldsymbol{J}_x(\boldsymbol{q})\dot{\boldsymbol{q}}, \\ \dot{\boldsymbol{r}} &= \boldsymbol{J}_r(\boldsymbol{q})\dot{\boldsymbol{q}}, \end{aligned} \tag{5.116}$$

where $\boldsymbol{J}_r(\boldsymbol{q})$ is the Jacobian matrix from joint space to Cartesian space, and $\boldsymbol{J}_x(\boldsymbol{q})$ is the Jacobian matrix from joint space to image space.

The multiple regional feedback controller is constructed using the region errors obtained from the gradient of potential energy. Differentiating $P_J(\boldsymbol{q})$ described by Eq. (5.93) with respect to $\boldsymbol{q}$ yields

$$\begin{aligned} \left(\frac{\partial P_J(\boldsymbol{q})}{\partial \boldsymbol{q}}\right)^T &= k_p \alpha_q \sum_{i=1}^{M} \left\{ k_{qE_i}[\max(0, h_{JE_i}(\boldsymbol{q}))]^{N-1} \left(\frac{\partial h_{JE_i}(\boldsymbol{q})}{\partial \boldsymbol{q}}\right)^T \right. \\ &\left. + \sum_{j=1}^{N_i} k_{qI_{ij}}[\min(0, h_{J_{I_{ij}}}(\boldsymbol{q}))]^{N-1} \left(\frac{\partial h_{J_{I_{ij}}}(\boldsymbol{q})}{\partial \boldsymbol{q}}\right)^T \right\} \triangleq k_p \alpha_q \Delta \varepsilon_q, \end{aligned} \tag{5.117}$$

where $\Delta\varepsilon_q$ denotes the joint-space region error, which drives the end effector away from the singular regions. From Eq. (5.117), it is seen that $\Delta\varepsilon_q$ naturally reduces to zero after the end effector leaves the singular positions where $h_{JE_i}(\boldsymbol{q}) \leq 0$ and $h_{J_{I_{ij}}}(\boldsymbol{q}) \geq 0$.

Differentiating $P_D(\boldsymbol{x})$ described by Eq. (5.95) with respect to $\boldsymbol{x}$ yields

$$\left(\frac{\partial P_D(\boldsymbol{x})}{\partial \boldsymbol{x}}\right)^T = k_p\alpha_x \sum_{i=1}^{M}\left\{\left(\frac{\partial P_i(\boldsymbol{x}_i)}{\partial \boldsymbol{x}}\right)^T \prod_{j=1}^{N_i} P_{I_{ij}}(\boldsymbol{x}_i) + P_i(\boldsymbol{x}_i)\sum_{j=1}^{N_i}\left[\left(\frac{\partial P_{I_{ij}}(\boldsymbol{x}_i)}{\partial \boldsymbol{x}}\right)^T \prod_{k\neq j}^{N_i} P_{I_{ik}}(\boldsymbol{x}_i)\right] - \sum_{j=1}^{N_i} k_{c_{ij}}\left(\frac{\partial P_{I_{ij}}(\boldsymbol{x}_i)}{\partial \boldsymbol{x}}\right)^T\right\} \triangleq k_p\alpha_x\Delta\varepsilon_x, \quad (5.118)$$

where $\Delta\varepsilon_x$ denotes the image-space region error moving the end effector to the desired trajectory, and it is activated only after the end effector enters the image-space region.

Differentiating $P_C(\boldsymbol{r})$ described by Eq. (5.115) with respect to $\boldsymbol{r}$ yields

$$\begin{aligned}\left(\frac{\partial P_C(\boldsymbol{r})}{\partial \boldsymbol{r}}\right)^T = k_p\alpha_r \sum_{i=1}^{M}\Bigg\{&\left[\left(\frac{\partial P_{C_{E_i}}(\boldsymbol{r}_i)}{\partial \boldsymbol{r}}\right)^T + \left(\frac{\partial P_{C_{EV_i}}(\boldsymbol{r}_i)}{\partial \boldsymbol{r}}\right)^T\right] P_{C_{V_i}}(\boldsymbol{r}_i)\prod_{j=1}^{N_i} P_{C_{I_{ij}}}(\boldsymbol{r}_i)\\ &+ [P_{C_{E_i}}(\boldsymbol{r}_i) + P_{C_{EV_i}}(\boldsymbol{r}_i)]\left(\frac{\partial P_{C_{V_i}}(\boldsymbol{r}_i)}{\partial \boldsymbol{r}}\right)^T \prod_{j=1}^{N_i} P_{C_{I_{ij}}}(\boldsymbol{r}_i)\\ &+ [P_{C_{E_i}}(\boldsymbol{r}_i) + P_{C_{EV_i}}(\boldsymbol{r}_i)] P_{C_{V_i}}(\boldsymbol{r}_i)\\ &\times \sum_{j=1}^{N_i}\left[\left(\frac{\partial P_{C_{I_{ij}}}(\boldsymbol{r}_i)}{\partial \boldsymbol{r}}\right)^T \prod_{k\neq j}^{N_i} P_{C_{I_{ik}}}(\boldsymbol{r}_i)\right]\\ &- \sum_{j=1}^{N_i} k_{p_{ij}}\left(\frac{\partial P_{C_{I_{ij}}}(\boldsymbol{r}_i)}{\partial \boldsymbol{r}}\right)^T\Bigg\} \triangleq k_p\alpha_r\Delta\varepsilon_r, \end{aligned} \quad (5.119)$$

where $(\frac{\partial P_{C_{E_i}}(\boldsymbol{r}_i)}{\partial \boldsymbol{r}})^T$, $(\frac{\partial P_{C_{I_{ij}}}(\boldsymbol{r}_i)}{\partial \boldsymbol{r}})^T$, $(\frac{\partial P_{C_{V_i}}(\boldsymbol{r}_i)}{\partial \boldsymbol{r}})^T$ and $(\frac{\partial P_{C_{EV_i}}(\boldsymbol{r}_i)}{\partial \boldsymbol{r}})^T$ are given as

$$\begin{aligned}\left(\frac{\partial P_{C_{E_i}}(\boldsymbol{r}_i)}{\partial \boldsymbol{r}}\right)^T = k_{CE_i}\{&\min\{0, [\min(0, h_{C_{E_i}}(\boldsymbol{r}_i))]^N\\ &- (\kappa_{C_{Ei}}^{n_{C_{Ei}}} - 1)^N\}\}^{N-1}\\ &\times [\min(0, h_{C_{E_i}}(\boldsymbol{r}_i))]^{N-1}\left(\frac{\partial h_{C_{E_i}}(\boldsymbol{r}_i)}{\partial \boldsymbol{r}}\right)^T, \end{aligned} \quad (5.120)$$

$$\left(\frac{\partial P_{CI_{ij}}(\boldsymbol{r}_i)}{\partial \boldsymbol{r}}\right)^T = k_{CI_{ij}} \min\left\{0, [\min(0, h_{CI_{ij}}(\boldsymbol{r}_i))]^N - \left(\frac{1}{\kappa_{CIij}^{n_{CIij}}} - 1\right)^N\right\}^{N-1} \times [\min(0, h_{CI_{ij}}(\boldsymbol{r}_i))]^{N-1}\left(\frac{\partial h_{CI_{ij}}(\boldsymbol{r}_i)}{\partial \boldsymbol{r}}\right)^T, \quad (5.121)$$

$$\left(\frac{\partial P_{CV_i}(\boldsymbol{r}_i)}{\partial \boldsymbol{r}}\right)^T = k_{CV_i} \min\left\{0, [\min(0, h_{CV_i}(\boldsymbol{r}_i))]^N - \left(\frac{1}{\kappa_{CVi}^{n_{CVi}}} - 1\right)^N\right\}^{N-1} \times [\min(0, h_{CV_i}(\boldsymbol{r}_i))]^{N-1}\left(\frac{\partial h_{CV_i}(\boldsymbol{r}_i)}{\partial \boldsymbol{r}}\right)^T, \quad (5.122)$$

$$\left(\frac{\partial P_{CEV_i}(\boldsymbol{r}_i)}{\partial \boldsymbol{r}}\right)^T = -k_{CEV_i}[\min(0, h_{CEV_i}(\boldsymbol{r}_i))]^{N-1}\left(\frac{\partial h_{CEV_i}(\boldsymbol{r}_i)}{\partial \boldsymbol{r}}\right)^T. \quad (5.123)$$

The Cartesian-space region error $\Delta\varepsilon_r$ defined in Eq. (5.119) is used to drive the end effector from joint-space regions to image-space regions. From Eq. (5.119), $\Delta\varepsilon_r$ reduces to zero when the end effector is inside the image-space region. Therefore, each region error acts in a corresponding local region, and the combination of region errors ensures the convergence of robot movement.

Similarly, since vision feedback is used only when the end effector is inside the image-space regions, the desired velocity is designated as

$$\dot{\boldsymbol{x}}_{di} = \boldsymbol{0}, \text{ if } f_{Er_i}(\boldsymbol{x}_i) \geq 0 \text{ or } f_{Ir_{ij}}(\boldsymbol{x}_i) \leq 0. \quad (5.124)$$

Equation (5.124) can also be satisfied by introducing weight factors.

In this section, we consider multiple regional feedback control with uncertain kinematics, where the Jacobian matrix $\boldsymbol{J}_x(\boldsymbol{q})$ is uncertain.

A new sliding vector is defined using the approximate Jacobian matrix as

$$\begin{aligned} s_m = \dot{\boldsymbol{q}} - \dot{\boldsymbol{q}}_{mr} = \dot{\boldsymbol{q}} - [\boldsymbol{I}_n - \hat{\boldsymbol{J}}_x^+(\boldsymbol{q}, \hat{\boldsymbol{\theta}}_k)\boldsymbol{A}\hat{\boldsymbol{J}}_x(\boldsymbol{q}, \hat{\boldsymbol{\theta}}_k)]^{-1} \\ \times [\hat{\boldsymbol{J}}_x^+(\boldsymbol{q}, \hat{\boldsymbol{\theta}}_k)\dot{\boldsymbol{x}}_r - \alpha_x \hat{\boldsymbol{J}}_x^T(\boldsymbol{q}, \hat{\boldsymbol{\theta}}_k)\Delta\varepsilon_x - \alpha_r \boldsymbol{J}_r^T(\boldsymbol{q})\Delta\varepsilon_r - \alpha_q \Delta\varepsilon_q], \quad (5.125) \end{aligned}$$

where $\boldsymbol{A}$ is the matrix defined in Eq. (5.68), $\hat{\boldsymbol{J}}_x(\boldsymbol{q}, \hat{\boldsymbol{\theta}}_k)$ represents the estimated matrix of $\boldsymbol{J}_x(\boldsymbol{q})$, where $\hat{\boldsymbol{\theta}}_k$ are approximate kinematic parameters, $\dot{\boldsymbol{x}}_r$ is the vector defined in Eq. (5.67), and

$$\dot{\boldsymbol{q}}_{mr} = [\boldsymbol{I}_n - \hat{\boldsymbol{J}}_x^{+}(\boldsymbol{q}, \hat{\boldsymbol{\theta}}_k)\boldsymbol{A}\hat{\boldsymbol{J}}_x(\boldsymbol{q}, \hat{\boldsymbol{\theta}}_k)]^{-1}[\hat{\boldsymbol{J}}_x^{+}(\boldsymbol{q}, \hat{\boldsymbol{\theta}}_k)\dot{\boldsymbol{x}}_r - \alpha_x \hat{\boldsymbol{J}}_x^{T}(\boldsymbol{q}, \hat{\boldsymbol{\theta}}_k)\Delta\varepsilon_x - \alpha_r \boldsymbol{J}_r^{T}(\boldsymbol{q})\Delta\varepsilon_r - \alpha_q \Delta\varepsilon_q], \tag{5.126}$$

is the reference vector. Note that the information of joint velocity $\dot{\boldsymbol{q}}$ is not required in the reference vector $\dot{\boldsymbol{q}}_{mr}$.

Multiple regional feedback control is achieved using the region errors in Eqs. (5.117)–(5.119) and the sliding vector in Eq. (5.125) as

$$\boldsymbol{\tau}_m = -[\boldsymbol{I}_n - \hat{\boldsymbol{J}}_x^{+}(\boldsymbol{q}, \hat{\boldsymbol{\theta}}_k)\boldsymbol{A}\hat{\boldsymbol{J}}_x(\boldsymbol{q}, \hat{\boldsymbol{\theta}}_k)]^{T}[k_p\alpha_x \hat{\boldsymbol{J}}_x^{T}(\boldsymbol{q}, \hat{\boldsymbol{\theta}}_k)\Delta\varepsilon_x + k_p\alpha_r \boldsymbol{J}_r^{T}(\boldsymbol{q})\Delta\varepsilon_r + k_p\alpha_q \Delta\varepsilon_q] - \boldsymbol{K}_s \boldsymbol{s}_m + \boldsymbol{Y}_d(\boldsymbol{q}, \dot{\boldsymbol{q}}, \dot{\boldsymbol{q}}_{mr}, \ddot{\boldsymbol{q}}_{mr})\hat{\boldsymbol{\theta}}_d. \tag{5.127}$$

When the end effector reaches the singular positions, $\Delta\varepsilon_x$ and $\Delta\varepsilon_r$ reduce to zero, and the joint-space region error $\Delta\varepsilon_q$ pushes the end effector out of the singular positions. After the end effector leaves the singular regions, the Cartesian-space region error $\Delta\varepsilon_r$ is activated to drive it toward the field of view. The image-space region error $\Delta\varepsilon_x$ is activated only within the field of view to enable the end effector to converge to the desired trajectory. In addition, note that the multiple regional feedback controller is continuous, since the region errors $\Delta\varepsilon_x$, $\Delta\varepsilon_r$, and $\Delta\varepsilon_q$ are continuous.

The estimated parameters $\hat{\boldsymbol{\theta}}_d$ are updated using the update law

$$\dot{\hat{\boldsymbol{\theta}}}_d = -\boldsymbol{L}_d \boldsymbol{Y}_d^{T}(\boldsymbol{q}, \dot{\boldsymbol{q}}, \dot{\boldsymbol{q}}_{mr}, \ddot{\boldsymbol{q}}_{mr})\boldsymbol{s}_m, \tag{5.128}$$

and the uncertain kinematic parameters $\hat{\boldsymbol{\theta}}_k$ are updated using the update law

$$\dot{\hat{\boldsymbol{\theta}}}_k = k_p\alpha_x \boldsymbol{L}_k \boldsymbol{Y}_k^{T}(\boldsymbol{q}, \dot{\boldsymbol{q}})\Delta\varepsilon_x. \tag{5.129}$$

From Eq. (1.5), the robot dynamic equation is stated as

$$\boldsymbol{M}(\boldsymbol{q})\ddot{\boldsymbol{q}} + \left(\tfrac{1}{2}\dot{\boldsymbol{M}}(\boldsymbol{q}) + \boldsymbol{S}(\boldsymbol{q}, \dot{\boldsymbol{q}})\right)\dot{\boldsymbol{q}} + \boldsymbol{g}(\boldsymbol{q}) = \boldsymbol{\tau}_m. \tag{5.130}$$

Using the sliding vector $\boldsymbol{s}_m$ in Eq. (5.125) and Property 1.3, Eq. (5.130) can be written as

$$\boldsymbol{M}(\boldsymbol{q})\dot{\boldsymbol{s}}_m + \left(\tfrac{1}{2}\dot{\boldsymbol{M}}(\boldsymbol{q}) + \boldsymbol{S}(\boldsymbol{q}, \dot{\boldsymbol{q}})\right)\boldsymbol{s}_m + \boldsymbol{Y}_d(\boldsymbol{q}, \dot{\boldsymbol{q}}, \dot{\boldsymbol{q}}_{mr}, \ddot{\boldsymbol{q}}_{mr})\boldsymbol{\theta}_d = \boldsymbol{\tau}_m. \tag{5.131}$$

The closed-loop equation of the system is obtained by substituting Eq. (5.127) into Eq. (5.131) to give

$$\boldsymbol{M}(\boldsymbol{q})\dot{\boldsymbol{s}}_m + \left(\tfrac{1}{2}\dot{\boldsymbol{M}}(\boldsymbol{q}) + \boldsymbol{S}(\boldsymbol{q}, \dot{\boldsymbol{q}})\right)\boldsymbol{s}_m + [\boldsymbol{I}_n - \hat{\boldsymbol{J}}_x^{+}(\boldsymbol{q}, \hat{\boldsymbol{\theta}}_k)\boldsymbol{A}\hat{\boldsymbol{J}}_x(\boldsymbol{q}, \hat{\boldsymbol{\theta}}_k)]^{T} \times [k_p\alpha_x \hat{\boldsymbol{J}}_x^{T}(\boldsymbol{q}, \hat{\boldsymbol{\theta}}_k)\Delta\varepsilon_x + k_p\alpha_r \boldsymbol{J}_r^{T}(\boldsymbol{q})\Delta\varepsilon_r + k_p\alpha_q \Delta\varepsilon_q] + \boldsymbol{K}_s \boldsymbol{s}_m + \boldsymbol{Y}_d(\boldsymbol{q}, \dot{\boldsymbol{q}}, \dot{\boldsymbol{q}}_{mr}, \ddot{\boldsymbol{q}}_{mr})\Delta\boldsymbol{\theta}_d = \boldsymbol{0}. \tag{5.132}$$

We can now state the following theorem.

Theorem 5.2 *The multiple regional feedback control law* (5.127) *and the update law* (5.128) *for the robot system* (5.130) *guarantee the global convergence of the tracking errors. That is,* $\boldsymbol{x} \rightarrow \boldsymbol{x}_d$, $\dot{\boldsymbol{x}} \rightarrow \dot{\boldsymbol{x}}_d$ *as* $t \rightarrow \infty$.

Proof Consider the following Lyapunov-like candidate:

$$V_m = \frac{1}{2}\boldsymbol{s}_m^T \boldsymbol{M}(\boldsymbol{q})\boldsymbol{s}_m + P_D(\boldsymbol{x}) + P_C(\boldsymbol{r}) + P_J(\boldsymbol{q}) + \frac{1}{2}\Delta\boldsymbol{\theta}_d^T \boldsymbol{L}_d^{-1}\Delta\boldsymbol{\theta}_d + \frac{1}{2}\Delta\boldsymbol{\theta}_k^T \boldsymbol{L}_k^{-1}\Delta\boldsymbol{\theta}_k. \tag{5.133}$$

Next, note that the derivative of $P_C(\boldsymbol{r})$ in Eq. (5.115) is given as

$$\begin{aligned}\dot{P}_C(\boldsymbol{r}) = k_p\alpha_r \Bigg\{ & \left[\dot{\boldsymbol{r}}^T \left(\frac{\partial P_{C_{E_i}}(\boldsymbol{r}_i)}{\partial \boldsymbol{r}}\right)^T + \dot{\boldsymbol{r}}^T \left(\frac{\partial P_{C_{EV_i}}(\boldsymbol{r}_i)}{\partial \boldsymbol{r}}\right)^T\right] P_{C_{V_i}}(\boldsymbol{r}_i)\prod_{j=1}^{N_i} P_{C_{I_{ij}}}(\boldsymbol{r}_i) \\ & + [P_{C_{E_i}}(\boldsymbol{r}_i) + P_{C_{EV_i}}(\boldsymbol{r}_i)]\dot{\boldsymbol{r}}^T \left(\frac{\partial P_{C_{V_i}}(\boldsymbol{r}_i)}{\partial \boldsymbol{r}}\right)^T \prod_{j=1}^{N_i} P_{C_{I_{ij}}}(\boldsymbol{r}_i) \\ & + [P_{C_{E_i}}(\boldsymbol{r}_i) + P_{C_{EV_i}}(\boldsymbol{r}_i)]P_{C_{V_i}}(\boldsymbol{r}_i)\sum_{j=1}^{N_i}\left[\dot{\boldsymbol{r}}^T \left(\frac{\partial P_{C_{I_{ij}}}(\boldsymbol{r}_i)}{\partial \boldsymbol{r}}\right)^T \prod_{k\neq j}^{N_i} P_{C_{I_{ik}}}(\boldsymbol{r}_i)\right] \\ & - \sum_{j=1}^{N_i}\dot{\boldsymbol{r}}^T \left(\frac{\partial P_{C_{I_{ij}}}(\boldsymbol{r}_i)}{\partial \boldsymbol{r}}\right)^T \Bigg\} = k_p\alpha_r\dot{\boldsymbol{r}}^T\Delta\boldsymbol{\varepsilon}_r. \end{aligned} \tag{5.134}$$

Differentiating Eq. (5.133) with respect to time and substituting Eq. (5.134) into its time derivative yields

$$\begin{aligned}\dot{V}_m = & \boldsymbol{s}_m^T \boldsymbol{M}(\boldsymbol{q})\dot{\boldsymbol{s}}_m + \frac{1}{2}\boldsymbol{s}_m^T \dot{\boldsymbol{M}}(\boldsymbol{q})\boldsymbol{s}_m + k_p\alpha_r\dot{\boldsymbol{r}}^T\Delta\boldsymbol{\varepsilon}_r + k_p\alpha_q\dot{\boldsymbol{q}}^T\Delta\boldsymbol{\varepsilon}_q \\ & + k_p\alpha_x\left\{\dot{\boldsymbol{x}}^T\Delta\boldsymbol{\varepsilon}_x - \sum_{i=1}^{M}\left[\dot{\boldsymbol{x}}_a^T\left(\frac{\partial P_{T_i}(\boldsymbol{x}_i)}{\partial \boldsymbol{x}}\right)^T\prod_{j=1}^{N_i} P_{I_{ij}}(\boldsymbol{x}_i)\right]\right\} \\ & - \dot{\hat{\boldsymbol{\theta}}}_d^T \boldsymbol{L}_d^{-1}\Delta\boldsymbol{\theta}_d - \dot{\hat{\boldsymbol{\theta}}}_k^T \boldsymbol{L}_k^{-1}\Delta\boldsymbol{\theta}_k. \end{aligned} \tag{5.135}$$

Substituting Eqs. (5.128) and (5.132) into Eq. (5.135) and using Property 1.2, we obtain

$$\begin{aligned}\dot{V}_m &= -\boldsymbol{s}_m^T \boldsymbol{K}_s \boldsymbol{s}_m - k_p \boldsymbol{s}_m^T [\boldsymbol{I}_n - \hat{\boldsymbol{J}}_x^+(\boldsymbol{q}, \hat{\boldsymbol{\theta}}_k) \boldsymbol{A} \hat{\boldsymbol{J}}_x(\boldsymbol{q}, \hat{\boldsymbol{\theta}}_k)]^T \\ &\quad \times [\alpha_x \hat{\boldsymbol{J}}_x^T(\boldsymbol{q}, \hat{\boldsymbol{\theta}}_k) \Delta\varepsilon_x + \alpha_r \boldsymbol{J}_r^T(\boldsymbol{q}) \Delta\varepsilon_r + \alpha_q \Delta\varepsilon_q] + k_p \alpha_r \dot{\boldsymbol{r}}^T \Delta\varepsilon_r \\ &\quad + k_p \alpha_q \dot{\boldsymbol{q}}^T \Delta\varepsilon_q + k_p \alpha_x \left\{ \dot{\boldsymbol{x}}^T \Delta\varepsilon_x - \sum_{i=1}^{M} \left[\dot{\boldsymbol{x}}_a^T \left(\frac{\partial P_{T_i}(\boldsymbol{x}_i)}{\partial \boldsymbol{x}} \right)^T \right.\right. \\ &\quad \left.\left. \times \prod_{j=1}^{N_i} P_{I_{ij}}(\boldsymbol{x}_i) \right] \right\} - \dot{\hat{\boldsymbol{\theta}}}_k^T \boldsymbol{L}_k^{-1} \Delta\boldsymbol{\theta}_k. \end{aligned} \tag{5.136}$$

Substituting Eqs. (5.125) and (5.129) into Eq. (5.136) yields

$$\begin{aligned}\dot{V}_m &= -\boldsymbol{s}_m^T \boldsymbol{K}_s \boldsymbol{s}_m - k_p [\alpha_x \hat{\boldsymbol{J}}_x^T(\boldsymbol{q}, \hat{\boldsymbol{\theta}}_k) \Delta\varepsilon_x + \alpha_r \boldsymbol{J}_r^T(\boldsymbol{q}) \Delta\varepsilon_r + \alpha_q \Delta\varepsilon_q]^T \\ &\quad \times [\alpha_x \hat{\boldsymbol{J}}_x^T(\boldsymbol{q}, \hat{\boldsymbol{\theta}}_k) \Delta\varepsilon_x + \alpha_r \boldsymbol{J}_r^T(\boldsymbol{q}) \Delta\varepsilon_r + \alpha_q \Delta\varepsilon_q] \\ &\quad + k_p \alpha_r \dot{\boldsymbol{x}}_a^T \hat{\boldsymbol{J}}_x^{+T}(\boldsymbol{q}, \hat{\boldsymbol{\theta}}_k) \boldsymbol{J}_r^T(\boldsymbol{q}) \Delta\varepsilon_r + k_p \alpha_q \dot{\boldsymbol{x}}_a^T \hat{\boldsymbol{J}}_x^{+T}(\boldsymbol{q}, \hat{\boldsymbol{\theta}}_k) \Delta\varepsilon_q \\ &\quad + k_p \alpha_x \sum_{i=1}^{M} \left\{ \dot{\boldsymbol{x}}_a^T \left(\frac{\partial P_{E_i}(\boldsymbol{x}_i)}{\partial \boldsymbol{x}} \right)^T \prod_{j=1}^{N_i} P_{I_{ij}}(\boldsymbol{x}_i) - \sum_{j=1}^{N_i} \left[k_{c_{ij}} \dot{\boldsymbol{x}}_a^T \left(\frac{\partial P_{I_{ij}}(\boldsymbol{x}_i)}{\partial \boldsymbol{x}} \right)^T \right] \right. \\ &\quad \left. + [P_{T_i}(\boldsymbol{x}_i) + P_{E_i}(\boldsymbol{x}_i)] \sum_{j=1}^{N_i} \left[\dot{\boldsymbol{x}}_a^T \left(\frac{\partial P_{I_{ij}}(\boldsymbol{x}_i)}{\partial \boldsymbol{x}} \right)^T \prod_{k \neq j}^{N_i} P_{I_{ik}}(\boldsymbol{x}_i) \right] \right\}. \end{aligned} \tag{5.137}$$

From Eqs. (5.124) and (5.63), note that

$$\dot{\boldsymbol{x}}_{ai} = \boldsymbol{0}, \text{ if } f_{Er_i}(\boldsymbol{x}_i) \geq 0 \text{ or } f_{Ir_{ij}}(\boldsymbol{x}_i) \leq 0. \tag{5.138}$$

The condition that $f_{Er_i}(\boldsymbol{x}_i) \geq 0$ or $f_{Ir_{ij}}(\boldsymbol{x}_i) \leq 0$ indicates that $h_{C_{E_i}}(\boldsymbol{r}_i) \leq 0$ and $h_{C_{I_{ij}}}(\boldsymbol{r}_i) \geq 0$, and $h_{J_{E_i}}(\boldsymbol{q}) \geq 0$ or $h_{J_{I_{ij}}}(\boldsymbol{q}) \leq 0$, where both the Cartesian-space region error $\Delta\varepsilon_r$ and the joint-space region error $\Delta\varepsilon_q$ are nonzero. Thus $\dot{\boldsymbol{x}}_{ai}$, $\Delta\varepsilon_r$, and $\Delta\varepsilon_q$ cannot be simultaneously nonzero.

Equation (5.138) also indicates that $\dot{\boldsymbol{x}}_a$ is nonzero where $f_{Er_i}(\boldsymbol{x}_i) < 0$ and $f_{Ir_{ij}}(\boldsymbol{x}_i) > 0$. The gradient of potential energy $P_{E_i}(\boldsymbol{x}_i)$ reduces to zero where $f_{Er_i}(\boldsymbol{x}_i) < 0$, while the gradient of potential energy $P_{I_{ij}}(\boldsymbol{x}_i)$ reduces to zero where $f_{Ir_{ij}}(\boldsymbol{x}_i) > 0$. Thus, $\boldsymbol{x}_{ai}$ and $(\frac{\partial P_{E_i}(\boldsymbol{x}_i)}{\partial \boldsymbol{x}})^T$, $\boldsymbol{x}_{ai}$, and $(\frac{\partial P_{I_{ij}}(\boldsymbol{x}_i)}{\partial \boldsymbol{x}})^T$ cannot be simultaneously nonzero.

Therefore, the last three terms in Eq. (5.137) are zero, and hence

$$\begin{aligned}\dot{V}_m &= -\boldsymbol{s}_m^T \boldsymbol{K}_s \boldsymbol{s}_m - k_p [\alpha_x \hat{\boldsymbol{J}}_x^T(\boldsymbol{q}, \hat{\boldsymbol{\theta}}_k) \Delta\varepsilon_x + \alpha_r \boldsymbol{J}_r^T(\boldsymbol{q}) \Delta\varepsilon_r + \alpha_q \Delta\varepsilon_q]^T \\ &\quad \times [\alpha_x \hat{\boldsymbol{J}}_x^T(\boldsymbol{q}, \hat{\boldsymbol{\theta}}_k) \Delta\varepsilon_x + \alpha_r \boldsymbol{J}_r^T(\boldsymbol{q}) \Delta\varepsilon_r + \alpha_q \Delta\varepsilon_q]. \end{aligned} \tag{5.139}$$

Since $V_m > 0$ and $\dot{V}_m \leq 0$, V_m is bounded. Hence, $\boldsymbol{s}_m$, $\Delta\boldsymbol{\theta}_d$, $\Delta\boldsymbol{\theta}_k$, $P_J(\boldsymbol{q})$, $P_D(\boldsymbol{x})$, and $P_C(\boldsymbol{r})$ are bounded. Hence, the region functions are bounded. The boundedness of region functions ensures the boundedness of $\boldsymbol{q}, \boldsymbol{x}$, and $\boldsymbol{r}$. Therefore, $\Delta\varepsilon_q$, $\Delta\varepsilon_x$, and

$\Delta\varepsilon_r$ are bounded. Since $\boldsymbol{x}$ is bounded, $\dot{\boldsymbol{x}}_a$ is bounded if $\dot{\boldsymbol{x}}_d$ is bounded. Hence, $\dot{\boldsymbol{q}}_{mr}$ is bounded. From Eq. (5.125), $\dot{\boldsymbol{q}}$ is bounded because $\boldsymbol{s}_m$ is bounded. The boundedness of $\dot{\boldsymbol{q}}$ guarantees the boundedness of $\dot{\boldsymbol{x}}$ and $\dot{\boldsymbol{r}}$, since $\boldsymbol{J}_x(\boldsymbol{q})$ and $\boldsymbol{J}_r(\boldsymbol{q})$ are bounded. Therefore, $\Delta\dot{\varepsilon}_q$, $\Delta\dot{\varepsilon}_r$, and $\Delta\dot{\varepsilon}_x$ are bounded. Then $\ddot{\boldsymbol{q}}_{mr}$ is bounded if $\ddot{\boldsymbol{x}}_d$ is bounded. From the closed-loop Eq. (5.132), we may conclude that $\dot{\boldsymbol{s}}_m$ is bounded. Thus, $\ddot{V}_m$ is bounded, since $\dot{\boldsymbol{s}}_m$, $\boldsymbol{s}_m$, $\Delta\dot{\varepsilon}_q$, $\Delta\varepsilon_q$, $\Delta\dot{\varepsilon}_x$, $\Delta\varepsilon_x$, $\Delta\dot{\varepsilon}_r$, $\Delta\varepsilon_r$ are bounded. Therefore, $\dot{V}_m$ is uniformly continuous. Applying Barbalat's lemma, we obtain that $\dot{V}_m \to 0$, which also indicates that

$$
\begin{aligned}
&\alpha_x \hat{\boldsymbol{J}}_x^T(\boldsymbol{q}, \hat{\boldsymbol{\theta}}_k)\Delta\varepsilon_x + \alpha_r \boldsymbol{J}_r^T(\boldsymbol{q})\Delta\varepsilon_r + \alpha_q \Delta\varepsilon_q \to \mathbf{0}\\
&\boldsymbol{s}_m \to \mathbf{0}.
\end{aligned} \tag{5.140}
$$

If the end effector is located at the singular positions, $\Delta\varepsilon_q \neq \mathbf{0}$, $\Delta\varepsilon_r = \mathbf{0}$, $\Delta\varepsilon_x = \mathbf{0}$, which contradicts Eq. (5.140). If the end effector is located at the Cartesian-space region, $\Delta\varepsilon_q = \mathbf{0}$, $\Delta\varepsilon_r \neq \mathbf{0}$, $\Delta\varepsilon_x = \mathbf{0}$, which also contradicts Eq. (5.140). Therefore, the end effector can settle down only within the image-space region. Then $\Delta\varepsilon_q = \mathbf{0}$ and $\Delta\varepsilon_r = \mathbf{0}$, and from Eq. (5.140), $\Delta\varepsilon_x = \mathbf{0}$, which means that $\boldsymbol{x} \to \boldsymbol{x}_d$ as $t \to \infty$. From the definition of $\dot{\boldsymbol{x}}_a$ in Eq. (5.63), $\boldsymbol{x} \to \boldsymbol{x}_d$ indicates that $\dot{\boldsymbol{x}}_a \to \dot{\boldsymbol{x}}_d$. From the definition of $\boldsymbol{s}_m$ in Eq. (5.125), $\boldsymbol{s}_m \to \mathbf{0}$, $\dot{\boldsymbol{x}}_a \to \dot{\boldsymbol{x}}_d$, $\Delta\varepsilon_x \to \mathbf{0}$, $\Delta\varepsilon_q \to \mathbf{0}$, $\Delta\varepsilon_r \to \mathbf{0}$ implies $\dot{\boldsymbol{x}} \to \dot{\boldsymbol{x}}_d$ as $t \to \infty$. $\triangle\triangle\triangle$

5.7 Summary and Notes

Singularities have been a long-standing problem in task-space control of robots. To overcome the problem of singularities in joint-space control, Nakamura and Hanafusa (1986) introduced a damped least-squares inverse Jacobian matrix that provided an approximate motion when the robot was in the neighborhood of the singular positions. Sciavicco and Siciliano (1988) proposed to use the Jacobian transpose instead of the inverse Jacobian matrix in kinematic control, and hence typical numerical instabilities were avoided. In addition, a large number of research works have been carried out on singularity avoidance using redundant robots (Nakamura et al. 1987; Nakamura 1991; Chiaverini 1997).

To deal with the problem of limited field of view, a number of approaches have been proposed in the literature of visual servoing to keep the image features within the field of view (Cowan et al. 2002; Mezouar and Chaumette 2002; Malis 2004; Gans and Hutchinson 2007; Chesi and Hashimoto 2010; Gans et al. 2011). Most visual servoing techniques are formulated as kinematic control methods without the consideration of the effects of robot dynamics. In addition, it is commonly assumed that the image features are observable at the beginning stage, but that is valid only locally within a limited field of view.

The aforementioned singularity problem and problem of limited field of view can be addressed using regional feedback. The concept of regional feedback was first brought up in Cheah and Li (2009). Inspired by human visually guided tasks, a setpoint controller that requires vision only when the end effector is near the desired position was proposed in Cheah and Li (2009). The proposed controller (Cheah and Li 2009) consists of a Cartesian-space region-reaching controller in the initial stages and a vision-based controller in the final stages, and it allows the end effector to start outside the field of view, transit smoothly from outside to inside the field of view, and move to the desired point at the end. Such results were extended to a regional feedback tracking controller (Cheah and Li 2010). An adaptive regional feedback tracking controller with uncertain kinematics, dynamics, and depth information was presented in Li and Cheah (2012c). In Cheah and Li (2011), a singularity-robust task-space controller was proposed for tracking control of a robot manipulator. Using the regional feedback information from joint space and Cartesian space, the proposed controller allows the end effector to start from a singular configuration or enter singular regions during the course of movement. The multiple task-space tracking controller was proposed in Li and Cheah (2012b), which smoothly integrates the regional feedback information from joint space, Cartesian space, and image space into a single controller. Moreover, it has been shown in Li and Cheah (2013) that the task-space control method using regional feedback provides a unified formulation to address various open issues in task-space control problems. Instead of designing multiple controllers in different regions and switching between them, the regional feedback control strategy guarantees the smooth transition from one feedback information to another without using any hard or discontinuous switching, and the combination of regional feedback ensures the global convergence of robot motion.

Chapter 6
Sensory Feedback Control of Complex Dynamical Robot Systems

Task-space sensory feedback control methods are also effective in dealing with various types of uncertainty in robot systems with more complicated dynamic behavior. In this chapter, sensory task-space control problems of robot systems with uncertain actuator dynamics are studied in Sects. 6.1 and 6.2. Two main types of actuator dynamics, electrically driven actuator systems and joint flexibility, are considered. In addition, force control problems with constraint uncertainty and multi-fingered robot control with uncertain contact points are treated in Sects. 6.3 and 6.4.

6.1 Rigid-Link Electrically Driven Robots

In this section, the problem of sensory task-space adaptive Jacobian control of rigid-link electrically driven (RLED) robots is considered. The dynamic equations of the rigid-link electrically driven robot manipulator with n degrees of freedom are described as

$$\boldsymbol{M}(\boldsymbol{q})\ddot{\boldsymbol{q}} + \left(\tfrac{1}{2}\dot{\boldsymbol{M}}(\boldsymbol{q}) + \boldsymbol{S}(\boldsymbol{q}, \dot{\boldsymbol{q}})\right)\dot{\boldsymbol{q}} + \boldsymbol{g}(\boldsymbol{q}) = \boldsymbol{K}_N \boldsymbol{I}, \tag{6.1}$$

$$\boldsymbol{L}\dot{\boldsymbol{I}} + \boldsymbol{R}\boldsymbol{I} + \boldsymbol{K}_E\dot{\boldsymbol{q}} = \boldsymbol{u}, \tag{6.2}$$

where $\boldsymbol{M}(\boldsymbol{q}) \in \Re^{n\times n}$ denotes a positive definite inertia matrix, $\boldsymbol{S}(\boldsymbol{q}, \dot{\boldsymbol{q}}) \in \Re^{n\times n}$ is a skew-symmetric matrix, $\boldsymbol{g}(\boldsymbol{q}) \in \Re^n$ denotes the gravitational force, $\boldsymbol{I} \in \Re^n$ is the vector of armature currents, $\boldsymbol{u} \in \Re^n$ denotes the vector of armature voltages, $\boldsymbol{L} \in \Re^{n\times n}$ represents the actuator inductances, $\boldsymbol{R} \in \Re^{n\times n}$ represents the actuator resistances, $\boldsymbol{K}_E \in \Re^{n\times n}$ is the matrix characterizing the voltage constant of the actuator, $\boldsymbol{K}_N \in \Re^{n\times n}$ is the motor torque transmission constant matrix, which characterizes the electromechanical conversion between current and torque, and $\boldsymbol{L}$, $\boldsymbol{R}$, $\boldsymbol{K}_E$, and $\boldsymbol{K}_N$ are positive definite constant diagonal matrices.

C.C. Cheah and X. Li, *Task-Space Sensory Feedback Control of Robot Manipulators*, Intelligent Systems, Control and Automation: Science and Engineering 73,
DOI 10.1007/978-981-287-062-9_6

The design of the sensory task-space adaptive Jacobian control scheme for the tracking control of RLED robots is based on the backstepping method, which provides a recursive method for analyzing the stability of a system. The main idea of backstepping is to start with a system that is to be stabilized with a fictitious control input. For the augmented system, a new control input is explicitly designed such that the fictitious control input is achieved.

Consider the RLED robot described by Eqs. (6.1) and (6.2). First, based on the manipulator subsystem dynamics (6.1), a desired armature current signal $\boldsymbol{I}_d$ is designed to ensure the convergence of the task-space tracking errors. Then, based on the actuator subsystem dynamics (6.2), a voltage control input $\boldsymbol{u}$ is designed to guarantee that the actual armature current $\boldsymbol{I}$ tracks the desired current signal $\boldsymbol{I}_d$. The overall system is a third-order system, and observers are constructed in the desired current signal design to avoid the use of accelerations in the voltage control input.

6.1.1 Desired Armature Current Design

Using a desired armature current $\boldsymbol{I}_d$, the manipulator subsystem dynamics (6.1) can be rewritten as

$$\boldsymbol{M}(\boldsymbol{q})\ddot{\boldsymbol{q}} + \left(\tfrac{1}{2}\dot{\boldsymbol{M}}(\boldsymbol{q})+\boldsymbol{S}(\boldsymbol{q},\dot{\boldsymbol{q}})\right)\dot{\boldsymbol{q}} + \boldsymbol{g}(\boldsymbol{q}) = \boldsymbol{K}_N\boldsymbol{I}_d + \boldsymbol{K}_N\tilde{\boldsymbol{I}}, \tag{6.3}$$

where $\tilde{\boldsymbol{I}} = \boldsymbol{I} - \boldsymbol{I}_d$ represents a current perturbation to the rigid-link robot dynamics. The subsystem (6.3) can be viewed as a system with a fictitious input $\boldsymbol{K}_N\boldsymbol{I}_d$ with a disturbance or perturbation $\boldsymbol{K}_N\tilde{\boldsymbol{I}}$.

Similarly, the Jacobian matrix $\boldsymbol{J}(\boldsymbol{q})$ is not exactly known, due to uncertainties in kinematics. We also consider the case that the exact values of the dynamic parameters in $\boldsymbol{M}(\boldsymbol{q})$, $\frac{1}{2}\dot{\boldsymbol{M}}(\boldsymbol{q})+\boldsymbol{S}(\boldsymbol{q},\dot{\boldsymbol{q}})$, $\boldsymbol{g}(\boldsymbol{q})$ and the actuator dynamic matrices $\boldsymbol{L}$, $\boldsymbol{R}$, $\boldsymbol{K}_E$, $\boldsymbol{K}_N$ are unknown. The objective is to design the fictitious input $\boldsymbol{I}_d$ so as to ensure the convergence of the task-space tracking errors in the presence of uncertainties in system kinematics, manipulator dynamics, and actuator dynamics.

In the presence of kinematic uncertainty, the approximate Jacobian matrix $\hat{\boldsymbol{J}}(\boldsymbol{q},\hat{\boldsymbol{\theta}}_k)$ is defined according to *Property* 1.5 as

$$\dot{\hat{\boldsymbol{x}}} = \hat{\boldsymbol{J}}(\boldsymbol{q},\hat{\boldsymbol{\theta}}_k)\dot{\boldsymbol{q}} = \boldsymbol{Y}_k(\boldsymbol{q},\dot{\boldsymbol{q}})\hat{\boldsymbol{\theta}}_k, \tag{6.4}$$

where $\dot{\hat{\boldsymbol{x}}} \in \Re^n$ denotes the estimated task-space velocity, and $\hat{\boldsymbol{\theta}}_k \in \Re^{n_k}$ is the estimated kinematic parameters vector, which is updated through a kinematic adaptation law.

Let $\boldsymbol{x}_d(t) \in \Re^n$ be the desired task-space trajectory, which is bounded and at least $\boldsymbol{x}_d(t) \in C^3$, and let $\hat{\boldsymbol{K}}_N \in \Re^{n\times n}$ be a fixed approximate motor torque transmission matrix. Then the desired armature current signal can be designated as

$$\boldsymbol{I}_d = \hat{\boldsymbol{K}}_N^{-1}[-\hat{\boldsymbol{J}}^T(\boldsymbol{q}, \hat{\boldsymbol{\theta}}_k)\boldsymbol{K}_p\Delta\boldsymbol{x} - \boldsymbol{K}_d(\dot{\hat{\boldsymbol{q}}}_o - \dot{\boldsymbol{q}}_r) + \boldsymbol{Y}_d(\boldsymbol{q}, \dot{\hat{\boldsymbol{q}}}_o, \dot{\boldsymbol{q}}_r, \ddot{\hat{\boldsymbol{q}}}_r)\hat{\boldsymbol{\theta}}_d + \boldsymbol{Y}_a(\boldsymbol{I}_o)\hat{\boldsymbol{\theta}}_a], \tag{6.5}$$

where $\hat{\boldsymbol{\theta}}_d \in \Re^{n_d}$ and $\hat{\boldsymbol{\theta}}_a \in \Re^{n_a}$ are vectors of estimated manipulator dynamic parameters and actuator dynamic parameters, which are updated through adaptation laws, $\Delta\boldsymbol{x} = \boldsymbol{x} - \boldsymbol{x}_d$, $\boldsymbol{K}_p \in \Re^{n\times n}$, $\boldsymbol{K}_d \in \Re^{n\times n}$ are positive definite diagonal constant matrices, $\dot{\hat{\boldsymbol{q}}}_o \in \Re^n$ is the observed signal of $\dot{\boldsymbol{q}}$, $\dot{\boldsymbol{q}}_r \in \Re^n$ is the reference velocity signal, and $\ddot{\hat{\boldsymbol{q}}}_r$ is a modified reference acceleration. The regressor $\boldsymbol{Y}_d(\boldsymbol{q}, \dot{\hat{\boldsymbol{q}}}_o, \dot{\boldsymbol{q}}_r, \ddot{\hat{\boldsymbol{q}}}_r) \in \Re^{n\times n_d}$ is defined such that

$$\boldsymbol{Y}_d(\boldsymbol{q}, \dot{\hat{\boldsymbol{q}}}_o, \dot{\boldsymbol{q}}_r, \ddot{\hat{\boldsymbol{q}}}_r)\hat{\boldsymbol{\theta}}_d = \hat{\boldsymbol{M}}(\boldsymbol{q})\ddot{\hat{\boldsymbol{q}}}_r + \left(\tfrac{1}{2}\dot{\hat{\boldsymbol{M}}}(\boldsymbol{q}) + \hat{\boldsymbol{S}}(\boldsymbol{q}, \dot{\hat{\boldsymbol{q}}}_o)\right)\dot{\boldsymbol{q}}_r + \hat{\boldsymbol{g}}(\boldsymbol{q}), \tag{6.6}$$

where $\hat{\boldsymbol{M}}(\boldsymbol{q})$, $\dot{\hat{\boldsymbol{M}}}(\boldsymbol{q})$, $\hat{\boldsymbol{S}}(\boldsymbol{q}, \dot{\hat{\boldsymbol{q}}}_o)$, and $\hat{\boldsymbol{g}}(\boldsymbol{q})$ are the approximate manipulator dynamic matrices and vector respectively using the estimated dynamic parameters in $\hat{\boldsymbol{\theta}}_d$. The regressor $\boldsymbol{Y}_a(\boldsymbol{I}_o) = \mathrm{diag}\{I_{o1}, \ I_{o2}, \ \ldots, \ I_{on}\}$ and I_{oi} denotes the ith element of the vector $\boldsymbol{I}_o \in \Re^n$, which is defined as

$$\boldsymbol{I}_o = -\hat{\boldsymbol{J}}^T(\boldsymbol{q}, \hat{\boldsymbol{\theta}}_k)\boldsymbol{K}_p\Delta\boldsymbol{x} - \boldsymbol{K}_d(\dot{\hat{\boldsymbol{q}}}_o - \dot{\boldsymbol{q}}_r) + \boldsymbol{Y}_d(\boldsymbol{q}, \dot{\hat{\boldsymbol{q}}}_o, \dot{\boldsymbol{q}}_r, \ddot{\hat{\boldsymbol{q}}}_r)\hat{\boldsymbol{\theta}}_d. \tag{6.7}$$

The reference velocity $\dot{\boldsymbol{q}}_r$ is defined as

$$\dot{\boldsymbol{q}}_r = \hat{\boldsymbol{J}}^{-1}(\boldsymbol{q}, \hat{\boldsymbol{\theta}}_k)[\dot{\boldsymbol{x}}_d - \alpha(\hat{\boldsymbol{x}}_o - \boldsymbol{x}_d)], \tag{6.8}$$

where $\alpha \in \Re$ is a positive constant, $\hat{\boldsymbol{J}}^{-1}(\boldsymbol{q}, \hat{\boldsymbol{\theta}}_k)$ is the inverse approximate Jacobian matrix, and $\hat{\boldsymbol{x}}_o \in \Re^n$ is the the observed signal of $\boldsymbol{x}$ updated through the following observer updating law:

$$\dot{\hat{\boldsymbol{x}}}_o = \hat{\boldsymbol{J}}(\boldsymbol{q}, \hat{\boldsymbol{\theta}}_k)\dot{\hat{\boldsymbol{q}}}_o + \boldsymbol{\Lambda}(\boldsymbol{x} - \hat{\boldsymbol{x}}_o), \tag{6.9}$$

where $\boldsymbol{\Lambda} \in \Re^{n\times n}$ is a positive definite diagonal constant matrix.

From Eq. (6.8), the time derivative of $\dot{\boldsymbol{q}}_r$ is obtained as

$$\ddot{\boldsymbol{q}}_r = \dot{\hat{\boldsymbol{J}}}^{-1}(\boldsymbol{q}, \hat{\boldsymbol{\theta}}_k)[\dot{\boldsymbol{x}}_d - \alpha(\hat{\boldsymbol{x}}_o - \boldsymbol{x}_d)] + \hat{\boldsymbol{J}}^{-1}(\boldsymbol{q}, \hat{\boldsymbol{\theta}}_k)[\ddot{\boldsymbol{x}}_d - \alpha(\dot{\hat{\boldsymbol{x}}}_o - \dot{\boldsymbol{x}}_d)]. \tag{6.10}$$

Since $\dot{\hat{\boldsymbol{J}}}^{-1}(\boldsymbol{q}, \hat{\boldsymbol{\theta}}_k) = -\hat{\boldsymbol{J}}^{-1}(\boldsymbol{q}, \hat{\boldsymbol{\theta}}_k)\dot{\hat{\boldsymbol{J}}}(\boldsymbol{q}, \hat{\boldsymbol{\theta}}_k)\hat{\boldsymbol{J}}^{-1}(\boldsymbol{q}, \hat{\boldsymbol{\theta}}_k)$, then using Eq. (6.8), $\ddot{\boldsymbol{q}}_r$ can be further written as

$$\begin{aligned}
\ddot{\boldsymbol{q}}_r &= -\hat{\boldsymbol{J}}^{-1}(\boldsymbol{q}, \hat{\boldsymbol{\theta}}_k)\dot{\hat{\boldsymbol{J}}}(\boldsymbol{q}, \hat{\boldsymbol{\theta}}_k)\hat{\boldsymbol{J}}^{-1}(\boldsymbol{q}, \hat{\boldsymbol{\theta}}_k)[\dot{\boldsymbol{x}}_d - \alpha(\hat{\boldsymbol{x}}_o - \boldsymbol{x}_d)] \\
&\quad + \hat{\boldsymbol{J}}^{-1}(\boldsymbol{q}, \hat{\boldsymbol{\theta}}_k)[\ddot{\boldsymbol{x}}_d - \alpha(\dot{\hat{\boldsymbol{x}}}_o - \dot{\boldsymbol{x}}_d)] \\
&= -\hat{\boldsymbol{J}}^{-1}(\boldsymbol{q}, \hat{\boldsymbol{\theta}}_k)\dot{\hat{\boldsymbol{J}}}(\boldsymbol{q}, \hat{\boldsymbol{\theta}}_k)\dot{\boldsymbol{q}}_r + \hat{\boldsymbol{J}}^{-1}(\boldsymbol{q}, \hat{\boldsymbol{\theta}}_k)[\ddot{\boldsymbol{x}}_d - \alpha(\dot{\hat{\boldsymbol{x}}}_o - \dot{\boldsymbol{x}}_d)] \\
&= -\hat{\boldsymbol{J}}^{-1}(\boldsymbol{q}, \hat{\boldsymbol{\theta}}_k)\boldsymbol{Z}_r(\boldsymbol{q}, \hat{\boldsymbol{\theta}}_k, \Delta\boldsymbol{x}, \tilde{\boldsymbol{x}})\dot{\boldsymbol{q}} + \hat{\boldsymbol{J}}^{-1}(\boldsymbol{q}, \hat{\boldsymbol{\theta}}_k)[\ddot{\boldsymbol{x}}_d - \alpha(\dot{\hat{\boldsymbol{x}}}_o - \dot{\boldsymbol{x}}_d)].
\end{aligned} \tag{6.11}$$

Note that the matrix $\boldsymbol{Z}_r(\boldsymbol{q}, \hat{\boldsymbol{\theta}}_k, \Delta\boldsymbol{x}, \tilde{\boldsymbol{x}}) \in \Re^{n\times n}$ is constructed as

$$\dot{\hat{\boldsymbol{J}}}(\boldsymbol{q}, \hat{\boldsymbol{\theta}}_k)\dot{\boldsymbol{q}}_r = \boldsymbol{Z}_r(\boldsymbol{q}, \hat{\boldsymbol{\theta}}_k, \Delta\boldsymbol{x}, \tilde{\boldsymbol{x}})\dot{\boldsymbol{q}}, \tag{6.12}$$

where $\tilde{\boldsymbol{x}} = \boldsymbol{x} - \hat{\boldsymbol{x}}_o$, and the main purpose is to extract the joint velocity $\dot{\boldsymbol{q}}$ from the term $\dot{\hat{\boldsymbol{J}}}(\boldsymbol{q}, \hat{\boldsymbol{\theta}}_k)\dot{\boldsymbol{q}}_r$. Note that $\dot{\boldsymbol{q}}_r$ is a function of $\hat{\boldsymbol{\theta}}_k$, $\hat{\boldsymbol{x}}_o$, $\boldsymbol{x}_d$, $\dot{\boldsymbol{x}}_d$, and $\dot{\hat{\boldsymbol{J}}}(\boldsymbol{q}, \hat{\boldsymbol{\theta}}_k)$ is a function of $\dot{\boldsymbol{q}}$ and $\dot{\hat{\boldsymbol{\theta}}}_k$.

Next, $\dot{\boldsymbol{q}}$ is replaced by the observed velocity $\dot{\hat{\boldsymbol{q}}}_o$, and the modified reference acceleration $\ddot{\hat{\boldsymbol{q}}}_r$ for $\ddot{\boldsymbol{q}}_r$ is thus defined as

$$\begin{aligned}
\ddot{\hat{\boldsymbol{q}}}_r &= -\hat{\boldsymbol{J}}^{-1}(\boldsymbol{q}, \hat{\boldsymbol{\theta}}_k)\boldsymbol{Z}_r(\boldsymbol{q}, \hat{\boldsymbol{\theta}}_k, \Delta\boldsymbol{x}, \tilde{\boldsymbol{x}})\dot{\hat{\boldsymbol{q}}}_o + \hat{\boldsymbol{J}}^{-1}(\boldsymbol{q}, \hat{\boldsymbol{\theta}}_k)[\ddot{\boldsymbol{x}}_d - \alpha(\dot{\hat{\boldsymbol{x}}}_o - \dot{\boldsymbol{x}}_d)] \\
&= -\hat{\boldsymbol{J}}^{-1}(\boldsymbol{q}, \hat{\boldsymbol{\theta}}_k)\boldsymbol{Z}_r(\boldsymbol{q}, \hat{\boldsymbol{\theta}}_k, \Delta\boldsymbol{x}, \tilde{\boldsymbol{x}})\dot{\boldsymbol{q}} + \hat{\boldsymbol{J}}^{-1}(\boldsymbol{q}, \hat{\boldsymbol{\theta}}_k)[\ddot{\boldsymbol{x}}_d \\
&\quad -\alpha(\dot{\hat{\boldsymbol{x}}}_o - \dot{\boldsymbol{x}}_d)] + \hat{\boldsymbol{J}}^{-1}(\boldsymbol{q}, \hat{\boldsymbol{\theta}}_k)\boldsymbol{Z}_r(\boldsymbol{q}, \hat{\boldsymbol{\theta}}_k, \Delta\boldsymbol{x}, \tilde{\boldsymbol{x}})(\dot{\boldsymbol{q}} - \dot{\hat{\boldsymbol{q}}}_o) \\
&= \ddot{\boldsymbol{q}}_r + \hat{\boldsymbol{J}}^{-1}(\boldsymbol{q}, \hat{\boldsymbol{\theta}}_k)\boldsymbol{Z}_r(\boldsymbol{q}, \hat{\boldsymbol{\theta}}_k, \Delta\boldsymbol{x}, \tilde{\boldsymbol{x}})\dot{\tilde{\boldsymbol{q}}},
\end{aligned} \tag{6.13}$$

where $\dot{\tilde{\boldsymbol{q}}} = \dot{\boldsymbol{q}} - \dot{\hat{\boldsymbol{q}}}_o$. Therefore, the joint velocity is not used in the modified reference acceleration $\ddot{\hat{\boldsymbol{q}}}_r$.

The observed signal $\dot{\hat{\boldsymbol{q}}}_o$ is updated as follows:

$$\begin{aligned}
\ddot{\hat{\boldsymbol{q}}}_o = \hat{\boldsymbol{M}}^{-1}(\boldsymbol{q}) \Big\{ & \hat{\boldsymbol{K}}_N \boldsymbol{I} - \big[\tfrac{1}{2}\dot{\hat{\boldsymbol{M}}}(\boldsymbol{q}) + \hat{\boldsymbol{S}}(\boldsymbol{q}, \dot{\boldsymbol{q}})\big]\dot{\hat{\boldsymbol{q}}}_o - \hat{\boldsymbol{g}}(\boldsymbol{q}) \\
& + \hat{\boldsymbol{J}}^T(\boldsymbol{q}, \hat{\boldsymbol{\theta}}_k)\boldsymbol{K}_x\tilde{\boldsymbol{x}} + \boldsymbol{K}_q\dot{\tilde{\boldsymbol{q}}} - \hat{\boldsymbol{K}}_N\boldsymbol{Y}_a(\boldsymbol{I})\hat{\hat{\boldsymbol{\theta}}}_a + \big[\boldsymbol{K}_d \\
& + \boldsymbol{Z}_r^T(\boldsymbol{q}, \hat{\boldsymbol{\theta}}_k, \Delta\boldsymbol{x}, \tilde{\boldsymbol{x}})\hat{\boldsymbol{J}}^{-T}(\boldsymbol{q}, \hat{\boldsymbol{\theta}}_k)\hat{\boldsymbol{M}}(\boldsymbol{q}) - \big(\tfrac{1}{2}\dot{\hat{\boldsymbol{M}}}(\boldsymbol{q}) + \hat{\boldsymbol{S}}(\boldsymbol{q}, \dot{\hat{\boldsymbol{q}}}_r)\big)\big]\boldsymbol{s}\Big\},
\end{aligned} \tag{6.14}$$

where $\boldsymbol{I} \in \Re^n$ is the armature current, which is measured with an ammeter or current probe, $\hat{\boldsymbol{M}}(\boldsymbol{q})$, $\frac{1}{2}\dot{\hat{\boldsymbol{M}}}(\boldsymbol{q}) + \hat{\boldsymbol{S}}(\boldsymbol{q}, \dot{\boldsymbol{q}})$, $\frac{1}{2}\dot{\hat{\boldsymbol{M}}}(\boldsymbol{q}) + \hat{\boldsymbol{S}}(\boldsymbol{q}, \dot{\hat{\boldsymbol{q}}}_r)$, and $\hat{\boldsymbol{g}}(\boldsymbol{q})$ use same estimated dynamics parameters in $\hat{\boldsymbol{\theta}}_d$ as in (6.6), $\boldsymbol{Y}_a(\boldsymbol{I}) = \mathrm{diag}\{I_1, \ldots, I_n\}$ is defined in the same way as $\boldsymbol{Y}_a(\boldsymbol{I}_o)$ but with $\boldsymbol{I}$ as the argument, and I_i denotes the ith element of the armature current $\boldsymbol{I}$, $\hat{\hat{\boldsymbol{\theta}}}_a \in \Re^n$ is an adaptive parameter vector updated through an adaptation law to be defined later, $\boldsymbol{s} \in \Re^n$ is an adaptive sliding vector defined using $\dot{\boldsymbol{q}}_r$ as

$$\boldsymbol{s} = \dot{\boldsymbol{q}} - \dot{\boldsymbol{q}}_r, \tag{6.15}$$

such that

$$\begin{aligned}
\hat{\boldsymbol{J}}(\boldsymbol{q}, \hat{\boldsymbol{\theta}}_k)\boldsymbol{s} &= \hat{\boldsymbol{J}}(\boldsymbol{q}, \hat{\boldsymbol{\theta}}_k)\dot{\boldsymbol{q}} - \dot{\boldsymbol{x}}_d + \alpha(\hat{\boldsymbol{x}}_o - \boldsymbol{x}_d) \\
&= \dot{\boldsymbol{x}} - \dot{\boldsymbol{x}}_d + \hat{\boldsymbol{J}}(\boldsymbol{q}, \hat{\boldsymbol{\theta}}_k)\dot{\boldsymbol{q}} - \dot{\boldsymbol{x}} + \alpha(\boldsymbol{x} - \boldsymbol{x}_d) - \alpha(\boldsymbol{x} - \hat{\boldsymbol{x}}_o) \\
&= \Delta\dot{\boldsymbol{x}} + \alpha\Delta\boldsymbol{x} - \boldsymbol{Y}_k(\boldsymbol{q}, \dot{\boldsymbol{q}})\Delta\boldsymbol{\theta}_k - \alpha\tilde{\boldsymbol{x}},
\end{aligned} \tag{6.16}$$

where $\Delta\boldsymbol{\theta}_k = \boldsymbol{\theta}_k - \hat{\boldsymbol{\theta}}_k$ and Eq. (6.4) have been used.

The control voltage input $\boldsymbol{u}$ utilizes the time derivative of the desired armature current $\boldsymbol{I}_d$, as will be seen in the next design step. Therefore, velocity information is not used in $\boldsymbol{I}_d$, so as to avoid the use of task-space and joint-space accelerations in the control input $\boldsymbol{u}$. The observed signals $\dot{\hat{\boldsymbol{q}}}_o$ and $\hat{\boldsymbol{x}}_o$ are thus designed to replace $\dot{\boldsymbol{q}}$ and $\boldsymbol{x}$ respectively, and the modified reference acceleration $\ddot{\hat{\boldsymbol{q}}}_r$ for $\ddot{\boldsymbol{q}}_r$ is introduced for the same purpose using the matrix $\boldsymbol{Z}_r(\boldsymbol{q}, \hat{\boldsymbol{\theta}}_k, \Delta\boldsymbol{x}, \tilde{\boldsymbol{x}})$ to extract $\dot{\boldsymbol{q}}$ out from $\dot{\hat{\boldsymbol{J}}}(\boldsymbol{q}, \hat{\boldsymbol{\theta}}_k)$ and replace it with the observed signal $\dot{\hat{\boldsymbol{q}}}_o$.

Substituting Eq. (6.5) into (6.3), the closed-loop equation of the manipulator subsystem is obtained as

$$\begin{aligned} \boldsymbol{M}(\boldsymbol{q})\ddot{\boldsymbol{q}} + \left[\tfrac{1}{2}\dot{\boldsymbol{M}}(\boldsymbol{q}) + \boldsymbol{S}(\boldsymbol{q}, \dot{\boldsymbol{q}})\right]\dot{\boldsymbol{q}} + \boldsymbol{g}(\boldsymbol{q}) + \hat{\boldsymbol{J}}^T(\boldsymbol{q}, \hat{\boldsymbol{\theta}}_k)\boldsymbol{K}_p\Delta\boldsymbol{x} \\ + \boldsymbol{K}_d(\dot{\hat{\boldsymbol{q}}}_o - \dot{\boldsymbol{q}}_r) - \boldsymbol{Y}_d(\boldsymbol{q}, \dot{\hat{\boldsymbol{q}}}_o, \dot{\boldsymbol{q}}_r, \ddot{\hat{\boldsymbol{q}}}_r)\hat{\boldsymbol{\theta}}_d \\ + (\boldsymbol{I}_n - \boldsymbol{K}_N\hat{\boldsymbol{K}}_N^{-1})\boldsymbol{I}_o - \boldsymbol{K}_N\hat{\boldsymbol{K}}_N^{-1}\boldsymbol{Y}_a(\boldsymbol{I}_o)\hat{\boldsymbol{\theta}}_a = \boldsymbol{K}_N\tilde{\boldsymbol{I}}, \end{aligned} \tag{6.17}$$

where $\boldsymbol{I}_n \in \Re^{n\times n}$ denotes the $n\times n$ identity matrix.

Substituting Eq. (6.13) into (6.6) and using *Property* 1.3, we have

$$\begin{aligned} \boldsymbol{Y}_d(\boldsymbol{q}, \dot{\hat{\boldsymbol{q}}}_o, \dot{\boldsymbol{q}}_r, \ddot{\hat{\boldsymbol{q}}}_r)\hat{\boldsymbol{\theta}}_d &= \hat{\boldsymbol{M}}(\boldsymbol{q})\ddot{\boldsymbol{q}}_r + \left(\tfrac{1}{2}\dot{\hat{\boldsymbol{M}}}(\boldsymbol{q}) + \hat{\boldsymbol{S}}(\boldsymbol{q}, \dot{\boldsymbol{q}})\right)\dot{\boldsymbol{q}}_r + \hat{\boldsymbol{g}}(\boldsymbol{q}) \\ &+ \hat{\boldsymbol{M}}(\boldsymbol{q})\hat{\boldsymbol{J}}^{-1}(\boldsymbol{q}, \hat{\boldsymbol{\theta}}_k)\boldsymbol{Z}_r(\boldsymbol{q}, \hat{\boldsymbol{\theta}}_k, \Delta\boldsymbol{x}, \tilde{\boldsymbol{x}})\dot{\tilde{\boldsymbol{q}}} - \left(\tfrac{1}{2}\dot{\hat{\boldsymbol{M}}}(\boldsymbol{q}) + \hat{\boldsymbol{S}}(\boldsymbol{q}, \dot{\hat{\boldsymbol{q}}}_r)\right)\dot{\tilde{\boldsymbol{q}}} \\ &= \boldsymbol{Y}_d(\boldsymbol{q}, \dot{\boldsymbol{q}}, \dot{\boldsymbol{q}}_r, \ddot{\boldsymbol{q}}_r)\hat{\boldsymbol{\theta}}_d + \hat{\boldsymbol{M}}(\boldsymbol{q})\hat{\boldsymbol{J}}^{-1}(\boldsymbol{q}, \hat{\boldsymbol{\theta}}_k)\boldsymbol{Z}_r(\boldsymbol{q}, \hat{\boldsymbol{\theta}}_k, \Delta\boldsymbol{x}, \tilde{\boldsymbol{x}})\dot{\tilde{\boldsymbol{q}}} \\ &- \left(\tfrac{1}{2}\dot{\hat{\boldsymbol{M}}}(\boldsymbol{q}) + \hat{\boldsymbol{S}}(\boldsymbol{q}, \dot{\hat{\boldsymbol{q}}}_r)\right)\dot{\tilde{\boldsymbol{q}}}, \end{aligned} \tag{6.18}$$

where $\boldsymbol{Y}_d(\boldsymbol{q}, \dot{\boldsymbol{q}}, \dot{\boldsymbol{q}}_r, \ddot{\boldsymbol{q}}_r)\hat{\boldsymbol{\theta}}_d$ is defined according to *Property* 1.3 as

$$\boldsymbol{Y}_d(\boldsymbol{q}, \dot{\boldsymbol{q}}, \dot{\boldsymbol{q}}_r, \ddot{\boldsymbol{q}}_r)\hat{\boldsymbol{\theta}}_d = \hat{\boldsymbol{M}}(\boldsymbol{q})\ddot{\boldsymbol{q}}_r + \left(\tfrac{1}{2}\dot{\hat{\boldsymbol{M}}}(\boldsymbol{q}) + \hat{\boldsymbol{S}}(\boldsymbol{q}, \dot{\boldsymbol{q}})\right)\dot{\boldsymbol{q}}_r + \hat{\boldsymbol{g}}(\boldsymbol{q}). \tag{6.19}$$

Differentiating Eq. (6.15), we have

$$\dot{\boldsymbol{s}} = \ddot{\boldsymbol{q}} - \ddot{\boldsymbol{q}}_r. \tag{6.20}$$

Substituting (6.15), (6.20) into (6.17) and using Eq. (6.18), the closed-loop dynamic equation is expressed in the following form:

$$\begin{aligned} \boldsymbol{M}(\boldsymbol{q})\dot{\boldsymbol{s}} + \left(\tfrac{1}{2}\dot{\hat{\boldsymbol{M}}}(\boldsymbol{q}) + \hat{\boldsymbol{S}}(\boldsymbol{q}, \dot{\boldsymbol{q}})\right)\boldsymbol{s} + \boldsymbol{Y}_d(\boldsymbol{q}, \dot{\boldsymbol{q}}, \dot{\boldsymbol{q}}_r, \ddot{\boldsymbol{q}}_r)\Delta\boldsymbol{\theta}_d \\ -\hat{\boldsymbol{M}}(\boldsymbol{q})\hat{\boldsymbol{J}}^{-1}(\boldsymbol{q}, \hat{\boldsymbol{\theta}}_k)\boldsymbol{Z}_r(\boldsymbol{q}, \hat{\boldsymbol{\theta}}_k, \Delta\boldsymbol{x}, \tilde{\boldsymbol{x}})\dot{\tilde{\boldsymbol{q}}} + \left[\tfrac{1}{2}\dot{\hat{\boldsymbol{M}}}(\boldsymbol{q}) + \hat{\boldsymbol{S}}(\boldsymbol{q}, \dot{\hat{\boldsymbol{q}}}_r)\right]\dot{\tilde{\boldsymbol{q}}} \\ +\hat{\boldsymbol{J}}^T(\boldsymbol{q}, \hat{\boldsymbol{\theta}}_k)\boldsymbol{K}_p\Delta\boldsymbol{x} + \boldsymbol{K}_d(\dot{\hat{\boldsymbol{q}}}_o - \dot{\boldsymbol{q}}_r) + \boldsymbol{Y}_a(\boldsymbol{I}_o)(\boldsymbol{\theta}_a - \boldsymbol{K}_N\hat{\boldsymbol{K}}_N^{-1}\hat{\boldsymbol{\theta}}_a) = \boldsymbol{K}_N\tilde{\boldsymbol{I}}, \end{aligned} \tag{6.21}$$

where *Property* 1.3 has been used, $\Delta\boldsymbol{\theta}_d = \boldsymbol{\theta}_d - \hat{\boldsymbol{\theta}}_d$, and $\boldsymbol{\theta}_a = [1 - \frac{k_N^1}{\hat{k}_N^1} \cdots 1 - \frac{k_N^n}{\hat{k}_N^n}]^T$ with k_N^i and $\hat{k}_N^i$ respectively the ith diagonal element of $\boldsymbol{K}_N$ and $\hat{\boldsymbol{K}}_N$.

Next, we consider the development of the closed-loop observer dynamics for $\dot{\hat{\boldsymbol{q}}}_o$. From Eq. (6.14), multiplying both sides by $\hat{\boldsymbol{M}}(\boldsymbol{q})$ yields

$$\begin{aligned}\hat{\boldsymbol{K}}_N\boldsymbol{I} - \hat{\boldsymbol{M}}(\boldsymbol{q})\ddot{\hat{\boldsymbol{q}}}_o - [\tfrac{1}{2}\dot{\hat{\boldsymbol{M}}}(\boldsymbol{q}) + \hat{\boldsymbol{S}}(\boldsymbol{q},\dot{\boldsymbol{q}})]\dot{\hat{\boldsymbol{q}}}_o - \hat{\boldsymbol{g}}(\boldsymbol{q}) + \hat{\boldsymbol{J}}^T(\boldsymbol{q},\hat{\boldsymbol{\theta}}_k)\boldsymbol{K}_x\tilde{\boldsymbol{x}} \\ +\boldsymbol{K}_q\dot{\tilde{\boldsymbol{q}}} + \big[\boldsymbol{K}_d + \boldsymbol{Z}_r^T(\boldsymbol{q},\hat{\boldsymbol{\theta}}_k,\Delta\boldsymbol{x},\tilde{\boldsymbol{x}})\hat{\boldsymbol{J}}^{-T}(\boldsymbol{q},\hat{\boldsymbol{\theta}}_k)\hat{\boldsymbol{M}}(\boldsymbol{q}) \\ -\big(\tfrac{1}{2}\dot{\hat{\boldsymbol{M}}}^T(\boldsymbol{q}) + \hat{\boldsymbol{S}}^T(\boldsymbol{q},\dot{\hat{\boldsymbol{q}}}_r)\big)\big]\boldsymbol{s} - \hat{\boldsymbol{K}}_N\boldsymbol{Y}_a(\boldsymbol{I})\hat{\hat{\boldsymbol{\theta}}}_a = \boldsymbol{0}.\end{aligned} \tag{6.22}$$

The first term of Eq. (6.22) can be written as

$$\hat{\boldsymbol{K}}_N\boldsymbol{I} = \boldsymbol{K}_N\boldsymbol{I} + \hat{\boldsymbol{K}}_N(\boldsymbol{I}_n - \hat{\boldsymbol{K}}_N^{-1}\boldsymbol{K}_N)\boldsymbol{I} = \boldsymbol{K}_N\boldsymbol{I} + \hat{\boldsymbol{K}}_N\boldsymbol{Y}_a(\boldsymbol{I})\boldsymbol{\theta}_a. \tag{6.23}$$

Substituting Eq. (6.23) into Eq. (6.22), we have

$$\begin{aligned}\boldsymbol{K}_N\boldsymbol{I} - \hat{\boldsymbol{M}}(\boldsymbol{q})\ddot{\hat{\boldsymbol{q}}}_o - [\tfrac{1}{2}\dot{\hat{\boldsymbol{M}}}(\boldsymbol{q}) + \hat{\boldsymbol{S}}(\boldsymbol{q},\dot{\boldsymbol{q}})]\dot{\hat{\boldsymbol{q}}}_o - \hat{\boldsymbol{g}}(\boldsymbol{q}) + \hat{\boldsymbol{J}}^T(\boldsymbol{q},\hat{\boldsymbol{\theta}}_k)\boldsymbol{K}_x\tilde{\boldsymbol{x}} \\ +\boldsymbol{K}_q\dot{\tilde{\boldsymbol{q}}} + \big[\boldsymbol{K}_d + \boldsymbol{Z}_r^T(\boldsymbol{q},\hat{\boldsymbol{\theta}}_k,\Delta\boldsymbol{x},\tilde{\boldsymbol{x}})\hat{\boldsymbol{J}}^{-T}(\boldsymbol{q},\hat{\boldsymbol{\theta}}_k)\hat{\boldsymbol{M}}(\boldsymbol{q}) \\ -\big(\tfrac{1}{2}\dot{\hat{\boldsymbol{M}}}^T(\boldsymbol{q}) + \hat{\boldsymbol{S}}^T(\boldsymbol{q},\dot{\hat{\boldsymbol{q}}}_r)\big)\big]\boldsymbol{s} + \hat{\boldsymbol{K}}_N\boldsymbol{Y}_a(\boldsymbol{I})(\boldsymbol{\theta}_a - \hat{\hat{\boldsymbol{\theta}}}_a) = \boldsymbol{0}.\end{aligned} \tag{6.24}$$

Then substituting $\boldsymbol{K}_N\boldsymbol{I}$ with the left-hand side of the manipulator subsystem dynamic Eq. (6.1) into Eq. (6.24), we obtain

$$\begin{aligned}\boldsymbol{M}(\boldsymbol{q})\ddot{\tilde{\boldsymbol{q}}} + \big[\boldsymbol{K}_q + \tfrac{1}{2}\dot{\boldsymbol{M}}(\boldsymbol{q}) + \boldsymbol{S}(\boldsymbol{q},\dot{\boldsymbol{q}})\big]\dot{\tilde{\boldsymbol{q}}} + [\boldsymbol{M}(\boldsymbol{q}) - \hat{\boldsymbol{M}}(\boldsymbol{q})]\ddot{\hat{\boldsymbol{q}}}_o \\ +\big(\tfrac{1}{2}\dot{\boldsymbol{M}}(\boldsymbol{q}) + \boldsymbol{S}(\boldsymbol{q},\dot{\boldsymbol{q}}) - \big(\tfrac{1}{2}\dot{\hat{\boldsymbol{M}}}(\boldsymbol{q}) + \hat{\boldsymbol{S}}(\boldsymbol{q},\dot{\boldsymbol{q}})\big)\big)\dot{\hat{\boldsymbol{q}}}_o + [\boldsymbol{g}(\boldsymbol{q}) - \hat{\boldsymbol{g}}(\boldsymbol{q})] \\ +\hat{\boldsymbol{J}}^T(\boldsymbol{q},\hat{\boldsymbol{\theta}}_k)\boldsymbol{K}_x\tilde{\boldsymbol{x}} + \big[\boldsymbol{K}_d + \boldsymbol{Z}_r^T(\boldsymbol{q},\hat{\boldsymbol{\theta}}_k,\Delta\boldsymbol{x},\tilde{\boldsymbol{x}})\hat{\boldsymbol{J}}^{-T}(\boldsymbol{q},\hat{\boldsymbol{\theta}}_k)\hat{\boldsymbol{M}}(\boldsymbol{q}) \\ -\big(\tfrac{1}{2}\dot{\hat{\boldsymbol{M}}}^T(\boldsymbol{q}) + \hat{\boldsymbol{S}}^T(\boldsymbol{q},\dot{\boldsymbol{q}}_r)\big)\big]\boldsymbol{s} + \hat{\boldsymbol{K}}_N\boldsymbol{Y}_a(\boldsymbol{I})(\boldsymbol{\theta}_a - \hat{\hat{\boldsymbol{\theta}}}_a) = \boldsymbol{0},\end{aligned} \tag{6.25}$$

where $\ddot{\tilde{\boldsymbol{q}}} = \ddot{\boldsymbol{q}} - \ddot{\hat{\boldsymbol{q}}}_o$. Next, note that

$$\begin{aligned}[\boldsymbol{M}(\boldsymbol{q}) - \hat{\boldsymbol{M}}(\boldsymbol{q})]\ddot{\hat{\boldsymbol{q}}}_o + \big[\tfrac{1}{2}\dot{\boldsymbol{M}}(\boldsymbol{q}) + \boldsymbol{S}(\boldsymbol{q},\dot{\boldsymbol{q}}) - \big(\tfrac{1}{2}\dot{\hat{\boldsymbol{M}}}(\boldsymbol{q}) + \hat{\boldsymbol{S}}(\boldsymbol{q},\dot{\boldsymbol{q}})\big)\big]\dot{\hat{\boldsymbol{q}}}_o \\ +[\boldsymbol{g}(\boldsymbol{q}) - \hat{\boldsymbol{g}}(\boldsymbol{q})] = \boldsymbol{Y}_d(\boldsymbol{q},\dot{\boldsymbol{q}},\dot{\hat{\boldsymbol{q}}}_o,\ddot{\hat{\boldsymbol{q}}}_o)\Delta\boldsymbol{\theta}_d,\end{aligned} \tag{6.26}$$

where the regressor $\boldsymbol{Y}_d(\boldsymbol{q},\dot{\boldsymbol{q}},\dot{\hat{\boldsymbol{q}}}_o,\ddot{\hat{\boldsymbol{q}}}_o)$ is defined similarly to $\boldsymbol{Y}_d(\boldsymbol{q},\dot{\boldsymbol{q}},\dot{\boldsymbol{q}}_r,\ddot{\boldsymbol{q}}_r)$ of Eq. (6.19).

Using Eq. (6.26), the closed-loop observer dynamic Eq. (6.25) can be simplified to

$$\begin{aligned}&\boldsymbol{M}(\boldsymbol{q})\ddot{\tilde{\boldsymbol{q}}} + \left[\boldsymbol{K}_q + \tfrac{1}{2}\dot{\boldsymbol{M}}(\boldsymbol{q}) + \boldsymbol{S}(\boldsymbol{q},\dot{\boldsymbol{q}})\right]\dot{\tilde{\boldsymbol{q}}} + \boldsymbol{Y}_d(\boldsymbol{q},\dot{\boldsymbol{q}},\dot{\hat{\boldsymbol{q}}}_o,\ddot{\hat{\boldsymbol{q}}}_o)\Delta\boldsymbol{\theta}_d \\ &+\hat{\boldsymbol{J}}^T(\boldsymbol{q},\hat{\boldsymbol{\theta}}_k)\boldsymbol{K}_x\tilde{\boldsymbol{x}} + \left[\boldsymbol{K}_d + \boldsymbol{Z}_r^T(\boldsymbol{q},\hat{\boldsymbol{\theta}}_k,\Delta\boldsymbol{x},\tilde{\boldsymbol{x}})\hat{\boldsymbol{J}}^{-T}(\boldsymbol{q},\hat{\boldsymbol{\theta}}_k)\hat{\boldsymbol{M}}(\boldsymbol{q})\right. \\ &\left.-\left(\tfrac{1}{2}\dot{\hat{\boldsymbol{M}}}^T(\boldsymbol{q}) + \hat{\boldsymbol{S}}^T(\boldsymbol{q},\dot{\boldsymbol{q}}_r)\right)\right]\boldsymbol{s} + \hat{\boldsymbol{K}}_N\boldsymbol{Y}_a(\boldsymbol{I})(\boldsymbol{\theta}_a - \hat{\hat{\boldsymbol{\theta}}}_a) = \boldsymbol{0}. \end{aligned} \tag{6.27}$$

With the closed-loop dynamics of the manipulator and observer, we can now proceed to show the stability of the closed-loop system with the desired armature current input. A Lyapunov function candidate V is defined as

$$\begin{aligned} V &= \tfrac{1}{2}\boldsymbol{s}^T\boldsymbol{M}(\boldsymbol{q})\boldsymbol{s} + \tfrac{1}{2}\Delta\boldsymbol{x}^T\boldsymbol{K}_p\Delta\boldsymbol{x} + \tfrac{1}{2}\dot{\tilde{\boldsymbol{q}}}^T\boldsymbol{M}(\boldsymbol{q})\dot{\tilde{\boldsymbol{q}}} + \tfrac{1}{2}\tilde{\boldsymbol{x}}^T\boldsymbol{K}_x\tilde{\boldsymbol{x}} + \tfrac{1}{2}\Delta\boldsymbol{\theta}_d^T\boldsymbol{L}_d^{-1}\Delta\boldsymbol{\theta}_d \\ &+\tfrac{1}{2}\Delta\boldsymbol{\theta}_k^T\boldsymbol{L}_k^{-1}\Delta\boldsymbol{\theta}_k + \tfrac{1}{2}(\hat{\boldsymbol{K}}_N\boldsymbol{K}_N^{-1}\boldsymbol{\theta}_a - \hat{\boldsymbol{\theta}}_a)^T\boldsymbol{L}_a^{-1}\boldsymbol{K}_N\hat{\boldsymbol{K}}_N^{-1}(\hat{\boldsymbol{K}}_N\boldsymbol{K}_N^{-1}\boldsymbol{\theta}_a - \hat{\boldsymbol{\theta}}_a) \\ &+\tfrac{1}{2}(\boldsymbol{\theta}_a - \hat{\hat{\boldsymbol{\theta}}}_a)^T\hat{\boldsymbol{K}}_N\bar{\boldsymbol{L}}_a^{-1}(\boldsymbol{\theta}_a - \hat{\hat{\boldsymbol{\theta}}}_a), \end{aligned} \tag{6.28}$$

where $\boldsymbol{K}_x$, $\boldsymbol{L}_d$, $\boldsymbol{L}_k$, $\boldsymbol{L}_a$, and $\bar{\boldsymbol{L}}_a$ are positive definite constant diagonal matrices.

Differentiating V with respect to time, we have

$$\begin{aligned} \dot{V} &= \boldsymbol{s}^T\boldsymbol{M}(\boldsymbol{q})\dot{\boldsymbol{s}} + \tfrac{1}{2}\boldsymbol{s}^T\dot{\boldsymbol{M}}(\boldsymbol{q})\boldsymbol{s} + \Delta\boldsymbol{x}^T\boldsymbol{K}_p\Delta\dot{\boldsymbol{x}} + \dot{\tilde{\boldsymbol{q}}}^T\boldsymbol{M}(\boldsymbol{q})\ddot{\tilde{\boldsymbol{q}}} \\ &+\tfrac{1}{2}\dot{\tilde{\boldsymbol{q}}}^T\dot{\boldsymbol{M}}(\boldsymbol{q})\dot{\tilde{\boldsymbol{q}}} + \tilde{\boldsymbol{x}}^T\boldsymbol{K}_x\dot{\tilde{\boldsymbol{x}}} - \Delta\boldsymbol{\theta}_d^T\boldsymbol{L}_d^{-1}\dot{\hat{\boldsymbol{\theta}}}_d - \Delta\boldsymbol{\theta}_k^T\boldsymbol{L}_k^{-1}\dot{\hat{\boldsymbol{\theta}}}_k \\ &-(\hat{\boldsymbol{K}}_N\boldsymbol{K}_N^{-1}\boldsymbol{\theta}_a - \hat{\boldsymbol{\theta}}_a)^T\boldsymbol{L}_a^{-1}\boldsymbol{K}_N\hat{\boldsymbol{K}}_N^{-1}\dot{\hat{\boldsymbol{\theta}}}_a - (\boldsymbol{\theta}_a - \hat{\hat{\boldsymbol{\theta}}}_a)^T\hat{\boldsymbol{K}}_N\bar{\boldsymbol{L}}_a^{-1}\dot{\hat{\hat{\boldsymbol{\theta}}}}_a, \end{aligned} \tag{6.29}$$

where $\dot{\tilde{\boldsymbol{x}}} = \dot{\boldsymbol{x}} - \dot{\hat{\boldsymbol{x}}}_o$.

Using the closed-loop manipulator and observer dynamic Eqs. (6.21) and (6.27), $\dot{V}$ is written as

$$\begin{aligned} \dot{V} &= -\boldsymbol{s}^T\boldsymbol{K}_d\boldsymbol{s} - \boldsymbol{s}^T\hat{\boldsymbol{J}}^T(\boldsymbol{q},\hat{\boldsymbol{\theta}}_k)\boldsymbol{K}_p\Delta\boldsymbol{x} - \boldsymbol{s}^T\boldsymbol{Y}_d(\boldsymbol{q},\dot{\boldsymbol{q}},\dot{\boldsymbol{q}}_r,\ddot{\boldsymbol{q}}_r)\Delta\boldsymbol{\theta}_d \\ &-\boldsymbol{s}^T\boldsymbol{Y}_a(\boldsymbol{I}_o)(\boldsymbol{\theta}_a - \boldsymbol{K}_N\hat{\boldsymbol{K}}_N^{-1}\hat{\boldsymbol{\theta}}_a) + \Delta\boldsymbol{x}^T\boldsymbol{K}_p\Delta\dot{\boldsymbol{x}} - \dot{\tilde{\boldsymbol{q}}}^T\boldsymbol{K}_q\dot{\tilde{\boldsymbol{q}}} - \dot{\tilde{\boldsymbol{q}}}^T\boldsymbol{Y}_d(\boldsymbol{q},\dot{\boldsymbol{q}},\dot{\hat{\boldsymbol{q}}}_o,\ddot{\hat{\boldsymbol{q}}}_o)\Delta\boldsymbol{\theta}_d \\ &-\dot{\tilde{\boldsymbol{q}}}^T\hat{\boldsymbol{J}}^T(\boldsymbol{q},\hat{\boldsymbol{\theta}}_k)\boldsymbol{K}_x\tilde{\boldsymbol{x}} - \dot{\tilde{\boldsymbol{q}}}^T\boldsymbol{Y}_a(\boldsymbol{I})\hat{\boldsymbol{K}}_N(\boldsymbol{\theta}_a - \hat{\hat{\boldsymbol{\theta}}}_a) - \Delta\boldsymbol{\theta}_d^T\boldsymbol{L}_d^{-1}\dot{\hat{\boldsymbol{\theta}}}_d - \Delta\boldsymbol{\theta}_k^T\boldsymbol{L}_k^{-1}\dot{\hat{\boldsymbol{\theta}}}_k \\ &-(\boldsymbol{\theta}_a - \boldsymbol{K}_N\hat{\boldsymbol{K}}_N^{-1}\hat{\boldsymbol{\theta}}_a)^T\boldsymbol{L}_a^{-1}\dot{\hat{\boldsymbol{\theta}}}_a - (\boldsymbol{\theta}_a - \hat{\hat{\boldsymbol{\theta}}}_a)^T\hat{\boldsymbol{K}}_N\bar{\boldsymbol{L}}_a^{-1}\dot{\hat{\hat{\boldsymbol{\theta}}}}_a + \tilde{\boldsymbol{x}}^T\boldsymbol{K}_x\dot{\tilde{\boldsymbol{x}}} + \boldsymbol{s}^T\boldsymbol{K}_N\tilde{\boldsymbol{I}}. \end{aligned} \tag{6.30}$$

Since $\dot{\tilde{\boldsymbol{q}}} = \dot{\boldsymbol{q}} - \dot{\hat{\boldsymbol{q}}}_o$, $\tilde{\boldsymbol{x}} = \boldsymbol{x} - \hat{\boldsymbol{x}}_o$, and from the definition of $\dot{\hat{\boldsymbol{x}}}$ and $\dot{\hat{\boldsymbol{x}}}_o$ in Eqs. (6.4) and (6.9) respectively, we have

$$\begin{aligned} \hat{\boldsymbol{J}}(\boldsymbol{q},\hat{\boldsymbol{\theta}}_k)\dot{\tilde{\boldsymbol{q}}} &= \hat{\boldsymbol{J}}(\boldsymbol{q},\hat{\boldsymbol{\theta}}_k)\dot{\boldsymbol{q}} - \hat{\boldsymbol{J}}(\boldsymbol{q},\hat{\boldsymbol{\theta}}_k)\dot{\hat{\boldsymbol{q}}}_o \\ &= \dot{\hat{\boldsymbol{x}}} - \dot{\hat{\boldsymbol{x}}}_o + \boldsymbol{\Lambda}\tilde{\boldsymbol{x}} = \dot{\boldsymbol{x}} - \dot{\hat{\boldsymbol{x}}}_o + \boldsymbol{\Lambda}\tilde{\boldsymbol{x}} + \dot{\hat{\boldsymbol{x}}} - \dot{\boldsymbol{x}} \\ &= \dot{\tilde{\boldsymbol{x}}} + \boldsymbol{\Lambda}\tilde{\boldsymbol{x}} + [\hat{\boldsymbol{J}}(\boldsymbol{q},\hat{\boldsymbol{\theta}}_k) - \boldsymbol{J}(\boldsymbol{q},\boldsymbol{\theta}_k)]\dot{\boldsymbol{q}} = \dot{\tilde{\boldsymbol{x}}} + \boldsymbol{\Lambda}\tilde{\boldsymbol{x}} - \boldsymbol{Y}_k(\boldsymbol{q},\dot{\boldsymbol{q}})\Delta\boldsymbol{\theta}_k. \end{aligned} \tag{6.31}$$

Substituting Eqs. (6.16) and (6.31) into Eq. (6.30), $\dot{V}$ can be written as

$$\dot{V} = -\boldsymbol{s}^T \boldsymbol{K}_d \boldsymbol{s} - \dot{\tilde{\boldsymbol{q}}}^T \boldsymbol{K}_q \dot{\tilde{\boldsymbol{q}}} - \alpha \Delta \boldsymbol{x}^T \boldsymbol{K}_p \Delta \boldsymbol{x} - \tilde{\boldsymbol{x}}^T \boldsymbol{\Lambda} \boldsymbol{K}_x \tilde{\boldsymbol{x}} + \alpha \tilde{\boldsymbol{x}}^T \boldsymbol{K}_p \Delta \boldsymbol{x}$$
$$+\Delta \boldsymbol{\theta}_k^T [\boldsymbol{Y}_k^T(\boldsymbol{q}, \dot{\boldsymbol{q}})(\boldsymbol{K}_p \Delta \boldsymbol{x} + \boldsymbol{K}_x \tilde{\boldsymbol{x}}) - \boldsymbol{L}_k^{-1} \dot{\hat{\boldsymbol{\theta}}}_k] - \Delta \boldsymbol{\theta}_d^T [\boldsymbol{Y}_d^T(\boldsymbol{q}, \dot{\boldsymbol{q}}, \dot{\boldsymbol{q}}_r, \ddot{\boldsymbol{q}}_r) \boldsymbol{s}$$
$$+\boldsymbol{Y}_d^T(\boldsymbol{q}, \dot{\boldsymbol{q}}, \dot{\hat{\boldsymbol{q}}}_o, \ddot{\hat{\boldsymbol{q}}}_o) \dot{\tilde{\boldsymbol{q}}} + \boldsymbol{L}_d^{-1} \dot{\hat{\boldsymbol{\theta}}}_d] - (\boldsymbol{\theta}_a - \boldsymbol{K}_N \hat{\boldsymbol{K}}_N^{-1} \hat{\boldsymbol{\theta}}_a)^T [\boldsymbol{Y}_a(\boldsymbol{I}_o) \boldsymbol{s} + \boldsymbol{L}_a^{-1} \dot{\hat{\boldsymbol{\theta}}}_a]$$
$$-(\boldsymbol{\theta}_a - \hat{\hat{\boldsymbol{\theta}}}_a)^T \hat{\boldsymbol{K}}_N [\boldsymbol{Y}_a(\boldsymbol{I}) \dot{\tilde{\boldsymbol{q}}} + \bar{\boldsymbol{L}}_a^{-1} \dot{\hat{\hat{\boldsymbol{\theta}}}}_a] + \boldsymbol{s}^T \boldsymbol{K}_N \tilde{\boldsymbol{I}}. \tag{6.32}$$

The adaptation laws for $\hat{\boldsymbol{\theta}}_k$, $\hat{\boldsymbol{\theta}}_d$, $\hat{\boldsymbol{\theta}}_a$, and $\hat{\hat{\boldsymbol{\theta}}}_a$ are designed as follows:

$$\dot{\hat{\boldsymbol{\theta}}}_k = \boldsymbol{L}_k \boldsymbol{Y}_k^T(\boldsymbol{q}, \dot{\boldsymbol{q}})(\boldsymbol{K}_p \Delta \boldsymbol{x} + \boldsymbol{K}_x \tilde{\boldsymbol{x}}), \tag{6.33}$$
$$\dot{\hat{\boldsymbol{\theta}}}_d = -\boldsymbol{L}_d [\boldsymbol{Y}_d^T(\boldsymbol{q}, \dot{\boldsymbol{q}}, \dot{\boldsymbol{q}}_r, \ddot{\boldsymbol{q}}_r) \boldsymbol{s} + \boldsymbol{Y}_d^T(\boldsymbol{q}, \dot{\boldsymbol{q}}, \dot{\hat{\boldsymbol{q}}}_o, \ddot{\hat{\boldsymbol{q}}}_o) \dot{\tilde{\boldsymbol{q}}}], \tag{6.34}$$
$$\dot{\hat{\boldsymbol{\theta}}}_a = -\boldsymbol{L}_a \boldsymbol{Y}_a(\boldsymbol{I}_o) \boldsymbol{s}, \tag{6.35}$$
$$\dot{\hat{\hat{\boldsymbol{\theta}}}}_a = -\bar{\boldsymbol{L}}_a \boldsymbol{Y}_a(\boldsymbol{I}) \dot{\tilde{\boldsymbol{q}}}. \tag{6.36}$$

On substituting Eqs. (6.33)–(6.36) into (6.32), $\dot{V}$ can be simplified as follows:

$$\begin{aligned}\dot{V} &= -\boldsymbol{s}^T \boldsymbol{K}_d \boldsymbol{s} - \dot{\tilde{\boldsymbol{q}}}^T \boldsymbol{K}_q \dot{\tilde{\boldsymbol{q}}} - \alpha \Delta \boldsymbol{x}^T \boldsymbol{K}_p \Delta \boldsymbol{x} - \tilde{\boldsymbol{x}}^T \boldsymbol{\Lambda} \boldsymbol{K}_x \tilde{\boldsymbol{x}} + \alpha \tilde{\boldsymbol{x}}^T \boldsymbol{K}_p \Delta \boldsymbol{x} + \boldsymbol{s}^T \boldsymbol{K}_N \tilde{\boldsymbol{I}} \\ &\le -\lambda_{\min}\{\boldsymbol{K}_d\} \|\boldsymbol{s}\|^2 - \lambda_{\min}\{\boldsymbol{K}_q\} \|\dot{\tilde{\boldsymbol{q}}}\|^2 - \alpha \lambda_{\min}\{\boldsymbol{K}_p\} \|\Delta \boldsymbol{x}\|^2 - \lambda_{\min}\{\boldsymbol{\Lambda} \boldsymbol{K}_x\} \|\tilde{\boldsymbol{x}}\|^2 \\ &\quad + \alpha \lambda_{\max}\{\boldsymbol{K}_p\} \|\tilde{\boldsymbol{x}}\| \|\Delta \boldsymbol{x}\| + \boldsymbol{s}^T \boldsymbol{K}_N \tilde{\boldsymbol{I}} \\ &\le -\lambda_{\min}\{\boldsymbol{K}_d\} \|\boldsymbol{s}\|^2 - \lambda_{\min}\{\boldsymbol{K}_q\} \|\dot{\tilde{\boldsymbol{q}}}\|^2 - [\|\Delta \boldsymbol{x}\| \ \|\tilde{\boldsymbol{x}}\|] \boldsymbol{P} [\|\Delta \boldsymbol{x}\| \ \|\tilde{\boldsymbol{x}}\|]^T + \boldsymbol{s}^T \boldsymbol{K}_N \tilde{\boldsymbol{I}},\end{aligned} \tag{6.37}$$

where $\lambda_{\min}\{\cdot\}$, $\lambda_{\max}\{\cdot\}$ denotes the minimum and maximum eigenvalue of a matrix respectively, and $\boldsymbol{P}$ is a matrix of the form

$$\boldsymbol{P} = \begin{bmatrix} \alpha \lambda_{\min}\{\boldsymbol{K}_p\} & -\frac{1}{2} \alpha \lambda_{\max}\{\boldsymbol{K}_p\} \\ -\frac{1}{2} \alpha \lambda_{\max}\{\boldsymbol{K}_p\} & \lambda_{\min}\{\boldsymbol{\Lambda} \boldsymbol{K}_x\} \end{bmatrix}. \tag{6.38}$$

The matrix $\boldsymbol{P}$ is positive definite if the following sufficient condition is satisfied:

$$4 \lambda_{\min}\{\boldsymbol{K}_p\} \lambda_{\min}\{\boldsymbol{\Lambda} \boldsymbol{K}_x\} > \alpha \lambda_{\max}^2\{\boldsymbol{K}_p\}. \tag{6.39}$$

Therefore,

$$\dot{V} \le -\lambda_{\min}\{\boldsymbol{K}_d\} \|\boldsymbol{s}\|^2 - \lambda_{\min}\{\boldsymbol{K}_q\} \|\dot{\tilde{\boldsymbol{q}}}\|^2$$
$$-\lambda_{\min}\{\boldsymbol{P}\} \|\Delta \boldsymbol{x}\|^2 - \lambda_{\min}\{\boldsymbol{P}\} \|\tilde{\boldsymbol{x}}\|^2 + \boldsymbol{s}^T \boldsymbol{K}_N \tilde{\boldsymbol{I}}. \tag{6.40}$$

Hence $\dot{V}$ is negative definite in $\boldsymbol{s}$, $\dot{\tilde{\boldsymbol{q}}}$, $\Delta \boldsymbol{x}$, and $\tilde{\boldsymbol{x}}$ when the armature current perturbation $\tilde{\boldsymbol{I}}$ is equal to $\boldsymbol{0}$.

A projection algorithm can be used for the adaptation of $\hat{\theta}_d$ to keep the estimated parameters within the range $\theta_{d\min} \le \hat{\theta}_d \le \theta_{d\max}$ (where $\theta_{d\max}$, $\theta_{d\min}$ are the upper and lower estimation bounds for $\hat{\theta}_d$) to ensure that the estimated inertia matrix $\hat{M}(q)$ is nonsingular.

We can now state the following lemma:

Lemma 6.1 *If the armature current perturbation $\tilde{I}$ equals $\mathbf{0}$ and the condition* (6.39) *is satisfied, then the desired current design* (6.5)*, the parameter adaptation laws* (6.33)–(6.36)*, together with the observers* (6.9) *and* (6.14)*, guarantee the convergence of $x - x_d \to \mathbf{0}$ and $\dot{x} - \dot{x}_d \to \mathbf{0}$ as $t \to \infty$. Moreover, $x - \hat{x}_o \to \mathbf{0}$ and $\dot{q} - \dot{\hat{q}}_o \to \mathbf{0}$ as $t \to \infty$.*

Proof From Eq. (6.28), it can be seen that V is positive definite in s, $\dot{\tilde{q}}$, Δx, $\tilde{x}$, $\Delta\theta_d$, $\Delta\theta_k$, $\hat{K}_N K_N^{-1}\theta_a - \hat{\theta}_a$, and $\theta_a - \hat{\bar{\theta}}_a$. Since $\dot{V}$ is negative definite in s, $\dot{\tilde{q}}$, Δx, and $\tilde{x}$, V is bounded as well the variables included in V. This implies that $\hat{\theta}_k$, $\hat{\theta}_d$, $\hat{\theta}_a$, $\hat{\bar{\theta}}_a$ are bounded, x and $\hat{x}_o$ are also bounded if x_d is bounded. Hence $\dot{q}_r$ in Eq. (6.8) is bounded if the approximate Jacobian matrix is nonsingular and $\dot{x}_d$ is bounded. From the boundedness of s, it can be seen from Eq. (6.15) that $\dot{q}$ is also bounded, which means that $\dot{x} = J(q)\dot{q}$ is bounded. Since $\Delta\dot{x} = \dot{x} - \dot{x}_d$, $\Delta\dot{x}$ is bounded if $\dot{x}_d$ is bounded. Then the boundedness of $\dot{\hat{q}}_o$ can be concluded, because $\dot{\tilde{q}} = \dot{q} - \dot{\hat{q}}_o$ is bounded. From Eq. (6.9), it can be seen now that $\dot{\hat{x}}_o$ is bounded, and hence $\dot{\tilde{x}} = \dot{x} - \dot{\hat{x}}_o$ is also bounded. From Eq. (6.33), $\dot{\hat{\theta}}_k$ is bounded, since $\dot{q}$, Δx, and $\tilde{x}$ are bounded. Then from Eq. (6.11), $\ddot{q}_r$ is bounded. Hence it can be seen from Eq. (6.5) that I_d is bounded, since all signals contained in it are bounded. In the absence of armature current perturbation, $\tilde{I} = \mathbf{0}$ or $I = I_d$, and hence from Eq. (6.14), we have that $\ddot{\hat{q}}_o$ is bounded. The assumption $\tilde{I} = \mathbf{0}$ will be removed in next design step, and it will be shown that the actual armature current I is still bounded and that it converges to I_d with the voltage control input. Then from Eq. (6.3), it can be seen that $\ddot{q}$ is bounded, whence $\ddot{\tilde{q}} = \ddot{q} - \ddot{\hat{q}}_o$ and $\dot{s} = \ddot{q} - \ddot{q}_r$ are also bounded. It has now been shown that $\dot{s}$, $\ddot{\tilde{q}}$, $\Delta\dot{x}$, and $\dot{\tilde{x}}$ are all bounded, which means that s, $\dot{\tilde{q}}$, Δx, and $\tilde{x}$ are uniformly continuous. Since V is bounded and $\dot{V}$ is negative definite in s, $\dot{\tilde{q}}$, Δx, and $\tilde{x}$ when the armature current perturbation $\tilde{I}$ is equal to $\mathbf{0}$, it is obvious that s, $\dot{\tilde{q}}$, Δx, $\tilde{x} \in L_2(0, \infty)$. Then we have s, $\dot{\tilde{q}}$, Δx, $\tilde{x} \to \mathbf{0}$ as $t \in \infty$. Since $\ddot{q}$ is bounded, $\ddot{x} = \dot{J}(q)\dot{q} + J(q)\ddot{q}$ is also bounded, and hence $\Delta\ddot{x} = \ddot{x} - \ddot{x}_d$ is bounded if $\ddot{x}_d$ is bounded. The boundedness of $\Delta\ddot{x}$ means that $\Delta\dot{x}$ is uniformly continuous; also, Δx has been shown to be bounded. Then according to Barbalat's lemma, $\Delta\dot{x} \to \mathbf{0}$ as $t \to \infty$, which completes the proof. △△△

6.1.2 Input-Control Voltage Design

In the previous subsection, it has been shown that the desired armature current I_d results in the convergence of the tracking errors if the armature current perturbation $\tilde{I}$

is zero. In this subsection, the control-voltage input $\boldsymbol{u}$ is designed so that the armature current perturbation $\tilde{\boldsymbol{I}}$ converges to zero.

Since the actuator dynamic coefficient matrices $\boldsymbol{L}$, $\boldsymbol{R}$, $\boldsymbol{K}_E$, and $\boldsymbol{K}_N$ are unknown, the adaptive approximate models $\hat{\boldsymbol{L}}$, $\hat{\boldsymbol{R}}$, $\hat{\boldsymbol{K}}_E$, and $\hat{\bar{\boldsymbol{K}}}_N$ are employed, and the input voltage $\boldsymbol{u}$ is designated as

$$\boldsymbol{u} = \hat{\boldsymbol{L}}\dot{\boldsymbol{I}}_d + \hat{\boldsymbol{R}}\boldsymbol{I}_d + \hat{\boldsymbol{K}}_E\dot{\boldsymbol{q}} - \boldsymbol{N}_N(\boldsymbol{s})\hat{\boldsymbol{\theta}}_{\bar{K}_N} - \boldsymbol{K}_I\tilde{\boldsymbol{I}}, \tag{6.41}$$

where we note that $\dot{\boldsymbol{I}}_d$ is used in the input voltage.

Next, the adaptation laws are design as follows:

$$\dot{\hat{\boldsymbol{\theta}}}_L = -\boldsymbol{L}_L\boldsymbol{N}_L(\dot{\boldsymbol{I}}_d)\tilde{\boldsymbol{I}}, \tag{6.42}$$
$$\dot{\hat{\boldsymbol{\theta}}}_R = -\boldsymbol{L}_R\boldsymbol{N}_R(\boldsymbol{I}_d)\tilde{\boldsymbol{I}}, \tag{6.43}$$
$$\dot{\hat{\boldsymbol{\theta}}}_{K_E} = -\boldsymbol{L}_E\boldsymbol{N}_E(\dot{\boldsymbol{q}})\tilde{\boldsymbol{I}}, \tag{6.44}$$
$$\dot{\hat{\boldsymbol{\theta}}}_{\bar{K}_N} = \boldsymbol{L}_N\boldsymbol{N}_N(\boldsymbol{s})\tilde{\boldsymbol{I}}, \tag{6.45}$$

where $\boldsymbol{L}_L, \boldsymbol{L}_R, \boldsymbol{L}_E, \boldsymbol{L}_N, \boldsymbol{K}_I$ are positive definite diagonal matrices; $\tilde{\boldsymbol{I}} = \boldsymbol{I} - \boldsymbol{I}_d$, and $\boldsymbol{I}_d$ is defined in (6.5); $\boldsymbol{N}_L(\dot{\boldsymbol{I}}_d) = \text{diag}(\dot{I}_{d1}, \ldots, \dot{I}_{dn})$, $\boldsymbol{N}_R(\boldsymbol{I}_d) = \text{diag}(I_{d1}, \ldots, I_{dn})$, $\boldsymbol{N}_E(\dot{\boldsymbol{q}}) = \text{diag}(\dot{q}_1, \ldots, \dot{q}_n)$, $\boldsymbol{N}_N(\boldsymbol{s}) = \text{diag}(s_1, \ldots, s_n)$, and hence $\hat{\boldsymbol{L}}\dot{\boldsymbol{I}}_d = \boldsymbol{N}_L(\dot{\boldsymbol{I}}_d)\hat{\boldsymbol{\theta}}_L$, $\hat{\boldsymbol{R}}\boldsymbol{I}_d = \boldsymbol{N}_R(\boldsymbol{I}_d)\hat{\boldsymbol{\theta}}_R$, $\hat{\boldsymbol{K}}_E\dot{\boldsymbol{q}} = \boldsymbol{N}_E(\dot{\boldsymbol{q}})\hat{\boldsymbol{\theta}}_{K_E}$, $\hat{\bar{\boldsymbol{K}}}_N\boldsymbol{Y} = \boldsymbol{N}_N(\boldsymbol{Y})\hat{\boldsymbol{\theta}}_{\bar{K}_N}$, where $\hat{\boldsymbol{\theta}}_L, \hat{\boldsymbol{\theta}}_R, \hat{\boldsymbol{\theta}}_{K_E}, \hat{\boldsymbol{\theta}}_{\bar{K}_N}$ are vectors of the diagonal elements in $\hat{\boldsymbol{L}}$, $\hat{\boldsymbol{R}}$, $\hat{\boldsymbol{K}}_E$, and $\hat{\bar{\boldsymbol{K}}}_N$ respectively. The adaptive approximate models $\hat{\boldsymbol{L}}$, $\hat{\boldsymbol{R}}$, $\hat{\boldsymbol{K}}_E$, and $\hat{\bar{\boldsymbol{K}}}_N$ are updated online by the updating laws (6.42)–(6.45) during the control process. Note that $\hat{\bar{\boldsymbol{K}}}_N$ is updated online by the updating law (6.45) during the control process; in comparison, note that $\hat{\boldsymbol{K}}_N$ used in the desired armature current design $\boldsymbol{I}_d$ is a constant matrix.

Substituting the input voltage (6.41) into the actuator dynamics (6.2) yields

$$\begin{aligned}\boldsymbol{L}(\dot{\boldsymbol{I}} - \dot{\boldsymbol{I}}_d) + \boldsymbol{R}(\boldsymbol{I} - \boldsymbol{I}_d) &= (\hat{\boldsymbol{L}} - \boldsymbol{L})\dot{\boldsymbol{I}}_d + (\hat{\boldsymbol{R}} - \boldsymbol{R})\boldsymbol{I}_d \\ &+(\hat{\boldsymbol{K}}_E - \boldsymbol{K}_E)\dot{\boldsymbol{q}} - \boldsymbol{N}_N(\boldsymbol{s})\hat{\boldsymbol{\theta}}_{\bar{K}_N} - \boldsymbol{K}_I\tilde{\boldsymbol{I}}.\end{aligned} \tag{6.46}$$

The above equation can be rewritten as follows:

$$\begin{aligned}\boldsymbol{L}\dot{\tilde{\boldsymbol{I}}} + (\boldsymbol{R} + \boldsymbol{K}_I)\tilde{\boldsymbol{I}} &= \boldsymbol{N}_L(\dot{\boldsymbol{I}}_d)(\hat{\boldsymbol{\theta}}_L - \boldsymbol{\theta}_L) + \boldsymbol{N}_R(\boldsymbol{I}_d)(\hat{\boldsymbol{\theta}}_R - \boldsymbol{\theta}_R) \\ &+\boldsymbol{N}_E(\dot{\boldsymbol{q}})(\hat{\boldsymbol{\theta}}_{K_E} - \boldsymbol{\theta}_{K_E}) - \boldsymbol{N}_N(\boldsymbol{s})\hat{\boldsymbol{\theta}}_{\bar{K}_N},\end{aligned} \tag{6.47}$$

where $\boldsymbol{\theta}_L$, $\boldsymbol{\theta}_R$, $\boldsymbol{\theta}_{K_E}$, and $\boldsymbol{\theta}_{K_N}$ denote vectors of the unknown parameters in $\boldsymbol{L}$, $\boldsymbol{R}$, $\boldsymbol{K}_E$, $\boldsymbol{K}_N$ respectively such that $\boldsymbol{L}\dot{\boldsymbol{I}}_d = \boldsymbol{N}_L(\dot{\boldsymbol{I}}_d)\boldsymbol{\theta}_L$, $\boldsymbol{R}\boldsymbol{I}_d = \boldsymbol{N}_R(\boldsymbol{I}_d)\boldsymbol{\theta}_R$, $\boldsymbol{K}_E\dot{\boldsymbol{q}} = \boldsymbol{N}_E(\dot{\boldsymbol{q}})\boldsymbol{\theta}_{K_E}$, $\boldsymbol{K}_N\boldsymbol{s} = \boldsymbol{N}_N(\boldsymbol{s})\boldsymbol{\theta}_{K_N}$.

Now consider another Lyapunov function candidate V' of the following form:

$$V' = V + \frac{1}{2}\tilde{I}^T L\tilde{I} + \frac{1}{2}(\theta_L - \hat{\theta}_L)^T L_L^{-1}(\theta_L - \hat{\theta}_L) + \frac{1}{2}(\theta_R - \hat{\theta}_R)^T L_R^{-1}(\theta_R - \hat{\theta}_R)$$
$$+\frac{1}{2}(\theta_{K_E} - \hat{\theta}_{K_E})^T L_E^{-1}(\theta_{K_E} - \hat{\theta}_{K_E}) + \frac{1}{2}(\theta_{K_N} - \hat{\theta}_{\bar{K}_N})^T L_N^{-1}(\theta_{K_N} - \hat{\theta}_{\bar{K}_N}), \tag{6.48}$$

where V is defined as in Eq. (6.28). Note that V' is positive definite in all signals that it contains.

Differentiating V' with respect to time and using Eqs. (6.42)–(6.44) and (6.47), we have

$$\dot{V}' = \dot{V} - \tilde{I}^T(R + K_I)\tilde{I} - \tilde{I}^T N_N(s)\hat{\theta}_{\bar{K}_N} - \dot{\hat{\theta}}_{\bar{K}_N}^T L_N^{-1}(\theta_{K_N} - \hat{\theta}_{\bar{K}_N}). \tag{6.49}$$

Substituting Eqs. (6.37) and (6.45) into equation (6.49) and using inequality (6.40), we obtain

$$\dot{V}' \leq -\lambda_{\min}\{K_d\}\|s\|^2 - \lambda_{\min}\{K_q\}\|\dot{\tilde{q}}\|^2 - \lambda_{\min}\{P\}\|\Delta x\|^2 - \lambda_{\min}\{P\}\|\tilde{x}\|^2$$
$$- \lambda_{\min}\{K_I\}\|\tilde{I}\|^2 \leq 0. \tag{6.50}$$

Note that the term $s^T K_N \tilde{I}$, which is contained in $\dot{V}$, has been canceled out by the adaptive parameter vector $\hat{\theta}_{\bar{K}_N}$ and its adaptation law (6.45). Hence, if the sufficient stability condition (6.39) is satisfied, $\dot{V}'$ is negative definite in s, $\dot{\tilde{q}}$, Δx, $\tilde{x}$, and $\tilde{I}$.

The following theorem can now be stated:

Theorem 6.1 *If the control input voltage* u *given by* (6.5), (6.9), (6.14), (6.33)–(6.35), *and* (6.41)–(6.45) *is applied to the rigid-link robot dynamics* (6.1), (6.2), *then the closed-loop system gives rise to the convergence of* $x \to x_d$, $\dot{x} \to \dot{x}_d$, *and* $I \to I_d$ *as* $t \to \infty$ *if the feedback gains are chosen as in condition* (6.39).

Proof Since V' is nonnegative and $\dot{V}'$ is negative definite in s, $\dot{\tilde{q}}$, Δx, $\tilde{x}$, and $\tilde{I}$, V' is bounded, which means that the variables $\hat{\theta}_k, \hat{\theta}_d, \hat{\theta}_a, \hat{\bar{\theta}}_a, \hat{\theta}_L, \hat{\theta}_R, \hat{\theta}_{K_E}, \hat{\theta}_{\bar{K}_N}, s, \tilde{x}, \Delta x$, $\dot{\tilde{q}}$ are bounded. It is noticeable that the current perturbation $\tilde{I}$ is also bounded, since $\tilde{I}$ is included in V'. As in the proof of Lemma 6.1, it is easy to verify that variables x, $\hat{x}_o$, $\dot{q}_r$, $\dot{q}$, $\dot{\hat{q}}_o$, $\dot{\hat{x}}_o$, and $\Delta\dot{x}$, $\dot{\tilde{x}}$ are all bounded, and hence $\ddot{q}_r$ is bounded from Eq. (6.11). From Eq. (6.5), it can be seen that I_d is bounded. Since $\tilde{I}$ is also bounded, the armature current I is also bounded. Then from equations (6.1) and (6.14) it is easily verified that $\ddot{q}$ and $\ddot{\hat{q}}_o$ are bounded. From the boundedness of $\ddot{q}$, $\ddot{q}_r$, and $\ddot{\hat{q}}_o$, we may conclude that $\ddot{\tilde{q}}$ and $\dot{s}$ are bounded. Now we have shown the boundedness of $\dot{s}$, $\ddot{\tilde{q}}$, $\Delta\dot{x}$, $\dot{\tilde{x}}$, and $\dot{\tilde{I}}$. Since V' is bounded and $\dot{V}'$ is negative definite in s, $\dot{\tilde{q}}$, Δx, $\tilde{x}$, and $\tilde{I}$, it is clear that $s, \dot{\tilde{q}}, \Delta x, \tilde{x}, \tilde{I} \in L_2(0, \infty)$. Then we have that $s, \dot{\tilde{q}}, \Delta x, \tilde{x}, \tilde{I} \to 0$ as $t \in \infty$. Since $\ddot{q}$ is bounded, $\ddot{x} = \dot{J}(q)\dot{q} + J(q)\ddot{q}$ is also bounded, and hence $\Delta\ddot{x} = \ddot{x} - \ddot{x}_d$ is bounded if $\ddot{x}_d$ is bounded. The boundedness of $\Delta\ddot{x}$ means that $\Delta\dot{x}$ is uniformly continuous; also, Δx has been shown to be bounded. Then according to Babalat's lemma, $\Delta\dot{x} \to 0$ as $t \to \infty$. Hence the proof is complete. △△△

6.2 Rigid-Link Flexible-Joint Robots

This section considers the sensory task-space control problem of a rigid-link robot manipulator with flexible joints. The dynamics of rigid-link flexible-joint (RLFJ) robots is described by the following equations:

$$\boldsymbol{M}(\boldsymbol{q})\ddot{\boldsymbol{q}} + \left(\tfrac{1}{2}\dot{\boldsymbol{M}}(\boldsymbol{q}) + \boldsymbol{S}(\boldsymbol{q},\dot{\boldsymbol{q}})\right)\dot{\boldsymbol{q}} + \boldsymbol{g}(\boldsymbol{q}) = \boldsymbol{K}_m(\boldsymbol{q}_m - \boldsymbol{q}), \tag{6.51}$$

$$\boldsymbol{J}_m\ddot{\boldsymbol{q}}_m + \boldsymbol{B}_m\dot{\boldsymbol{q}}_m + \boldsymbol{K}_m(\boldsymbol{q}_m - \boldsymbol{q}) = \boldsymbol{u}, \tag{6.52}$$

where $\boldsymbol{q} \in \Re^n$ denotes the vector of robot link angles, $\boldsymbol{q}_m \in \Re^n$ denotes the vector of motor rotor shaft positions, $\boldsymbol{M}(\boldsymbol{q}) \in \Re^{n\times n}$ and $\boldsymbol{J}_m \in \Re^{n\times n}$ denote positive definite link and actuator inertia matrices respectively, $\boldsymbol{S}(\boldsymbol{q},\dot{\boldsymbol{q}}) \in \Re^{n\times n}$ is a skew-symmetric matrix, $\boldsymbol{g}(\boldsymbol{q}) \in \Re^n$ denotes the gravitational force, $\boldsymbol{B}_m \in \Re^{n\times n}$ is the positive definite damping matrix, $\boldsymbol{K}_m \in \Re^{n\times n}$ denotes the joint stiffness constant matrix, which is diagonal and positive definite, $\boldsymbol{u} \in \Re^n$ is the vector of actuator control torque inputs.

The overall system is of fourth order, and the backstepping method is used to design the control input for the RLFJ robot. For simplicity, we consider only the point-to-point control or regulation problem of the RLFJ robot. Based on the backstepping method, a desired actuator position trajectory denoted by $\boldsymbol{q}_{md}$ is first formulated based on the link subsystem (6.51), which leads to the convergence of the task-space position to the desired task-space position $\boldsymbol{x}_d$. Then a control torque input $\boldsymbol{u}$ is designed based on the joint subsystem (6.52), so that $\boldsymbol{q}_m$ tracks the desired actuator position $\boldsymbol{q}_{md}$.

6.2.1 Desired Actuator Position Trajectory Design

Using the desired actuator position trajectory $\boldsymbol{q}_{md}$, the link subsystem dynamics (6.51) can be rewritten as

$$\boldsymbol{M}(\boldsymbol{q})\ddot{\boldsymbol{q}} + \left(\tfrac{1}{2}\dot{\boldsymbol{M}}(\boldsymbol{q}) + \boldsymbol{S}(\boldsymbol{q},\dot{\boldsymbol{q}})\right)\dot{\boldsymbol{q}} + \boldsymbol{g}(\boldsymbol{q}) = \boldsymbol{K}_m(\boldsymbol{q}_{md} - \boldsymbol{q}) + \boldsymbol{K}_m\tilde{\boldsymbol{q}}_m, \tag{6.53}$$

where $\tilde{\boldsymbol{q}}_m = \boldsymbol{q}_m - \boldsymbol{q}_{md}$. The above subsystem (6.53) can be deemed as being controlled by $\boldsymbol{K}_m(\boldsymbol{q}_{md} - \boldsymbol{q})$ with an actuator rotor perturbation $\tilde{\boldsymbol{q}}_m$, which will be dealt with by the control torque input $\boldsymbol{u}$ in next design step.

To simplify the system expression, $\boldsymbol{q}_{md}$ is defined with an auxiliary variable $\boldsymbol{q}'_{md}$ as

$$\boldsymbol{q}_{md} = \boldsymbol{q}'_{md} + \boldsymbol{q}. \tag{6.54}$$

Then Eq. (6.53) can be written as

$$\boldsymbol{M}(\boldsymbol{q})\ddot{\boldsymbol{q}} + \left(\tfrac{1}{2}\dot{\boldsymbol{M}}(\boldsymbol{q}) + \boldsymbol{S}(\boldsymbol{q},\dot{\boldsymbol{q}})\right)\dot{\boldsymbol{q}} + \boldsymbol{g}(\boldsymbol{q}) = \boldsymbol{K}_m\boldsymbol{q}'_{md} + \boldsymbol{K}_m\tilde{\boldsymbol{q}}_m. \tag{6.55}$$

In the presence of uncertainty in the robot's kinematics and dynamics, the exact models of the Jacobian matrix $\boldsymbol{J}(\boldsymbol{q})$, gravitational force $\boldsymbol{g}(\boldsymbol{q})$, and joint stiffness matrix $\boldsymbol{K}_m$ are unavailable.

The approximate Jacobian matrix $\hat{\boldsymbol{J}}(\boldsymbol{q}, \hat{\boldsymbol{\theta}}_k)$ is defined according to *Property* 1.5 as

$$\dot{\hat{\boldsymbol{x}}} = \hat{\boldsymbol{J}}(\boldsymbol{q}, \hat{\boldsymbol{\theta}}_k)\dot{\boldsymbol{q}} = \boldsymbol{Y}_k(\boldsymbol{q}, \dot{\boldsymbol{q}})\hat{\boldsymbol{\theta}}_k, \tag{6.56}$$

where $\dot{\hat{\boldsymbol{x}}}$ denotes the estimated task-space velocity. The approximate gravitational force $\hat{\boldsymbol{g}}(\boldsymbol{q})$ is defined according to *Property* 1.4 as

$$\hat{\boldsymbol{g}}(\boldsymbol{q}) = \boldsymbol{Y}_g(\boldsymbol{q})\hat{\boldsymbol{\theta}}_g. \tag{6.57}$$

The adaptive dynamic and kinematic parameter vectors $\hat{\boldsymbol{\theta}}_k$ and $\hat{\boldsymbol{\theta}}_g$ are updated through adaptation laws given by (6.62) and (6.60) in the following, and $\hat{\boldsymbol{K}}_m$ is the approximate joint stiffness matrix, which is fixed and not updated by any adaptation law.

The fictitious signal $\boldsymbol{q}'_{md}$ is designed as follows using the approximate joint stiffness matrix, gravitational force, and adaptive Jacobian matrix:

$$\boldsymbol{q}'_{md} = -\hat{\boldsymbol{K}}_m^{-1}(\hat{\boldsymbol{J}}^T(\boldsymbol{q}, \hat{\boldsymbol{\theta}}_k)\boldsymbol{K}_p\boldsymbol{s}(\Delta\boldsymbol{x}) + \boldsymbol{K}_v\boldsymbol{v} - \boldsymbol{Y}_g(\boldsymbol{q})\hat{\boldsymbol{\theta}}_g - \boldsymbol{Y}(\boldsymbol{q}, \Delta\boldsymbol{x}, \hat{\boldsymbol{\theta}}_g, \hat{\boldsymbol{\theta}}_k)\hat{\varphi}), \tag{6.58}$$

$$\dot{\boldsymbol{v}} = -\boldsymbol{\Lambda}_1\boldsymbol{v} + \dot{\boldsymbol{q}} + \alpha\hat{\boldsymbol{J}}^T(\boldsymbol{q}, \hat{\boldsymbol{\theta}}_k)\boldsymbol{s}(\Delta\boldsymbol{x}), \tag{6.59}$$

$$\dot{\hat{\boldsymbol{\theta}}}_g = -\boldsymbol{L}_g\boldsymbol{Y}_g^T(\boldsymbol{q})(\dot{\boldsymbol{q}} + \alpha\hat{\boldsymbol{J}}^T(\boldsymbol{q}, \hat{\boldsymbol{\theta}}_k)\boldsymbol{s}(\Delta\boldsymbol{x})), \tag{6.60}$$

$$\dot{\hat{\varphi}} = -\boldsymbol{L}_a\boldsymbol{Y}(\boldsymbol{q}, \Delta\boldsymbol{x}, \hat{\phi}, \hat{\boldsymbol{\theta}}_k)(\dot{\boldsymbol{q}} + \alpha\hat{\boldsymbol{J}}^T(\boldsymbol{q}, \hat{\boldsymbol{\theta}}_k)\boldsymbol{s}(\Delta\boldsymbol{x})), \tag{6.61}$$

where $\boldsymbol{s}(\cdot)$ denotes the saturation function (see Sect. 2.3.2), $\boldsymbol{K}_p = k_p\boldsymbol{I}_m \in \Re^{m\times m}$, where $\boldsymbol{I}_m$ is the identity matrix, and $\boldsymbol{K}_v \in \Re^{n\times n}$ are positive definite diagonal feedback gains, $\boldsymbol{\Lambda}_1$, $\boldsymbol{L}_a \in \Re^{n\times n}$, $\boldsymbol{L}_g \in \Re^{n_g\times n_g}$ are positive definite diagonal adaptation gain matrices, and $\alpha \in \Re$ is a positive constant, $\boldsymbol{v}$, $\hat{\varphi} \in \Re^n$, $\hat{\boldsymbol{\theta}}_g \in \Re^{n_g}$ are updated through adaptation laws (6.59), (6.61) and (6.60) respectively. The adaptive kinematic parameter vector $\hat{\boldsymbol{\theta}}_k$ used in $\hat{\boldsymbol{J}}(\boldsymbol{q}, \hat{\boldsymbol{\theta}}_k)$ is updated through the following projection adaptation law:

$$\dot{\hat{\boldsymbol{\theta}}}_k = \boldsymbol{\Phi}\boldsymbol{L}_k\boldsymbol{Y}_k^T(\boldsymbol{q}, \dot{\boldsymbol{q}})\boldsymbol{s}(\Delta\boldsymbol{x}), \tag{6.62}$$

where $\boldsymbol{L}_k \in \Re^{n_k\times n_k}$ is a positive definite diagonal matrix, $\boldsymbol{\Phi} \in \Re^{n_k\times n_k}$ is the diagonal projection matrix defined as

$$\Phi_i = \begin{cases} 0, & \text{if } \hat{\theta}_{ki} \geq \theta_{\max i}, \text{ and } \dot{\hat{\theta}}_{ki} \geq 0, \\ 0, & \text{if } \hat{\theta}_{ki} \leq \theta_{\min i}, \text{ and } \dot{\hat{\theta}}_{ki} \leq 0, \\ 1, & \text{otherwise}, \end{cases} \tag{6.63}$$

where Φ_i is the ith diagonal element of $\boldsymbol{\Phi}$, and $\boldsymbol{\theta}_{\min}$ and $\boldsymbol{\theta}_{\max}$ denote the estimated upper and lower bounds of the unknown $\boldsymbol{\theta}_k$. The purpose of the projection law is to guarantee the boundedness of the adaptive kinematic parameter vector $\hat{\boldsymbol{\theta}}_k$ within a specified range $[\boldsymbol{\theta}_{\min}, \boldsymbol{\theta}_{\max}]$.

The regressor $\boldsymbol{Y}(\boldsymbol{q}, \Delta\boldsymbol{x}, \hat{\boldsymbol{\theta}}_g, \hat{\boldsymbol{\theta}}_k)$ is defined as $\boldsymbol{Y}(\boldsymbol{q}, \Delta\boldsymbol{x}, \hat{\boldsymbol{\theta}}_g, \hat{\boldsymbol{\theta}}_k) = \mathrm{diag}\{\boldsymbol{Y}_g(\boldsymbol{q})\hat{\boldsymbol{\theta}}_g - \hat{\boldsymbol{J}}^T(\boldsymbol{q}, \hat{\boldsymbol{\theta}}_k)\boldsymbol{K}_p\boldsymbol{s}(\Delta\boldsymbol{x})\}$ such that

$$(\boldsymbol{I}_n - \boldsymbol{K}_m\hat{\boldsymbol{K}}_m^{-1})(\boldsymbol{Y}_g(\boldsymbol{q})\hat{\boldsymbol{\theta}}_g - \hat{\boldsymbol{J}}^T(\boldsymbol{q}, \hat{\boldsymbol{\theta}}_k)\boldsymbol{K}_p\boldsymbol{s}(\Delta\boldsymbol{x})) = \boldsymbol{Y}(\boldsymbol{q}, \Delta\boldsymbol{x}, \hat{\boldsymbol{\theta}}_g, \hat{\boldsymbol{\theta}}_k)\phi_k, \quad (6.64)$$

where $\phi_k = (1 - \frac{k_m^1}{\hat{k}_m^1}, \ldots, 1 - \frac{k_m^n}{\hat{k}_m^n})^T$ is an unknown parameter vector, with k_m^i and $\hat{k}_m^i$ the ith diagonal elements of $\boldsymbol{K}_m$ and $\hat{\boldsymbol{K}}_m$ respectively.

Substituting $\boldsymbol{q}'_{md}$ defined by (6.58) into (6.55) and using Eq. (6.64), the closed-loop equation of the link subsystem is obtained as

$$\begin{aligned}&\boldsymbol{M}(\boldsymbol{q})\ddot{\boldsymbol{q}} + \left(\tfrac{1}{2}\dot{\boldsymbol{M}}(\boldsymbol{q}) + \boldsymbol{S}(\boldsymbol{q}, \dot{\boldsymbol{q}})\right)\dot{\boldsymbol{q}} + \boldsymbol{K}_m\hat{\boldsymbol{K}}_m^{-1}\boldsymbol{K}_v\boldsymbol{v} + k_p\hat{\boldsymbol{J}}^T(\boldsymbol{q}, \hat{\boldsymbol{\theta}}_k)\boldsymbol{s}(\Delta\boldsymbol{x})\\ &+\boldsymbol{Y}_g(\boldsymbol{q})(\boldsymbol{\theta}_g - \hat{\boldsymbol{\theta}}_g) + \boldsymbol{Y}(\boldsymbol{q}, \Delta\boldsymbol{x}, \hat{\boldsymbol{\theta}}_g, \hat{\boldsymbol{\theta}}_k)(\phi_k - \boldsymbol{K}_m\hat{\boldsymbol{K}}_m^{-1}\hat{\varphi}) = \boldsymbol{K}_m\tilde{\boldsymbol{q}}_m, \quad (6.65)\end{aligned}$$

where we note that $\boldsymbol{g}(\boldsymbol{q}) = \boldsymbol{Y}_g(\boldsymbol{q})\boldsymbol{\theta}_g$ has been used and $\boldsymbol{Y}(\boldsymbol{q}, \Delta\boldsymbol{x}, \hat{\boldsymbol{\theta}}_g, \hat{\boldsymbol{\theta}}_k)$, $\boldsymbol{K}_m$, $\hat{\boldsymbol{K}}_m^{-1}$ are diagonal matrices.

To analyze the stability of the closed-loop link subsystem with the desired actuator position trajectory $\boldsymbol{q}_{md} = \boldsymbol{q}'_{md} + \boldsymbol{q}$, a Lyapunov function candidate V_1 is chosen as

$$\begin{aligned}V_1 &= \tfrac{1}{2}\dot{\boldsymbol{q}}^T\boldsymbol{M}(\boldsymbol{q})\dot{\boldsymbol{q}} + \alpha\dot{\boldsymbol{q}}^T\boldsymbol{M}(\boldsymbol{q})\hat{\boldsymbol{J}}^T(\boldsymbol{q}, \hat{\boldsymbol{\theta}}_k)\boldsymbol{s}(\Delta\boldsymbol{x}) + \sum_{i=1}^{m} k_p S_i(\Delta x_i)\\ &+\tfrac{1}{2}\boldsymbol{v}^T\boldsymbol{K}_m\hat{\boldsymbol{K}}_m^{-1}\boldsymbol{K}_v\boldsymbol{v} + \tfrac{1}{2}(\boldsymbol{\theta}_g - \hat{\boldsymbol{\theta}}_g)^T\boldsymbol{L}_g^{-1}(\boldsymbol{\theta}_g - \hat{\boldsymbol{\theta}}_g) + \tfrac{1}{2}(\boldsymbol{\theta}_k - \hat{\boldsymbol{\theta}}_k)^T\boldsymbol{L}_k^{-1}k_p(\boldsymbol{\theta}_k - \hat{\boldsymbol{\theta}}_k)\\ &+\tfrac{1}{2}(\hat{\boldsymbol{K}}_m\boldsymbol{K}_m^{-1}\phi_k - \hat{\varphi})^T\boldsymbol{K}_m\hat{\boldsymbol{K}}_m^{-1}\boldsymbol{L}_a^{-1}(\hat{\boldsymbol{K}}_m\boldsymbol{K}_m^{-1}\phi_k - \hat{\varphi}). \quad (6.66)\end{aligned}$$

Using inequality $S_i(\Delta x_i) \geq \hat{c}_i s_i{}^2(\Delta x_i)$, we have

$$\begin{aligned}&\tfrac{1}{4}\dot{\boldsymbol{q}}^T\boldsymbol{M}(\boldsymbol{q})\dot{\boldsymbol{q}} + \alpha\dot{\boldsymbol{q}}^T\boldsymbol{M}(\boldsymbol{q})\hat{\boldsymbol{J}}^T(\boldsymbol{q}, \hat{\boldsymbol{\theta}}_J)\boldsymbol{s}(\Delta\boldsymbol{x}) + \sum_{i=1}^{m} k_p S_i(\Delta x_i)\\ &= \tfrac{1}{4}(\dot{\boldsymbol{q}} + 2\alpha\hat{\boldsymbol{J}}^T(\boldsymbol{q}, \hat{\boldsymbol{\theta}}_J)\boldsymbol{s}(\Delta\boldsymbol{x}))^T\boldsymbol{M}(\boldsymbol{q})(\dot{\boldsymbol{q}} + 2\alpha\hat{\boldsymbol{J}}^T(\boldsymbol{q}, \hat{\boldsymbol{\theta}}_J)\boldsymbol{s}(\Delta\boldsymbol{x}))\\ &-\alpha^2\boldsymbol{s}^T(\Delta\boldsymbol{x})\hat{\boldsymbol{J}}(\boldsymbol{q}, \hat{\boldsymbol{\theta}}_J)\boldsymbol{M}(\boldsymbol{q})\hat{\boldsymbol{J}}^T(\boldsymbol{q}, \hat{\boldsymbol{\theta}}_J)\boldsymbol{s}(\Delta\boldsymbol{x}) + \sum_{i=1}^{m} k_p S_i(\Delta x_i)\\ &\geq \sum_{i=1}^{m}(k_p\hat{c}_i - \alpha^2\lambda_{JM})s_i^2(\Delta x_i), \quad (6.67)\end{aligned}$$

where $\hat{c}_i$ is a positive constant such that $S_i(\Delta x_i) \geq \hat{c}_i {s_i}^2(\Delta x_i)$, $\lambda_{JM} \triangleq \lambda_{\max}[\hat{\boldsymbol{J}}(\boldsymbol{q}, \hat{\boldsymbol{\theta}}_k)\boldsymbol{M}(\boldsymbol{q})\hat{\boldsymbol{J}}^T(\boldsymbol{q}, \hat{\boldsymbol{\theta}}_k)]$, and $\lambda_{\max}[\cdot]$ denotes the maximum eigenvalue of a matrix. It is easy to see that α can be chosen small enough or k_p can be chosen large enough that the following inequality holds:

$$k_p\hat{c}_i - \alpha^2\lambda_{JM} > 0. \tag{6.68}$$

If condition (6.68) is satisfied, then on substituting inequalities (6.67) and (6.68) into Eq. (6.66), we obtain

$$\begin{aligned} V_1 \geq{}& \tfrac{1}{4}\dot{\boldsymbol{q}}^T\boldsymbol{M}(\boldsymbol{q})\dot{\boldsymbol{q}} + \sum_{i=1}^{m}(k_p\hat{c}_i - \alpha^2\lambda_{JM})s_i^2(\Delta x_i) + \tfrac{1}{2}\boldsymbol{v}^T\boldsymbol{K}_m\hat{\boldsymbol{K}}_m^{-1}\boldsymbol{K}_v\boldsymbol{v} \\ &+\tfrac{1}{2}(\boldsymbol{\theta}_g - \hat{\boldsymbol{\theta}}_g)^T\boldsymbol{L}_g^{-1}(\boldsymbol{\theta}_g - \hat{\boldsymbol{\theta}}_g) + \tfrac{1}{2}(\boldsymbol{\theta}_k - \hat{\boldsymbol{\theta}}_k)^T\boldsymbol{L}_k^{-1}k_p(\boldsymbol{\theta}_k - \hat{\boldsymbol{\theta}}_k) \\ &+\tfrac{1}{2}(\hat{\boldsymbol{K}}_m\boldsymbol{K}_m^{-1}\boldsymbol{\phi}_k - \hat{\boldsymbol{\varphi}})^T\boldsymbol{L}_a^{-1}\boldsymbol{K}_m\hat{\boldsymbol{K}}_m^{-1}(\hat{\boldsymbol{K}}_m\boldsymbol{K}_m^{-1}\boldsymbol{\phi}_k - \hat{\boldsymbol{\varphi}}) > 0. \end{aligned} \tag{6.69}$$

Hence V_1 is positive definite if condition (6.68) is met. Differentiating V_1 with respect to time and substituting Eqs. (6.56), (6.59)–(6.62), (6.65) into it, we have

$$\begin{aligned} \dot{V}_1 ={}& -\boldsymbol{v}^T\boldsymbol{K}_m\hat{\boldsymbol{K}}_m^{-1}\boldsymbol{K}_v\boldsymbol{\Lambda}_1\boldsymbol{v} - \alpha k_p\boldsymbol{s}^T(\Delta\boldsymbol{x})\hat{\boldsymbol{J}}(\boldsymbol{q}, \hat{\boldsymbol{\theta}}_k)\hat{\boldsymbol{J}}^T(\boldsymbol{q}, \hat{\boldsymbol{\theta}}_k)\boldsymbol{s}(\Delta\boldsymbol{x}) \\ &+k_p[\boldsymbol{s}^T(\Delta\boldsymbol{x})\boldsymbol{Y}_k(\boldsymbol{q}, \dot{\boldsymbol{q}})(\boldsymbol{\theta}_k - \hat{\boldsymbol{\theta}}_k) - \boldsymbol{s}^T(\Delta\boldsymbol{x})\boldsymbol{Y}_k(\boldsymbol{q}, \dot{\boldsymbol{q}})\boldsymbol{\Phi}(\boldsymbol{\theta}_k - \hat{\boldsymbol{\theta}}_k)] \\ &-\alpha\{\boldsymbol{s}^T(\Delta\boldsymbol{x})\hat{\boldsymbol{J}}(\boldsymbol{q}, \hat{\boldsymbol{\theta}}_k)[-\dot{\boldsymbol{M}}(\boldsymbol{q}) + \tfrac{1}{2}\dot{\boldsymbol{M}}(\boldsymbol{q}) + \boldsymbol{S}(\boldsymbol{q}, \dot{\boldsymbol{q}})]\dot{\boldsymbol{q}} \\ &-\dot{\boldsymbol{s}}^T(\Delta\boldsymbol{x})\hat{\boldsymbol{J}}(\boldsymbol{q}, \hat{\boldsymbol{\theta}}_k)\boldsymbol{M}(\boldsymbol{q})\dot{\boldsymbol{q}} - \boldsymbol{s}^T(\Delta\boldsymbol{x})\dot{\hat{\boldsymbol{J}}}(\boldsymbol{q}, \hat{\boldsymbol{\theta}}_k)\boldsymbol{M}(\boldsymbol{q})\dot{\boldsymbol{q}}\}. \end{aligned} \tag{6.70}$$

The position perturbation $\tilde{\boldsymbol{q}}_m$ will be considered in the analysis of overall closed-loop systems in next step, and it will be shown that the stability of the overall system can be ensured.

Since $\boldsymbol{s}(\Delta\boldsymbol{x})$ is bounded and $\hat{\boldsymbol{\theta}}_k$ is also bounded by the projection algorithm, there exist constants $c_0 > 0$ and $c_1 > 0$ such that

$$\begin{aligned} &\alpha|\boldsymbol{s}^T(\Delta\boldsymbol{x})\hat{\boldsymbol{J}}(\boldsymbol{q}, \hat{\boldsymbol{\theta}}_k)\big[-\dot{\boldsymbol{M}}(\boldsymbol{q}) + \tfrac{1}{2}\dot{\boldsymbol{M}}(\boldsymbol{q}) + \boldsymbol{S}(\boldsymbol{q}, \dot{\boldsymbol{q}})\big]\dot{\boldsymbol{q}} - \boldsymbol{s}^T(\Delta\boldsymbol{x})\dot{\hat{\boldsymbol{J}}}(\boldsymbol{q}, \hat{\boldsymbol{\theta}}_k)\boldsymbol{M}(\boldsymbol{q})\dot{\boldsymbol{q}} \\ &\quad-\dot{\boldsymbol{s}}^T(\Delta\boldsymbol{x})\hat{\boldsymbol{J}}(\boldsymbol{q}, \hat{\boldsymbol{\theta}}_k)\boldsymbol{M}(\boldsymbol{q})\dot{\boldsymbol{q}}| \geq -\alpha c_0\|\dot{\boldsymbol{q}}\|^2 - \alpha c_1\|\boldsymbol{s}(\Delta\boldsymbol{x})\|^2. \end{aligned} \tag{6.71}$$

In addition, it can be easily verified that

$$\boldsymbol{s}^T(\Delta\boldsymbol{x})\boldsymbol{Y}_k(\boldsymbol{q}, \dot{\boldsymbol{q}})(\boldsymbol{\theta}_k - \hat{\boldsymbol{\theta}}_k) - \boldsymbol{s}^T(\Delta\boldsymbol{x})\boldsymbol{Y}_k(\boldsymbol{q}, \dot{\boldsymbol{q}})\boldsymbol{\Phi}(\boldsymbol{\theta}_k - \hat{\boldsymbol{\theta}}_k) < 0. \tag{6.72}$$

Hence substituting inequalities (6.71) and (6.72) into (6.70), we obtain

$$\dot{V}_1 \leq -\lambda_1\|\dot{\boldsymbol{q}}\|^2 - \lambda_v\|\boldsymbol{v}\|^2 - \alpha k_p\lambda_2\|\boldsymbol{s}(\Delta\boldsymbol{x})\|^2, \tag{6.73}$$

where

$$\lambda_1=-\alpha c_0,\ \lambda_v=\lambda_{\min}[\boldsymbol{K}_m\hat{\boldsymbol{K}}_m^{-1}\boldsymbol{K}_v\boldsymbol{\Lambda}_1],\ \lambda_2=\lambda_{\min}[\hat{\boldsymbol{J}}(\boldsymbol{q},\hat{\boldsymbol{\theta}}_k)\hat{\boldsymbol{J}}^T(\boldsymbol{q},\hat{\boldsymbol{\theta}}_k)]-\tfrac{c_1}{k_p}. \tag{6.74}$$

Therefore, if $\boldsymbol{K}_p$ is chosen large enough, we have

$$\lambda_{\min}\Big[\hat{\boldsymbol{J}}(\boldsymbol{q},\hat{\boldsymbol{\theta}}_k)\hat{\boldsymbol{J}}^T(\boldsymbol{q},\hat{\boldsymbol{\theta}}_k)\Big]-\tfrac{c_1}{k_p}>0. \tag{6.75}$$

Now the following lemma can be stated:

Lemma 6.2 *If the feedback gains $\boldsymbol{K}_p$, α are chosen to satisfy conditions* (6.68) *and* (6.75), *then the closed-loop system described by equations* (6.65) *and* (6.59)–(6.62) *gives rise to the convergence of $\boldsymbol{x}\to\boldsymbol{x}_d$ and $\dot{\boldsymbol{q}}\to\boldsymbol{0}$ as $t\to\infty$ without the presence of actuator position perturbation, i.e., $\tilde{\boldsymbol{q}}_m=\boldsymbol{0}$.*

Proof If condition (6.68) is satisfied, V_1 is positive definite in $\dot{\boldsymbol{q}}$, $\boldsymbol{v}$, $\boldsymbol{s}(\Delta\boldsymbol{x})$, $\boldsymbol{\theta}_g-\hat{\boldsymbol{\theta}}_g$, $\hat{\boldsymbol{K}}_m\boldsymbol{K}_m^{-1}\phi_k-\hat{\varphi}$, and $\boldsymbol{\theta}_k-\hat{\boldsymbol{\theta}}_k$. Since the time derivative of V_1 is nonpositive with condition (6.75) satisfied, V_1 is bounded, and hence $\dot{\boldsymbol{q}}$, $\boldsymbol{v}$, $\boldsymbol{s}(\Delta\boldsymbol{x})$, $\hat{\boldsymbol{\theta}}_g$, $\hat{\varphi}$ are all bounded. From inequality (6.73), we can conclude also that $\dot{\boldsymbol{q}},\boldsymbol{v},\boldsymbol{s}(\Delta\boldsymbol{x})\in L_2(0,\infty)$. From Eqs. (6.59) and (6.65), it is easy to see that $\dot{\boldsymbol{v}}$ and $\ddot{\boldsymbol{q}}$ are bounded, and hence $\boldsymbol{v}$ and $\dot{\boldsymbol{q}}$ are uniformly continuous. Noting that $\boldsymbol{s}(\Delta\boldsymbol{x})$ is differentiable and $\dot{\boldsymbol{x}}=\boldsymbol{J}(\boldsymbol{q})\dot{\boldsymbol{q}}$ is bounded, we conclude that $\boldsymbol{s}(\Delta\boldsymbol{x})$ is also uniformly continuous. We obtain that $\dot{\boldsymbol{q}}$, $\boldsymbol{v}$, and $\boldsymbol{s}(\Delta\boldsymbol{x})$ converge to $\boldsymbol{0}$ as $t\to\infty$, which means that $\Delta\boldsymbol{x}=\boldsymbol{x}-\boldsymbol{x}_d$ also converges to $\boldsymbol{0}$ as $t\to\infty$. The proof is complete. △△△

6.2.2 Control Torque Input Design

It has been shown in the previous subsection that the regulation task can be achieved with the desired actuator position trajectory $\boldsymbol{q}_{md}$. Next, a control torque input $\boldsymbol{u}$ is designed to deal with the actuator position perturbation $\tilde{\boldsymbol{q}}_m$ and ensure the convergence of the task-space position error.

Based on the actuator dynamics defined by Eq. (6.52), a control torque with a combination of adaptive terms and a PD feedback part is designed as

$$\boldsymbol{u}=\hat{\boldsymbol{J}}_m\ddot{\boldsymbol{q}}_{md}+\hat{\boldsymbol{B}}_m\dot{\boldsymbol{q}}_{md}+\hat{\boldsymbol{K}}_m'(\boldsymbol{q}_{md}-\boldsymbol{q})-\boldsymbol{K}_{mp}\tilde{\boldsymbol{q}}_m-\boldsymbol{K}_{mv}\dot{\tilde{\boldsymbol{q}}}_m, \tag{6.76}$$

where $\hat{\boldsymbol{J}}_m,\hat{\boldsymbol{B}}_m,\hat{\boldsymbol{K}}_m'\in\Re^{n\times n}$ are the approximate models of $\boldsymbol{J}_m,\boldsymbol{B}_m,\boldsymbol{K}_m$ respectively, and $\boldsymbol{K}_{mp}$ and $\boldsymbol{K}_{mv}$ are two positive definite control gain matrices. Note from the definition of $\boldsymbol{q}_{md}$ and $\boldsymbol{q}_{md}'$ in Eqs. (6.54) and (6.58) that the control torque design (6.76) requires the use of acceleration information, since $\ddot{\boldsymbol{q}}_{md}$ is employed.

To avoid the use of accelerations in the control input, an observed signal $\hat{\boldsymbol{q}}_{md}$ for $\boldsymbol{q}_{md}$ is used in the control torque input as

$$\boldsymbol{u} = \hat{\boldsymbol{J}}_m\ddot{\hat{\boldsymbol{q}}}_{md} + \hat{\boldsymbol{B}}_m\dot{\hat{\boldsymbol{q}}}_{md} + \hat{\boldsymbol{K}}'_m(\hat{\boldsymbol{q}}_{md} - \boldsymbol{q}) - \boldsymbol{K}_{mp}\Delta\boldsymbol{q}_m - \boldsymbol{K}_{mv}\Delta\dot{\boldsymbol{q}}_m, \quad (6.77)$$

where $\Delta\boldsymbol{q}_m = \boldsymbol{q}_m - \hat{\boldsymbol{q}}_{md}$, $\Delta\dot{\boldsymbol{q}}_m = \dot{\boldsymbol{q}}_m - \dot{\hat{\boldsymbol{q}}}_{md}$. The approximate models $\hat{\boldsymbol{J}}_m$, $\hat{\boldsymbol{B}}_m$, $\hat{\boldsymbol{K}}'_m$ are updated by the adaptation laws

$$\dot{\hat{\boldsymbol{\theta}}}_{J_m} = -\boldsymbol{L}_{J_m}\boldsymbol{N}_{J_m}(\ddot{\hat{\boldsymbol{q}}}_{md})(\Delta\dot{\boldsymbol{q}}_m + \beta\Delta\boldsymbol{q}_m), \quad (6.78)$$

$$\dot{\hat{\boldsymbol{\theta}}}_{B_m} = -\boldsymbol{L}_{B_m}\boldsymbol{N}_{B_m}(\dot{\hat{\boldsymbol{q}}}_{md})(\Delta\dot{\boldsymbol{q}}_m + \beta\Delta\boldsymbol{q}_m), \quad (6.79)$$

$$\dot{\hat{\boldsymbol{\theta}}}_{K'_m} = -\boldsymbol{L}_{K_m}\boldsymbol{N}_{K_m}(\hat{\boldsymbol{q}}_{md} - \boldsymbol{q})(\Delta\dot{\boldsymbol{q}}_m + \beta\Delta\boldsymbol{q}_m), \quad (6.80)$$

where $\hat{\boldsymbol{\theta}}_{J_m}$, $\hat{\boldsymbol{\theta}}_{B_m}$, $\hat{\boldsymbol{\theta}}_{K'_m} \in \Re^n$ are adaptive actuator dynamic parameter vectors, $\boldsymbol{N}_{J_m}(\ddot{\hat{\boldsymbol{q}}}_{md})$, $\boldsymbol{N}_{B_m}(\dot{\hat{\boldsymbol{q}}}_{md})$, $\boldsymbol{N}_{K_m}(\hat{\boldsymbol{q}}_{md} - \boldsymbol{q}) \in \Re^{n\times n}$ are regressors defined as $\boldsymbol{N}_{J_m}(\ddot{\hat{\boldsymbol{q}}}_{md})$ $= \mathrm{diag}\{\ddot{\hat{\boldsymbol{q}}}_{md}\}$, $\boldsymbol{N}_{B_m}(\dot{\hat{\boldsymbol{q}}}_{md}) = \mathrm{diag}\{\dot{\hat{\boldsymbol{q}}}_{md}\}$, $\boldsymbol{N}_{K_m}(\hat{\boldsymbol{q}}_{md} - \boldsymbol{q}) = \mathrm{diag}\{\hat{\boldsymbol{q}}_{md} - \boldsymbol{q}\}$, $\beta \in \Re$ is a positive control gain, $\boldsymbol{L}_{J_m}$, $\boldsymbol{L}_{B_m}$, $\boldsymbol{L}_{K_m} \in \Re^{n\times n}$ are positive definite diagonal adaptation gain matrices. Note that the approximate model $\hat{\boldsymbol{K}}'_m$ used in the control torque design is updated through an adaptation law, while the approximate model $\hat{\boldsymbol{K}}_m$ in Eq. (6.58) of the desired actuator position design is fixed.

Substituting (6.77) into (6.52), the closed-loop dynamic equation of the actuator subsystem is obtained as

$$\begin{aligned}
&\boldsymbol{J}_m\Delta\ddot{\boldsymbol{q}}_m + (\boldsymbol{B}_m + \boldsymbol{K}_{mv})\Delta\dot{\boldsymbol{q}}_m + (\boldsymbol{K}_m + \boldsymbol{K}_{mp})\Delta\boldsymbol{q}_m \\
&= (\hat{\boldsymbol{J}}_m - \boldsymbol{J}_m)\ddot{\hat{\boldsymbol{q}}}_{md} + (\hat{\boldsymbol{B}}_m - \boldsymbol{B}_m)\dot{\hat{\boldsymbol{q}}}_{md} + (\hat{\boldsymbol{K}}'_m - \boldsymbol{K}_m)(\hat{\boldsymbol{q}}_{md} - \boldsymbol{q}) \\
&= -\boldsymbol{N}_{J_m}(\ddot{\hat{\boldsymbol{q}}}_{md})\Delta\boldsymbol{\theta}_{J_m} - \boldsymbol{N}_{B_m}(\dot{\hat{\boldsymbol{q}}}_{md})\Delta\boldsymbol{\theta}_{B_m} - \boldsymbol{N}_{K_m}(\hat{\boldsymbol{q}}_{md} - \boldsymbol{q})\Delta\boldsymbol{\theta}_{K_m},
\end{aligned} \quad (6.81)$$

where $\Delta\boldsymbol{\theta}_{J_m} = \boldsymbol{\theta}_{J_m} - \hat{\boldsymbol{\theta}}_{J_m}$, $\Delta\boldsymbol{\theta}_{B_m} = \boldsymbol{\theta}_{B_m} - \hat{\boldsymbol{\theta}}_{B_m}$, $\Delta\boldsymbol{\theta}_{K_m} = \boldsymbol{\theta}_{K_m} - \hat{\boldsymbol{\theta}}_{K'_m}$.

The observer dynamics for $\hat{\boldsymbol{q}}_{md}$ is designed in the following form:

$$\begin{cases}
\dot{\hat{\boldsymbol{q}}}_{md} = \boldsymbol{\eta} + \boldsymbol{\Lambda}_2\boldsymbol{e}_m, \\
\dot{\boldsymbol{\eta}} = \ddot{\hat{\boldsymbol{q}}}_{md} - \boldsymbol{\Lambda}_2\dot{\boldsymbol{e}}_m = \hat{\boldsymbol{J}}_{mo}^{-1}(\boldsymbol{u} - \hat{\boldsymbol{B}}_{mo}\boldsymbol{\eta} - \hat{\boldsymbol{K}}_{mo}(\boldsymbol{q}_m + \hat{\boldsymbol{q}}_{md} - \boldsymbol{q} - \boldsymbol{q}_{md}) + \boldsymbol{K}_s\boldsymbol{z}),
\end{cases} \quad (6.82)$$

where

$$\boldsymbol{z} = \dot{\boldsymbol{q}}_m - \boldsymbol{\eta} = \Delta\dot{\boldsymbol{q}}_m + \boldsymbol{\Lambda}_2\boldsymbol{e}_m, \quad (6.83)$$

and $\boldsymbol{e}_m = \boldsymbol{q}_{md} - \hat{\boldsymbol{q}}_{md}$ is the observation error, $\boldsymbol{\eta} \in \Re^n$ is an auxiliary variable, $\boldsymbol{\Lambda}_2 \in \Re^{n\times n}$ is a positive definite diagonal constant matrix, $\hat{\boldsymbol{J}}_{mo}$, $\hat{\boldsymbol{B}}_{mo}$, $\hat{\boldsymbol{K}}_{mo} \in \Re^{n\times n}$ are the approximate models for $\boldsymbol{J}_m$, $\boldsymbol{B}_m$, $\boldsymbol{K}_m$ respectively, and $\hat{\boldsymbol{\theta}}_{J_{mo}}$, $\hat{\boldsymbol{\theta}}_{B_{mo}}$, $\hat{\boldsymbol{\theta}}_{K_{mo}} \in \Re^n$

are the vectors of estimated parameters updated through adaptation laws defined as follows:

$$\dot{\hat{\boldsymbol{\theta}}}_{J_{mo}} = -\boldsymbol{L}_{J_{mo}} \boldsymbol{N}_{J_{mo}}(\dot{\boldsymbol{\eta}})\boldsymbol{z}, \tag{6.84}$$

$$\dot{\hat{\boldsymbol{\theta}}}_{B_{mo}} = -\boldsymbol{L}_{B_{mo}} \boldsymbol{N}_{B_{mo}}(\boldsymbol{\eta})\boldsymbol{z}, \tag{6.85}$$

$$\dot{\hat{\boldsymbol{\theta}}}_{K_{mo}} = -\boldsymbol{L}_{K_{mo}} \boldsymbol{N}_{K_{mo}}(\boldsymbol{q}_m + \hat{\boldsymbol{q}}_{md} - \boldsymbol{q} - \boldsymbol{q}_{md})\boldsymbol{z}, \tag{6.86}$$

where $\boldsymbol{L}_{J_{mo}}, \boldsymbol{L}_{B_{mo}}, \boldsymbol{L}_{K_{mo}} \in \Re^{n\times n}$ are positive definite constant matrices, $\boldsymbol{N}_{J_{mo}}(\dot{\boldsymbol{\eta}}) = \mathrm{diag}\{\dot{\boldsymbol{\eta}}\}$, $\boldsymbol{N}_{B_{mo}}(\boldsymbol{\eta}) = \mathrm{diag}\{\boldsymbol{\eta}\}$, $\boldsymbol{N}_{K_{mo}}(\boldsymbol{q}_m + \hat{\boldsymbol{q}}_{md} - \boldsymbol{q} - \boldsymbol{q}_{md}) = \mathrm{diag}\{\boldsymbol{q}_m + \hat{\boldsymbol{q}}_{md} - \boldsymbol{q} - \boldsymbol{q}_{md}\}$.

Multiplying both sides of Eq. (6.82) by $\hat{\boldsymbol{J}}_{mo}$ and substituting $\boldsymbol{u}$ with the right-hand side of Eq. (6.52), the closed-loop observer dynamics is obtained as

$$\begin{aligned}
&\boldsymbol{J}_m\dot{\boldsymbol{z}} + (\boldsymbol{B}_m + \boldsymbol{K}_s)\boldsymbol{z} + \boldsymbol{K}_m\boldsymbol{e}_m \\
&= (\hat{\boldsymbol{J}}_{mo} - \boldsymbol{J}_m)\dot{\boldsymbol{\eta}} + (\hat{\boldsymbol{B}}_{mo} - \boldsymbol{B}_m)\boldsymbol{\eta} + (\hat{\boldsymbol{K}}_{mo} - \boldsymbol{K}_m)(\boldsymbol{q}_m + \hat{\boldsymbol{q}}_{md} - \boldsymbol{q} - \boldsymbol{q}_{md}) \\
&= -\boldsymbol{N}_{J_{mo}}(\dot{\boldsymbol{\eta}})\Delta\boldsymbol{\theta}_{J_{mo}} - \boldsymbol{N}_{B_{mo}}(\boldsymbol{\eta})\Delta\boldsymbol{\theta}_{B_{mo}} - \boldsymbol{N}_{K_{mo}}(\boldsymbol{q}_m + \hat{\boldsymbol{q}}_{md} - \boldsymbol{q} - \boldsymbol{q}_{md})\Delta\boldsymbol{\theta}_{K_{mo}},
\end{aligned} \tag{6.87}$$

where $\Delta\boldsymbol{\theta}_{J_{mo}} = \boldsymbol{\theta}_{J_m} - \hat{\boldsymbol{\theta}}_{J_{mo}}$, $\Delta\boldsymbol{\theta}_{B_{mo}} = \boldsymbol{\theta}_{B_m} - \hat{\boldsymbol{\theta}}_{B_{mo}}$, $\Delta\boldsymbol{\theta}_{K_{mo}} = \boldsymbol{\theta}_{K_m} - \hat{\boldsymbol{\theta}}_{K_{mo}}$.

A Lyapunov-like function candidate V_2 is chosen for the actuator–observer integrated closed-loop subsystem as

$$\begin{aligned}
V_2 &= \tfrac{1}{2}(\Delta\dot{\boldsymbol{q}}_m + \beta\Delta\boldsymbol{q}_m)^T \boldsymbol{J}_m(\Delta\dot{\boldsymbol{q}}_m + \beta\Delta\boldsymbol{q}_m) + \tfrac{1}{2}\Delta\boldsymbol{q}_m^T(\beta\boldsymbol{B}_m + \beta\boldsymbol{K}_{mv} + \boldsymbol{K}_m + \\
&\boldsymbol{K}_{mp} - \beta^2\boldsymbol{J}_m)\Delta\boldsymbol{q}_m + \tfrac{1}{2}\boldsymbol{z}^T\boldsymbol{J}_m\boldsymbol{z} + \tfrac{1}{2}\Delta\boldsymbol{\theta}_{J_m}^T\boldsymbol{L}_{J_m}^{-1}\Delta\boldsymbol{\theta}_{J_m} + \tfrac{1}{2}\Delta\boldsymbol{\theta}_{B_m}^T\boldsymbol{L}_{B_m}^{-1}\Delta\boldsymbol{\theta}_{B_m} + \tfrac{1}{2}\Delta\boldsymbol{\theta}_{J_{mo}}^T\boldsymbol{L}_{J_{mo}}^{-1}\Delta\boldsymbol{\theta}_{J_{mo}} \\
&+ \tfrac{1}{2}\Delta\boldsymbol{\theta}_{K_m}^T\boldsymbol{L}_{K_m}^{-1}\Delta\boldsymbol{\theta}_{K_m} + \tfrac{1}{2}\Delta\boldsymbol{\theta}_{B_{mo}}^T\boldsymbol{L}_{B_{mo}}^{-1}\Delta\boldsymbol{\theta}_{B_{mo}} + \tfrac{1}{2}\Delta\boldsymbol{\theta}_{K_{mo}}^T\boldsymbol{L}_{K_{mo}}^{-1}\Delta\boldsymbol{\theta}_{K_{mo}},
\end{aligned} \tag{6.88}$$

where the control gain $\boldsymbol{K}_{mv}$ should be chosen large enough that

$$\lambda_{\min}[\boldsymbol{B}_m + \boldsymbol{K}_{mv}] > \beta\lambda_{\max}[\boldsymbol{J}_m] \tag{6.89}$$

and hence V_2 is positive definite.

Differentiating V_2 with respect to time and using Eqs. (6.78)–(6.81) and (6.84)–(6.87), one has

$$\begin{aligned}
\dot{V}_2 &= -\Delta\dot{\boldsymbol{q}}_m^T(\boldsymbol{B}_m + \boldsymbol{K}_{mv} - \beta\boldsymbol{J}_m)\Delta\dot{\boldsymbol{q}}_m - \beta\Delta\boldsymbol{q}_m^T(\boldsymbol{K}_m + \boldsymbol{K}_{mp})\Delta\boldsymbol{q}_m \\
&\quad - \boldsymbol{z}^T(\boldsymbol{B}_m + \boldsymbol{K}_s)\boldsymbol{z} - \Delta\dot{\boldsymbol{q}}_m^T\boldsymbol{K}_m\boldsymbol{e}_m - \boldsymbol{e}_m^T\boldsymbol{\Lambda}_2\boldsymbol{K}_m\boldsymbol{e}_m \\
&\le -(\lambda_{\min}[\boldsymbol{B}_m + \boldsymbol{K}_{mv}] - \beta\lambda_{\max}[\boldsymbol{J}_m])\|\Delta\dot{\boldsymbol{q}}_m\|^2 - \beta(\lambda_{\min}[\boldsymbol{K}_m] \\
&\quad + \lambda_{\min}[\boldsymbol{K}_{mp}])\|\Delta\boldsymbol{q}_m\|^2 - \lambda_{\min}[\boldsymbol{B}_m + \boldsymbol{K}_s]\|\boldsymbol{z}\|^2 - \lambda_{\min}[\boldsymbol{\Lambda}_2\boldsymbol{K}_m]\|\boldsymbol{e}_m\|^2 \\
&\quad + \lambda_{\max}[\boldsymbol{K}_m]\|\Delta\dot{\boldsymbol{q}}_m\|\|\boldsymbol{e}_m\|.
\end{aligned} \tag{6.90}$$

Now consider the overall Lyapunov-like function candidate V as

$$V = V_1 + V_2, \tag{6.91}$$

where V_1 is as defined previously in Eq. (6.66). It is clear that V is positive definite in all the signals contained if conditions (6.68) and (6.89) are satisfied.

Next, the time derivative of V is obtained using inequalities (6.73) and (6.90):

$$\begin{aligned}\dot{V} \leq & -\lambda_1\|\dot{\boldsymbol{q}}\|^2 - \lambda_v\|\boldsymbol{v}\|^2 - \alpha k_p\lambda_2\|\boldsymbol{s}(\Delta\boldsymbol{x})\|^2 + (\dot{\boldsymbol{q}} + \alpha\hat{\boldsymbol{J}}^T(\boldsymbol{q},\hat{\boldsymbol{\theta}}_J)\boldsymbol{s}(\Delta\boldsymbol{x}))^T\boldsymbol{K}_m\tilde{\boldsymbol{q}}_m \\ & - (\lambda_{\min}[\boldsymbol{B}_m + \boldsymbol{K}_{mv}] - \beta\lambda_{\max}[\boldsymbol{J}_m])\|\Delta\dot{\boldsymbol{q}}_m\|^2 - \beta(\lambda_{\min}[\boldsymbol{K}_m] \\ & + \lambda_{\min}[\boldsymbol{K}_{mp}])\|\Delta\boldsymbol{q}_m\|^2 - \lambda_{\min}[\boldsymbol{B}_m + \boldsymbol{K}_s]\|\boldsymbol{z}\|^2 \\ & - \lambda_{\min}[\boldsymbol{\Lambda}_2\boldsymbol{K}_m]\|\boldsymbol{e}_m\|^2 + \lambda_{\max}[\boldsymbol{K}_m]\|\Delta\dot{\boldsymbol{q}}_m\|\|\boldsymbol{e}_m\|. \end{aligned} \tag{6.92}$$

Note the presence of the term $(\dot{\boldsymbol{q}} + \alpha\hat{\boldsymbol{J}}^T(\boldsymbol{q},\hat{\boldsymbol{\theta}}_k)\boldsymbol{s}(\Delta\boldsymbol{x}))^T\boldsymbol{K}_m\tilde{\boldsymbol{q}}_m$ in the first line of the above inequality due to the perturbation term $\tilde{\boldsymbol{q}}_m$ in Eq. (6.65).

Since $\tilde{\boldsymbol{q}}_m = (\boldsymbol{q}_m - \hat{\boldsymbol{q}}_{md}) - (\boldsymbol{q}_{md} - \hat{\boldsymbol{q}}_{md}) = \Delta\boldsymbol{q}_m - \boldsymbol{e}_m$, inequality (6.92) can be further written as

$$\begin{aligned}\dot{V} \leq & -\lambda_1\|\dot{\boldsymbol{q}}\|^2 - \lambda_v\|\boldsymbol{v}\|^2 - \alpha k_p\lambda_2\|\boldsymbol{s}(\Delta\boldsymbol{x})\|^2 + \lambda_{\max}[\boldsymbol{K}_m]\|\dot{\boldsymbol{q}}\|(\|\Delta\boldsymbol{q}_m\| + \|\boldsymbol{e}_m\|) \\ & + \alpha\lambda_{\max}[\boldsymbol{K}_m]b_{\hat{J}}\|\boldsymbol{s}(\Delta\boldsymbol{x})\|(\|\Delta\boldsymbol{q}_m\| + \|\boldsymbol{e}_m\|) - (\lambda_{\min}[\boldsymbol{B}_m + \boldsymbol{K}_{mv}] \\ & - \beta\lambda_{\max}[\boldsymbol{J}_m])\|\Delta\dot{\boldsymbol{q}}_m\|^2 - \beta(\lambda_{\min}[\boldsymbol{K}_m] + \lambda_{\min}[\boldsymbol{K}_{mp}])\|\Delta\boldsymbol{q}_m\|^2 \\ & - \lambda_{\min}[\boldsymbol{B}_m + \boldsymbol{K}_s]\|\boldsymbol{z}\|^2 - \lambda_{\min}[\boldsymbol{\Lambda}_2\boldsymbol{K}_m]\|\boldsymbol{e}_m\|^2 + \lambda_{\max}[\boldsymbol{K}_m]\|\Delta\dot{\boldsymbol{q}}_m\|\|\boldsymbol{e}_m\| \\ = & -\lambda_1\|\dot{\boldsymbol{q}}\|^2 - \lambda_v\|\boldsymbol{v}\|^2 - \alpha k_p\lambda_2\|\boldsymbol{s}(\Delta\boldsymbol{x})\|^2 - \lambda_{\min}[\boldsymbol{B}_m + \boldsymbol{K}_s]\|\boldsymbol{z}\|^2 \\ & - \lambda_{\min}[\Lambda_2\boldsymbol{K}_m]\|\boldsymbol{e}_m\|^2 - \beta(\lambda_{\min}[\boldsymbol{K}_m] + \lambda_{\min}[\boldsymbol{K}_{mp}])\|\Delta\boldsymbol{q}_m\|^2 \\ & - (\lambda_{\min}[\boldsymbol{B}_m + \boldsymbol{K}_{mv}] - \beta\lambda_{\max}[\boldsymbol{J}_m])\|\Delta\dot{\boldsymbol{q}}_m\|^2 + \lambda_m\|\dot{\boldsymbol{q}}\|\|\Delta\boldsymbol{q}_m\| \\ & + \lambda_m\|\dot{\boldsymbol{q}}\|\|\boldsymbol{e}_m\| + \alpha\lambda_m b_{\hat{J}}\|\boldsymbol{s}(\Delta\boldsymbol{x})\|\|\Delta\boldsymbol{q}_m\| \\ & + \alpha\lambda_m b_{\hat{J}}\|\boldsymbol{s}(\Delta\boldsymbol{x})\|\|\boldsymbol{e}_m\| + \lambda_m\|\Delta\dot{\boldsymbol{q}}_m\|\|\boldsymbol{e}_m\| \\ = & -\left[\, \|\dot{\boldsymbol{q}}\| \;\; \|\boldsymbol{s}(\Delta\boldsymbol{x})\| \;\; \|\Delta\boldsymbol{q}_m\| \;\; \|\Delta\dot{\boldsymbol{q}}_m\| \;\; \|\boldsymbol{e}_m\| \,\right]\boldsymbol{A} \\ & \times \left[\, \|\dot{\boldsymbol{q}}\| \;\; \|\boldsymbol{s}(\Delta\boldsymbol{x})\| \;\; \|\Delta\boldsymbol{q}_m\| \;\; \|\Delta\dot{\boldsymbol{q}}_m\| \;\; \|\boldsymbol{e}_m\| \,\right]^T \\ & - \lambda_v\|\boldsymbol{v}\|^2 - \lambda_{\min}[\boldsymbol{B}_m + \boldsymbol{K}_s]\|\boldsymbol{z}\|^2, \end{aligned} \tag{6.93}$$

where

$$\boldsymbol{A} = \begin{bmatrix} \lambda_1 & 0 & \frac{1}{2}\lambda_m & 0 & \frac{1}{2}\lambda_m \\ 0 & \alpha k_p\lambda_2 & \frac{1}{2}\alpha\lambda_m b_{\hat{J}} & 0 & \frac{1}{2}\alpha\lambda_m b_{\hat{J}} \\ \frac{1}{2}\lambda_m & \frac{1}{2}\alpha\lambda_m b_{\hat{J}} & \beta\psi_1 & 0 & 0 \\ 0 & 0 & 0 & \psi_2 & \frac{1}{2}\lambda_m \\ \frac{1}{2}\lambda_m & \frac{1}{2}\alpha\lambda_m b_{\hat{J}} & 0 & \frac{1}{2}\lambda_m & \lambda_{\min}[\boldsymbol{\Lambda}_2\boldsymbol{K}_m] \end{bmatrix}, \tag{6.94}$$

with $\lambda_m = \lambda_{\max}[\boldsymbol{K}_m]$, $\psi_1 = \lambda_{\min}[\boldsymbol{K}_m] + \lambda_{\min}[\boldsymbol{K}_{mp}]$, and $\psi_2 = \lambda_{\min}[\boldsymbol{B}_m + \boldsymbol{K}_{mv}] - \beta\lambda_{\max}[\boldsymbol{J}_m]$.

If the matrix $\boldsymbol{A}$ is positive definite, then inequality (6.93) can be written as

$$\dot{V} \leq -\lambda_v \|\boldsymbol{v}\|^2 - \lambda_{\min}[\boldsymbol{B}_m + \boldsymbol{K}_s]\|\boldsymbol{z}\|^2 - \lambda_{\min}[\boldsymbol{A}](\|\dot{\boldsymbol{q}}\|^2 + \|\boldsymbol{s}(\Delta\boldsymbol{x})\|^2 + \|\Delta\boldsymbol{q}_m\|^2 + \|\Delta\dot{\boldsymbol{q}}_m\|^2 + \|\boldsymbol{e}_m\|^2), \tag{6.95}$$

so that $\dot{V}$ is negative definite in $\boldsymbol{v}$, $\boldsymbol{z}$, $\dot{\boldsymbol{q}}$, $\boldsymbol{s}(\Delta\boldsymbol{x})$, $\Delta\boldsymbol{q}_m$, $\Delta\dot{\boldsymbol{q}}_m$, and $\boldsymbol{e}_m$.

According to Sylvester's theorem, the symmetric matrix $\boldsymbol{A}$ is positive definite if and only if all its principal minors A_i $(i = 1, 2, \ldots, 5)$ are strictly positive, where a principal minor is defined as $A_1 = a_{11}$, $A_2 = a_{11}a_{22} - a_{12}a_{21}$, $\ldots$, $A_5 = \det(\boldsymbol{A})$.

Next we should derive conditions to guarantee the positive definiteness of matrix $\boldsymbol{A}$. Since λ_1, λ_2 are positive if condition (6.75) is satisfied, it is easy to see that A_1 and A_2 are positive, and A_3 can be calculated as

$$A_3 = \tfrac{\alpha}{4}[4\beta k_p \lambda_1 \lambda_2 \psi_1 - \alpha\lambda_m^2 b_{\hat{J}}^2 \lambda_1 - \lambda_m^2 k_p \lambda_2]. \tag{6.96}$$

By choosing α small enough and k_p, ψ_1 large enough (i.e., $\boldsymbol{K}_p$, $\boldsymbol{K}_{mp}$ large enough) that

$$4\beta\lambda_1\lambda_2\psi_1 - \tfrac{\alpha}{k_p}\lambda_m^2 b_{\hat{J}}^2 \lambda_1 - \lambda_m^2 \lambda_2 > 0, \tag{6.97}$$

the positivity of A_3 can be achieved. From the structure of $\boldsymbol{A}$, it can be seen that A_4 is simply $A_4 = A_3 \cdot \psi_2$. From (6.89), we have

$$\psi_2 = \lambda_{\min}[\boldsymbol{B}_m + \boldsymbol{K}_{mv}] - \beta\lambda_{\max}[\boldsymbol{J}_m] > 0. \tag{6.98}$$

Hence A_4 is positive.

The determinant $\det(\boldsymbol{A}) = A_5$ can be calculated as

$$A_5 = \tfrac{\alpha}{4} k_p \left(\psi_2 \lambda_{\min}[\boldsymbol{\Lambda}_2 \boldsymbol{K}_m] + \beta\psi_1\psi_2 - \tfrac{1}{4}\lambda_m^2\right)\Big\{4\lambda_1\lambda_2\beta\psi_1\Big(1 - \frac{\beta\psi_1\psi_2}{\psi_2\lambda_{\min}[\boldsymbol{\Lambda}_2\boldsymbol{K}_m] + \beta\psi_1\psi_2 - \frac{1}{4}\lambda_m^2}\Big) - \lambda_m^2\left(\tfrac{\alpha}{k_p} b_{\hat{J}} + \lambda_2\right)\Big\}, \tag{6.99}$$

where $\boldsymbol{\Lambda}_2$ can be chosen large enough that

$$1 - \frac{\beta\psi_1\psi_2}{\psi_2\lambda_{\min}[\boldsymbol{\Lambda}_2\boldsymbol{K}_m] + \beta\psi_1\psi_2 - \frac{1}{4}\lambda_m^2} > 0. \tag{6.100}$$

Then α can be chosen small enough and k_p, ψ_1 can be chosen large enough (i.e., $\boldsymbol{K}_p$, $\boldsymbol{K}_{mp}$ large enough) to satisfy

$$4\lambda_1\lambda_2\beta\psi_1\left(1 - \frac{\beta\psi_1\psi_2}{\psi_2\lambda_{\min}[\boldsymbol{\Lambda}_2\boldsymbol{K}_m] + \beta\psi_1\psi_2 - \frac{1}{4}\lambda_m^2}\right) - \lambda_m^2\left(\frac{\alpha}{k_p} b_{\hat{J}} + \lambda_2\right) > 0. \tag{6.101}$$

Using sufficiently large $\boldsymbol{\Lambda}_2$ and ψ_1, it can be easily achieved that

$$\psi_2\lambda_{\min}[\boldsymbol{\Lambda}_2\boldsymbol{K}_m] + \beta\psi_1\psi_2 - \tfrac{1}{4}\lambda_m^2 > 0. \tag{6.102}$$

Hence

$$A_5 = \tfrac{\alpha}{4}k_p\big(\psi_2\lambda_{\min}[\boldsymbol{\Lambda}_2\boldsymbol{K}_m] + \beta\psi_1\psi_2 - \tfrac{1}{4}\lambda_m^2\big)\Big\{4\lambda_1\lambda_2\beta\psi_1\Big(1 - \frac{\beta\psi_1\psi_2}{\psi_2\lambda_{\min}[\boldsymbol{\Lambda}_2\boldsymbol{K}_m]+\beta\psi_1\psi_2-\frac{1}{4}\lambda_m^2}\Big) - \lambda_m^2\big(\tfrac{\alpha}{k_p}b_{\hat{J}} + \lambda_2\big)\Big\} > 0. \tag{6.103}$$

Conditions (6.89), (6.97), (6.103) can be satisfied by simply choosing α small enough and $\boldsymbol{K}_p$, $\boldsymbol{K}_{mp}$, $\boldsymbol{K}_{mv}$, $\boldsymbol{\Lambda}_2$ large enough.

We can now state the following theorem:

Theorem 6.2 *The rigid-link flexible-joint robot system described by equations* (6.65), (6.81) *together with the parameter adaptation laws* (6.59)–(6.62), (6.78)–(6.80), *and* (6.84)–(6.86) *gives rise to the convergence of* $\boldsymbol{x} \to \boldsymbol{x}_d$ *and* $\dot{\boldsymbol{q}} \to \boldsymbol{0}$ *as* $t \to \infty$ *if the feedback gains* α, $\boldsymbol{K}_p$, $\boldsymbol{K}_{mp}$, $\boldsymbol{K}_{mv}$, *and* $\boldsymbol{\Lambda}_2$ *are chosen to satisfy conditions* (6.68), (6.75), (6.89), (6.97), *and* (6.103).

Proof Since V is positive definite in $\dot{\boldsymbol{q}}, \boldsymbol{s}(\Delta\boldsymbol{x}), \boldsymbol{v}, \boldsymbol{\theta}_g - \hat{\boldsymbol{\theta}}_g, \hat{\boldsymbol{K}}_m\boldsymbol{K}_m^{-1}\phi_k - \hat{\varphi}, \boldsymbol{\theta}_k - \hat{\boldsymbol{\theta}}_k$, $\Delta\dot{\boldsymbol{q}}_m + \beta\Delta\boldsymbol{q}_m, \Delta\boldsymbol{q}_m, \boldsymbol{z}, \Delta\boldsymbol{\theta}_{J_m}, \Delta\boldsymbol{\theta}_{B_m}, \Delta\boldsymbol{\theta}_{K_m}, \Delta\boldsymbol{\theta}_{J_{mo}}, \Delta\boldsymbol{\theta}_{B_{mo}}, \Delta\boldsymbol{\theta}_{K_{mo}}$ with conditions (6.68) (6.89) satisfied, and $\dot{V}$ is negative definite in $\boldsymbol{v}$, $\boldsymbol{z}$, $\dot{\boldsymbol{q}}$, $\boldsymbol{s}(\Delta\boldsymbol{x})$, $\Delta\boldsymbol{q}_m$, $\Delta\dot{\boldsymbol{q}}_m$, and $\boldsymbol{e}_m$ with conditions (6.75), (6.97), and (6.103) satisfied, the variables included in V are all bounded. Since $\tilde{\boldsymbol{q}}_m = \Delta\boldsymbol{q}_m - \boldsymbol{e}_m$, $\tilde{\boldsymbol{q}}_m$ is also bounded. From Eq. (6.65), it can be seen that $\ddot{\boldsymbol{q}}$ is also bounded. Since $\boldsymbol{s}(\Delta\boldsymbol{x})$ is differentiable and $\Delta\dot{\boldsymbol{x}} = \dot{\boldsymbol{x}} = \boldsymbol{J}(\boldsymbol{q})\dot{\boldsymbol{q}}$ is bounded, the time derivative $\dot{\boldsymbol{s}}(\Delta\boldsymbol{x})$ is also bounded. Now that $\ddot{\boldsymbol{q}}$ and $\dot{\boldsymbol{s}}(\Delta\boldsymbol{x})$ are bounded, $\boldsymbol{s}(\Delta\boldsymbol{x})$ and $\dot{\boldsymbol{q}}$ are uniformly continuous. From inequality (6.95), it is easy to verify that $\boldsymbol{s}(\Delta\boldsymbol{x}), \dot{\boldsymbol{q}} \in L_2(0,\infty)$. Then it can be concluded that $\boldsymbol{s}(\Delta\boldsymbol{x}) \to \boldsymbol{0}$ and $\dot{\boldsymbol{q}} \to \boldsymbol{0}$ as $t \to \infty$. Since $\boldsymbol{s}(\Delta\boldsymbol{x}) = \boldsymbol{0}$ means that $\boldsymbol{x} = \boldsymbol{x}_d$, the proof is complete. △△△

6.3 Constrained Robots

This section considers a constrained robot system in which the end effector is in contact with a constraint surface, as shown in Fig. 6.1. In this system, it is necessary to control not only the position but also the interaction force between the robot end effector and environment. In many applications of constrained robots, it is difficult to obtain an exact model of the constraint surface. In the presence of constraint uncertainty, the robot control system needs to adapt to the uncertainty in external parameters due to the environment, in addition to the uncertainties in internal parameters of the robot kinematics and dynamics.

When the constraint surface is uncertain, the desired path on the surface cannot be obtained in Cartesian space and consequently in joint space. Using cameras, the

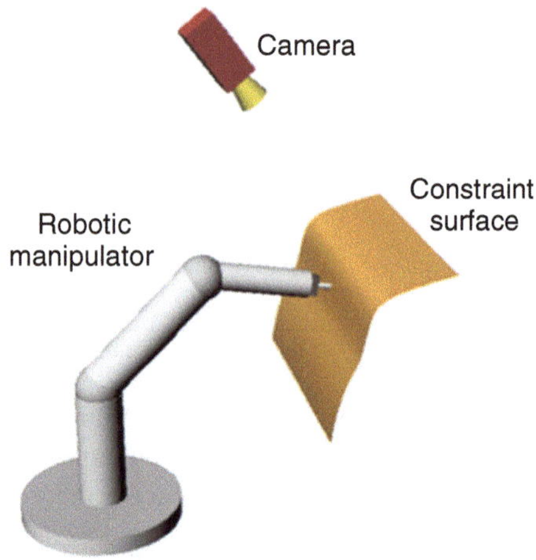

Fig. 6.1 A constrained robot system

desired path is obtained in image space using features such as edges and corners. In the absence of obvious features, markers can be used on the constraint surface.

6.3.1 Dynamic Model of a Constrained Robot System

The equations of motion of a constrained robot with n degrees of freedom can be expressed in joint coordinates as

$$\boldsymbol{M}(\boldsymbol{q})\ddot{\boldsymbol{q}} + \left(\tfrac{1}{2}\dot{\boldsymbol{M}}(\boldsymbol{q}) + \boldsymbol{S}(\boldsymbol{q},\dot{\boldsymbol{q}})\right)\dot{\boldsymbol{q}} + \boldsymbol{g}(\boldsymbol{q}) + \boldsymbol{J}_m^T(\boldsymbol{q})\boldsymbol{B}_t\dot{\boldsymbol{r}} = \boldsymbol{\tau} + \boldsymbol{J}_m^T(\boldsymbol{q})\boldsymbol{f}, \quad (6.104)$$

where $\boldsymbol{M}(\boldsymbol{q}) \in \Re^{n\times n}$ is the inertia matrix, $\boldsymbol{S}(\boldsymbol{q},\dot{\boldsymbol{q}})\dot{\boldsymbol{q}} = \frac{1}{2}\dot{\boldsymbol{M}}(\boldsymbol{q})\dot{\boldsymbol{q}} - \frac{1}{2}\{\frac{\partial}{\partial \boldsymbol{q}}\dot{\boldsymbol{q}}^T\boldsymbol{M}(\boldsymbol{q})\dot{\boldsymbol{q}}\}^T$ is a skew-symmetric matrix, $\boldsymbol{g}(\boldsymbol{q}) \in \Re^n$ is the gravitational force, $\boldsymbol{f} \in \Re^n$ is a contact force vector normal to the constraint surface, $\boldsymbol{B}_t\dot{\boldsymbol{r}}$ denotes the contact friction arising in the direction of the manipulator endpoint movement $\dot{\boldsymbol{r}}$ and $\boldsymbol{B}_t \in \Re^{m\times m}$ is the contact friction matrix, $\boldsymbol{\tau} \in \Re^n$ is the applied joint torque to the robot.

A constraint surface is defined as an algebraic term

$$\Psi(\boldsymbol{r}) = 0, \quad (6.105)$$

where $\Psi(\boldsymbol{r}) : \Re^m \rightarrow \Re$ is a given scalar function, and $\boldsymbol{r} \in \Re^m$ denotes the position of the end effector in Cartesian space. Differentiating Eq. (6.105) with respect to time yields the following velocity constraint:

$$\frac{\partial\Psi(\boldsymbol{r})}{\partial\boldsymbol{r}}\dot{\boldsymbol{r}} = 0. \quad (6.106)$$

The contact force normal to the constraint surface is then given by

$$f = d(r)\lambda, \tag{6.107}$$

where $\lambda \in \Re$ is defined as the magnitude of the contact force, and $d(r) = \frac{(\frac{\partial \Psi(r)}{\partial r})}{\|\frac{\partial \Psi(r)}{\partial r}\|} \in \Re^m$ is a unit vector that denotes the normal direction to the constraint surface. Hence, using Eq. (6.107), the robot dynamics (6.104) can be represented as

$$M(q)\ddot{q} + \left(\tfrac{1}{2}\dot{M}(q) + S(q, \dot{q})\right)\dot{q} + g(q) + J_m^T(q)B_t\dot{r} = \tau + D^T(q)\lambda, \tag{6.108}$$

where $D(q) = \frac{(\partial\Psi(r)/\partial r)^T}{\|\partial\Psi(r)/\partial r\|} J_m(q)$, $J_m(q) \in \Re^{m\times n}$ is the Jacobian matrix from joint space to Cartesian space, and $D(q)$ is the Jacobian of the constraint function such that

$$D(q)\dot{q} = \mathbf{0}. \tag{6.109}$$

6.3.2 Adaptive Vision and Force-Tracking Control

The dynamic model as described by Eq. (6.108) is linear in a set of physical parameters $\theta_d = (\theta_{d1}, \ldots, \theta_{dn_d})^T$:

$$M(q)\ddot{q} + \left(\tfrac{1}{2}\dot{M}(q) + S(q, \dot{q})\right)\dot{q} + g(q) + J_m^T(q)B_t J_m(q)\dot{q} = Y_d(q, \dot{q}, \dot{q}, \ddot{q})\theta_d, \tag{6.110}$$

where $Y_d(\cdot) \in \Re^{n\times n_d}$ is the dynamic regressor matrix.

When the kinematics is uncertain, the parameters of the Jacobian matrix are uncertain, and the velocity of the end effector in image space is estimated as

$$\dot{\hat{x}} = \hat{J}(q, \hat{\theta}_k)\dot{q} = Y_k(q, \dot{q})\hat{\theta}_k, \tag{6.111}$$

where $\dot{\hat{x}}$ is the estimated velocity of the end effector in image space, and $\hat{J}(q, \hat{\theta}_k)$ denotes the approximate Jacobian matrix. The estimated parameters $\hat{\theta}_k$ are updated by a parameter adaptation law.

The constraint surface is uncertain, and $J_m(q)\hat{d}(\hat{r})$ is expressed as

$$J_m^T(q)\hat{d}(\hat{r}) = Y_f(q)\theta_f, \tag{6.112}$$

where $\theta_f = (\theta_{f1}, \ldots, \theta_{fn_f})^T$ is a vector of parameters of $J_m(q)$ and $Y_f(q) \in \Re^{n\times n_f}$, $\hat{d}(\hat{r})$ is a fixed estimate of $d(r)$ that is not updated. Since $\hat{d}(\hat{r})$ is uncertain, the use of $\hat{d}(\hat{r})$ results in an estimation error $J_m(q)(d(r) - \hat{d}(\hat{r}))$, which is expressed as

$$J_m^T(q)(d(r) - \hat{d}(\hat{r})) = Y_{md}(q)\theta_{md}, \tag{6.113}$$

where $\boldsymbol{r} = \boldsymbol{h}(\boldsymbol{q})$, $\hat{\boldsymbol{r}} = \hat{\boldsymbol{h}}(\boldsymbol{q})$, $\boldsymbol{\theta}_{md} = (\theta_{md1}, \ldots, \theta_{mdn_m})^T$ denotes the unknown lumped parameters of $\boldsymbol{J}_m^T(\boldsymbol{q})$ and $\boldsymbol{d}(\boldsymbol{r}) - \hat{\boldsymbol{d}}(\hat{\boldsymbol{r}})$, and $\boldsymbol{Y}_{md}(\boldsymbol{q}) \in \Re^{n \times n_m}$. The unknown lumped parameters are updated by a parameter adaptation law.

Let us define a vector $\dot{\boldsymbol{x}}_r \in \Re^m$ as

$$\dot{\boldsymbol{x}}_r = (\dot{\boldsymbol{x}}_d - \alpha \Delta \boldsymbol{x}) + \beta(\hat{\boldsymbol{J}}_m(\boldsymbol{q}, \hat{\boldsymbol{\theta}}_f)\hat{\boldsymbol{J}}^+(\boldsymbol{q}, \hat{\boldsymbol{\theta}}_k))^{-1}\boldsymbol{R}(\boldsymbol{x})\hat{\boldsymbol{d}}(\hat{\boldsymbol{r}})\Delta F, \quad (6.114)$$

where α and β are positive constants, $\boldsymbol{x}_d(t) \in \Re^m$ is the desired motion trajectory in image space and $\dot{\boldsymbol{x}}_d(t) \in \Re^m$ is the desired speed trajectory, $\Delta \boldsymbol{x} = \boldsymbol{x} - \boldsymbol{x}_d$ is the image-tracking error, $\hat{\boldsymbol{J}}^+(\boldsymbol{q}, \hat{\boldsymbol{\theta}}_k) = \hat{\boldsymbol{J}}^T(\boldsymbol{q}, \hat{\boldsymbol{\theta}}_k)(\hat{\boldsymbol{J}}(\boldsymbol{q}, \hat{\boldsymbol{\theta}}_k)\hat{\boldsymbol{J}}^T(\boldsymbol{q}, \hat{\boldsymbol{\theta}}_k))^{-1}$ is the pseudoinverse of the Jacobian matrix, $\hat{\boldsymbol{J}}_m(\boldsymbol{q}, \hat{\boldsymbol{\theta}}_f)$ is an estimate of $\boldsymbol{J}_m(\boldsymbol{q})$, $\Delta F = \int_0^t (\lambda(\sigma) - \lambda_d(\sigma))d\sigma$, $\lambda_d(t)$ is the desired force trajectory, and $\boldsymbol{R}(\boldsymbol{x})$ is a rotation matrix, which will be defined later.

Differentiating Eq. (6.114) with respect to time yields

$$\begin{aligned}\ddot{\boldsymbol{x}}_r = (\ddot{\boldsymbol{x}}_d - \alpha \Delta \dot{\boldsymbol{x}}) + \beta(\hat{\boldsymbol{J}}_m(\boldsymbol{q}, \hat{\boldsymbol{\theta}}_f)\hat{\boldsymbol{J}}^+(\boldsymbol{q}, \hat{\boldsymbol{\theta}}_k))^{-1}\boldsymbol{R}(\boldsymbol{x})\hat{\boldsymbol{d}}(\hat{\boldsymbol{r}})\Delta\lambda \\ + \beta(\hat{\boldsymbol{J}}_m(\boldsymbol{q}, \hat{\boldsymbol{\theta}}_f)\hat{\boldsymbol{J}}^+(\boldsymbol{q}, \hat{\boldsymbol{\theta}}_k))^{-1}\boldsymbol{R}(\boldsymbol{x})\dot{\hat{\boldsymbol{d}}}(\hat{\boldsymbol{r}}, \dot{\hat{\boldsymbol{r}}})\Delta F \\ + \beta(\hat{\boldsymbol{J}}_m(\boldsymbol{q}, \hat{\boldsymbol{\theta}}_f)\hat{\boldsymbol{J}}^+(\boldsymbol{q}, \hat{\boldsymbol{\theta}}_k))^{-1}\dot{\boldsymbol{R}}(\boldsymbol{x})\hat{\boldsymbol{d}}(\hat{\boldsymbol{r}})\Delta \boldsymbol{F} + \beta\dot{\hat{\boldsymbol{J}}}_s\boldsymbol{R}(\boldsymbol{x})\hat{\boldsymbol{d}}(\hat{\boldsymbol{r}})\Delta F, \quad (6.115)\end{aligned}$$

where $\dot{\hat{\boldsymbol{J}}}_s = \frac{\mathrm{d}}{\mathrm{d}t}((\hat{\boldsymbol{J}}_m(\boldsymbol{q}, \hat{\boldsymbol{\theta}}_f)\hat{\boldsymbol{J}}^+(\boldsymbol{q}, \hat{\boldsymbol{\theta}}_k))^{-1})$.

To prove the stability of the vision-force tracking system, an adaptive vision space sliding vector is defined as

$$\hat{\boldsymbol{s}}_x = \dot{\hat{\boldsymbol{x}}} - \dot{\boldsymbol{x}}_r = \hat{\boldsymbol{J}}(\boldsymbol{q}, \hat{\boldsymbol{\theta}}_k)\dot{\boldsymbol{q}} - \dot{\boldsymbol{x}}_r. \quad (6.116)$$

Differentiating the above equation with respect to time yields

$$\dot{\hat{\boldsymbol{s}}}_x = \ddot{\hat{\boldsymbol{x}}} - \ddot{\boldsymbol{x}}_r = \hat{\boldsymbol{J}}(\boldsymbol{q}, \hat{\boldsymbol{\theta}}_k)\ddot{\boldsymbol{q}} + \dot{\hat{\boldsymbol{J}}}(\boldsymbol{q}, \hat{\boldsymbol{\theta}}_k)\dot{\boldsymbol{q}} - \ddot{\boldsymbol{x}}_r. \quad (6.117)$$

Next, let

$$\dot{\boldsymbol{q}}_r = \hat{\boldsymbol{J}}^+(\boldsymbol{q}, \hat{\boldsymbol{\theta}}_k)\dot{\boldsymbol{x}}_r + (\boldsymbol{I}_n - \hat{\boldsymbol{J}}^+(\boldsymbol{q}, \hat{\boldsymbol{\theta}}_k)\hat{\boldsymbol{J}}(\boldsymbol{q}, \hat{\boldsymbol{\theta}}_k))\psi, \quad (6.118)$$

where $\psi \in \Re^n$ is minus the gradient of the convex function to be optimized. From Eq. (6.118), we have

$$\begin{aligned}\ddot{\boldsymbol{q}}_r = \hat{\boldsymbol{J}}^+(\boldsymbol{q}, \hat{\boldsymbol{\theta}}_k)\ddot{\boldsymbol{x}}_r + \dot{\hat{\boldsymbol{J}}}^+(\boldsymbol{q}, \hat{\boldsymbol{\theta}}_k)\dot{\boldsymbol{x}}_r + (\boldsymbol{I}_n - \hat{\boldsymbol{J}}^+(\boldsymbol{q}, \hat{\boldsymbol{\theta}}_k)\hat{\boldsymbol{J}}(\boldsymbol{q}, \hat{\boldsymbol{\theta}}_k))\dot{\psi} \\ - (\dot{\hat{\boldsymbol{J}}}^+(\boldsymbol{q}, \hat{\boldsymbol{\theta}}_k)\hat{\boldsymbol{J}}(\boldsymbol{q}, \hat{\boldsymbol{\theta}}_k) + \hat{\boldsymbol{J}}^+(\boldsymbol{q}, \hat{\boldsymbol{\theta}}_k)\dot{\hat{\boldsymbol{J}}}(\boldsymbol{q}, \hat{\boldsymbol{\theta}}_k))\psi. \quad (6.119)\end{aligned}$$

Hence, an adaptive sliding vector in joint space is introduced as

$$s = \dot{q} - \dot{q}_r, \tag{6.120}$$

and

$$\dot{s} = \ddot{q} - \ddot{q}_r. \tag{6.121}$$

Multiplying both side of Eq. (6.120) by $\hat{J}(q, \hat{\theta}_k)$ and using Eq. (6.116), one has

$$\hat{J}(q, \hat{\theta}_k)s = \hat{J}(q, \hat{\theta}_k)\dot{q} - \dot{x}_r = \hat{s}_x. \tag{6.122}$$

Substituting Eq. (6.120) into Eq. (6.108) yields

$$\begin{aligned} M(q)\dot{s} + \left(\tfrac{1}{2}\dot{M}(q) + S(q, \dot{q})\right) s + J_m^T(q)B_t J_m(q)s + M(q)\ddot{q}_r \\ + \left(\tfrac{1}{2}\dot{M}(q) + S(q, \dot{q})\right) \dot{q}_r + g(q) + J_m^T(q)B_t J_m(q)\dot{q}_r = \tau + D^T(q)\lambda. \end{aligned} \tag{6.123}$$

The last three terms on the left-hand side of Eq. (6.123) are linear in a set of dynamic parameters θ_d and can be expressed as

$$\begin{aligned} M(q)\ddot{q}_r + \left(\tfrac{1}{2}\dot{M}(q) + S(q, \dot{q})\right) \dot{q}_r + g(q) \\ + J_m^T(q)B_t J_m(q)\dot{q}_r = Y_d(q, \dot{q}, \dot{q}_r, \ddot{q}_r)\theta_d. \end{aligned} \tag{6.124}$$

Then the dynamics Eq. (6.123) can be written as

$$\begin{aligned} M(q)\dot{s} + \left(\tfrac{1}{2}\dot{M}(q) + S(q, \dot{q})\right) s + J_m^T(q)B_t J_m(q)s \\ + Y_d(q, \dot{q}, \dot{q}_r, \ddot{q}_r)\theta_d = \tau + D^T(q)\lambda. \end{aligned} \tag{6.125}$$

The vision and force-tracking controller is designed as

$$\begin{aligned} u = -\hat{J}^T(q, \hat{\theta}_k)K(\Delta\dot{\hat{x}} + \alpha\Delta x) + Y_d(q, \dot{q}, \dot{q}_r, \ddot{q}_r)\hat{\theta}_d \\ - Y_{md}(q)\hat{\theta}_{md}\lambda - \hat{J}_m^T(q, \hat{\theta}_f)\hat{d}(\hat{r})\lambda + \hat{J}_m^T(q, \hat{\theta}_f)R(x)\hat{d}(\hat{r})(\Delta\lambda + \gamma\Delta F), \end{aligned} \tag{6.126}$$

where $\Delta\dot{\hat{x}} = \dot{\hat{x}} - \dot{x}_d$, $K \in \Re^{m \times m}$ is a positive definite matrix, and γ is a positive constant. The estimated parameters $\hat{\theta}_d, \hat{\theta}_f, \hat{\theta}_{md}, \hat{\theta}_k$ are updated by

$$\dot{\hat{\theta}}_d = -L_d Y_d^T(q, \dot{q}, \dot{q}_r, \ddot{q}_r)s, \tag{6.127}$$

$$\dot{\hat{\theta}}_f = L_f Y_f^T(q)s\lambda, \tag{6.128}$$

$$\dot{\hat{\theta}}_{md} = L_{md} Y_{md}^T(q)s\lambda, \tag{6.129}$$

$$\dot{\hat{\theta}}_k = 2L_k Y_k^T(q, \dot{q})K(\Delta\dot{x} + \alpha\Delta x), \tag{6.130}$$

where $\boldsymbol{L}_d \in \Re^{n_d \times n_d}$, $\boldsymbol{L}_f \in \Re^{n_f \times n_f}$, $\boldsymbol{L}_{md} \in \Re^{n_m \times n_m}$, and $\boldsymbol{L}_k \in \Re^{n_k \times n_k}$ are symmetric positive definite gain matrices of the update laws. The rotation matrix $\boldsymbol{R}(\boldsymbol{x})$ is designed so that

$$\bar{\boldsymbol{s}}_x^T \boldsymbol{R}(\boldsymbol{x})\hat{\boldsymbol{d}}(\hat{\boldsymbol{r}}) = 0, \tag{6.131}$$

where

$$\begin{aligned} \bar{\boldsymbol{s}}_x &= \left\{\beta(\Delta\dot{\boldsymbol{x}} + \alpha\Delta\boldsymbol{x})^T \boldsymbol{K}(\hat{\boldsymbol{J}}_m(\boldsymbol{q}, \hat{\boldsymbol{\theta}}_f)\hat{\boldsymbol{J}}^+(\boldsymbol{q}, \hat{\boldsymbol{\theta}}_k))^{-1}\Delta\boldsymbol{F} + \boldsymbol{s}_m^T(\Delta\lambda + \gamma\Delta F)\right\}^T \\ \boldsymbol{s}_m &= \hat{\boldsymbol{J}}_m(\boldsymbol{q}, \hat{\boldsymbol{\theta}}_f)\left\{\dot{\boldsymbol{q}} - \hat{\boldsymbol{J}}^+(\boldsymbol{q}, \hat{\boldsymbol{\theta}}_k)(\dot{\boldsymbol{x}}_d - \alpha\Delta\boldsymbol{x}) - (\boldsymbol{I}_n - \hat{\boldsymbol{J}}^+(\boldsymbol{q}, \hat{\boldsymbol{\theta}}_k)\hat{\boldsymbol{J}}(\boldsymbol{q}, \hat{\boldsymbol{\theta}}_k))\boldsymbol{\psi}\right\}. \end{aligned} \tag{6.132}$$

The role of the rotation matrix is to rotate $\hat{\boldsymbol{d}}(\hat{\boldsymbol{r}})$ so that $\boldsymbol{R}(\boldsymbol{x})\hat{\boldsymbol{d}}(\hat{\boldsymbol{r}})$ is orthonormal to $\bar{\boldsymbol{s}}_x$. In general, $\bar{\boldsymbol{s}}_x$ can be partitioned as $\bar{\boldsymbol{s}}_x = (\bar{\boldsymbol{s}}_{xp}, \bar{\boldsymbol{s}}_{xo})^T$, where $\bar{\boldsymbol{s}}_{xp}$ is the position vector and $\bar{\boldsymbol{s}}_{xo}$ is the orientation vector. Therefore, $\hat{\boldsymbol{d}}(\hat{\boldsymbol{r}})$ can also be partitioned as $(\hat{\boldsymbol{d}}_p(\hat{\boldsymbol{r}}), \hat{\boldsymbol{d}}_o(\hat{\boldsymbol{r}}))^T$. Next, let $\boldsymbol{n}_p$ be a unit vector normal to both the vectors $\bar{\boldsymbol{s}}_{xp}$ and $\hat{\boldsymbol{d}}_p(\hat{\boldsymbol{r}})$, as illustrated in Fig. 6.2.

The rotation matrix $\boldsymbol{R}(\boldsymbol{x})$ is then introduced as

$$\boldsymbol{R}(\boldsymbol{x}) = \begin{bmatrix} \boldsymbol{R}_p & 0 \\ 0 & \boldsymbol{R}_o \end{bmatrix}, \tag{6.133}$$

where the rotation matrix $\boldsymbol{R}_p$ is to rotate the vector $\bar{\boldsymbol{s}}_{xp}$ about the axis $\boldsymbol{n}_p$ so that the vector $\bar{\boldsymbol{s}}_{xp}^T\boldsymbol{R}_p$ is perpendicular to the vector $\hat{\boldsymbol{d}}_p(\hat{\boldsymbol{r}})$, as shown in Fig. 6.2, ϕ_p is the angle between $\bar{\boldsymbol{s}}_{xp}^T\boldsymbol{R}_p$ and $\hat{\boldsymbol{d}}_p(\hat{\boldsymbol{r}})$, which can be determined from the angle ϕ between $\bar{\boldsymbol{s}}_{xp}$ and $\hat{\boldsymbol{d}}_p(\hat{\boldsymbol{r}})$ (see Fig. 6.2). The rotation matrix $\boldsymbol{R}_o$ can be similarly designed. However, since the constraint surface is usually independent of $\boldsymbol{r}_o$, the rotation matrix $\boldsymbol{R}_o$ can be set as the identity matrix. Note that when $\bar{\boldsymbol{s}}_x$ reduces to zero, $\boldsymbol{R}(\boldsymbol{x})$ can be set to any value, because Eq. (6.131) is always satisfied when $\bar{\boldsymbol{s}}_x = \boldsymbol{0}$.

Fig. 6.2 Rotation matrix

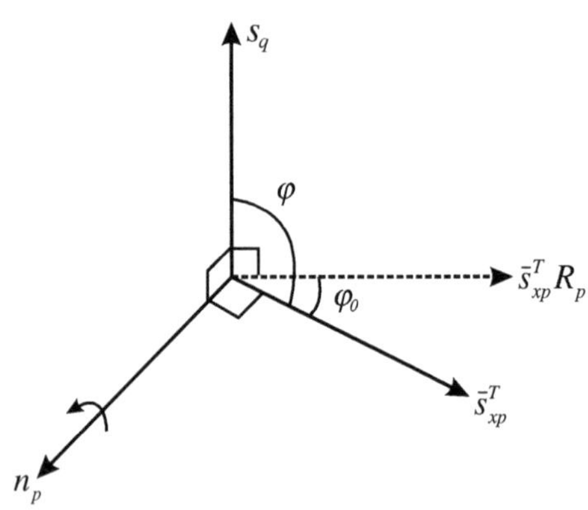

Substituting Eqs. (6.126) into Eq. (6.125), the closed-loop equation is obtained as

$$\begin{aligned} &\boldsymbol{M}(\boldsymbol{q})\dot{\boldsymbol{s}} + \left(\tfrac{1}{2}\dot{\boldsymbol{M}}(\boldsymbol{q}) + \boldsymbol{S}(\boldsymbol{q},\dot{\boldsymbol{q}})\right)\boldsymbol{s} + \boldsymbol{J}_m^T(\boldsymbol{q})\boldsymbol{B}_t\boldsymbol{J}_m(\boldsymbol{q})\boldsymbol{s} \\ &+\hat{\boldsymbol{J}}^T(\boldsymbol{q},\hat{\boldsymbol{\theta}}_k)\boldsymbol{K}(\Delta\dot{\hat{\boldsymbol{x}}} + \alpha\Delta\boldsymbol{x}) + \boldsymbol{Y}_d(\boldsymbol{q},\dot{\boldsymbol{q}},\dot{\boldsymbol{q}}_r,\ddot{\boldsymbol{q}}_r)\Delta\boldsymbol{\theta}_d \\ = &\boldsymbol{Y}_f(\boldsymbol{q})\Delta\boldsymbol{\theta}_f\lambda + \boldsymbol{Y}_{md}(\boldsymbol{q})\Delta\boldsymbol{\theta}_{md}\lambda + \hat{\boldsymbol{J}}_m^T(\boldsymbol{q},\hat{\boldsymbol{\theta}}_f)\boldsymbol{R}(\boldsymbol{x})\hat{\boldsymbol{d}}(\hat{\boldsymbol{r}})(\Delta\lambda + \gamma\Delta F), \end{aligned} \tag{6.134}$$

where $\Delta\boldsymbol{\theta}_d = \boldsymbol{\theta}_d - \hat{\boldsymbol{\theta}}_d$, and $\Delta\boldsymbol{\theta}_f = \boldsymbol{\theta}_f - \hat{\boldsymbol{\theta}}_f$.

To carry out the stability analysis, a Lyapunov-like function candidate V is defined as

$$\begin{aligned} V = &\tfrac{1}{2}\boldsymbol{s}^T\boldsymbol{M}(\boldsymbol{q})\boldsymbol{s} + \alpha\Delta\boldsymbol{x}^T\boldsymbol{K}\Delta\boldsymbol{x} + \tfrac{1}{2}\Delta\boldsymbol{\theta}_d^T\boldsymbol{L}_d^{-1}\Delta\boldsymbol{\theta}_d \\ &+\tfrac{1}{2}\Delta\boldsymbol{\theta}_f^T\boldsymbol{L}_f^{-1}\Delta\boldsymbol{\theta}_f + \tfrac{1}{2}\Delta\boldsymbol{\theta}_k^T\boldsymbol{L}_k^{-1}\Delta\boldsymbol{\theta}_k + \tfrac{1}{2}\Delta\boldsymbol{\theta}_{md}^T\boldsymbol{L}_{md}^{-1}\Delta\boldsymbol{\theta}_{md} + \tfrac{1}{2}\beta\Delta F^2, \end{aligned} \tag{6.135}$$

where $\Delta\boldsymbol{\theta}_k = \boldsymbol{\theta}_k - \hat{\boldsymbol{\theta}}_k$. Differentiating V with respect to time and substituting Eqs. (6.127)–(6.129), and (6.134) into it, and using Eq. (6.122), one has

$$\begin{aligned} \dot{V} = &-\boldsymbol{s}^T(\boldsymbol{J}_m^T(\boldsymbol{q})\boldsymbol{B}_t\boldsymbol{J}_m(\boldsymbol{q}))\boldsymbol{s} - \hat{\boldsymbol{s}}_x^T\boldsymbol{K}(\Delta\dot{\hat{\boldsymbol{x}}} + \alpha\Delta\boldsymbol{x}) + 2\alpha\Delta\boldsymbol{x}^T\boldsymbol{K}\Delta\dot{\boldsymbol{x}} \\ &+\boldsymbol{s}^T\hat{\boldsymbol{J}}_m^T(\boldsymbol{q},\hat{\boldsymbol{\theta}}_f)\boldsymbol{R}(\boldsymbol{x})\hat{\boldsymbol{d}}(\hat{\boldsymbol{r}})(\Delta\lambda + \gamma\Delta F) - \Delta\boldsymbol{\theta}_k^T\boldsymbol{L}_k^{-1}\dot{\hat{\boldsymbol{\theta}}}_k + \beta\Delta F\Delta\lambda. \end{aligned} \tag{6.136}$$

From Eqs. (6.120), (6.114), and (6.132), note that

$$\begin{aligned} \hat{\boldsymbol{J}}_m(\boldsymbol{q},\hat{\boldsymbol{\theta}}_f)\boldsymbol{s} = &\hat{\boldsymbol{J}}_m(\boldsymbol{q},\hat{\boldsymbol{\theta}}_f)\dot{\boldsymbol{q}} - \hat{\boldsymbol{J}}_m(\boldsymbol{q},\hat{\boldsymbol{\theta}}_f)\hat{\boldsymbol{J}}^+(\boldsymbol{q},\hat{\boldsymbol{\theta}}_k)(\dot{\boldsymbol{x}}_d - \alpha\Delta\boldsymbol{x}) - \beta\boldsymbol{R}(\boldsymbol{x})\hat{\boldsymbol{d}}(\hat{\boldsymbol{r}})\Delta F \\ &-\hat{\boldsymbol{J}}_m(\boldsymbol{q},\hat{\boldsymbol{\theta}}_f)(\boldsymbol{I}_n - \hat{\boldsymbol{J}}^+(\boldsymbol{q},\hat{\boldsymbol{\theta}}_k)\hat{\boldsymbol{J}}(\boldsymbol{q},\hat{\boldsymbol{\theta}}_k))\boldsymbol{\psi} = \boldsymbol{s}_m - \beta\boldsymbol{R}(\boldsymbol{x})\hat{\boldsymbol{d}}(\hat{\boldsymbol{r}})\Delta F. \end{aligned} \tag{6.137}$$

Next, substituting Eqs. (6.116), (6.114), and (6.137) into Eq. (6.136) and using Eq. (6.132), one has

$$\begin{aligned} \dot{V} = &-\boldsymbol{s}^T(\boldsymbol{J}_m^T(\boldsymbol{q})\boldsymbol{B}_t\boldsymbol{J}_m(\boldsymbol{q}))\boldsymbol{s} - (\Delta\dot{\hat{\boldsymbol{x}}} + \alpha\Delta\boldsymbol{x})^T\boldsymbol{K}(\Delta\dot{\hat{\boldsymbol{x}}} + \alpha\Delta\boldsymbol{x}) \\ &+ \beta(\Delta\dot{\hat{\boldsymbol{x}}} + \alpha\Delta\boldsymbol{x})^T\boldsymbol{K}(\hat{\boldsymbol{J}}_m(\boldsymbol{q},\hat{\boldsymbol{\theta}}_f)\hat{\boldsymbol{J}}^+(\boldsymbol{q},\hat{\boldsymbol{\theta}}_k))^{-1}\boldsymbol{R}(\boldsymbol{x})\hat{\boldsymbol{d}}(\hat{\boldsymbol{r}})\Delta F - \beta\Delta F(\Delta\lambda + \gamma\Delta F) \\ &+ \boldsymbol{s}_m^T\boldsymbol{R}(\boldsymbol{x})\hat{\boldsymbol{d}}(\hat{\boldsymbol{r}})(\Delta\boldsymbol{\lambda} + \gamma\Delta\boldsymbol{F}) + 2\alpha\Delta\boldsymbol{x}^T\boldsymbol{K}\Delta\dot{\boldsymbol{x}} - \Delta\boldsymbol{\theta}_k^T\boldsymbol{L}_k^{-1}\dot{\hat{\boldsymbol{\theta}}}_k + \beta\Delta F\Delta\lambda \\ = &-\boldsymbol{s}^T(\boldsymbol{J}_m^T(\boldsymbol{q})\boldsymbol{B}_t\boldsymbol{J}_m(\boldsymbol{q}))\boldsymbol{s} - (\Delta\dot{\hat{\boldsymbol{x}}} + \alpha\Delta\boldsymbol{x})^T\boldsymbol{K}(\Delta\dot{\hat{\boldsymbol{x}}} + \alpha\Delta\boldsymbol{x}) + \bar{\boldsymbol{s}}_x^T\boldsymbol{R}(\boldsymbol{x})\hat{\boldsymbol{d}}(\hat{\boldsymbol{r}}) \\ &- \beta\Delta F(\Delta\lambda + \gamma\Delta F) + 2\alpha\Delta\boldsymbol{x}^T\boldsymbol{K}\Delta\dot{\boldsymbol{x}} - \Delta\boldsymbol{\theta}_k^T\boldsymbol{L}_k^{-1}\dot{\hat{\boldsymbol{\theta}}}_k + \beta\Delta F\Delta\lambda. \end{aligned} \tag{6.138}$$

Using Eq. (6.131) on the above equation yields

$$\begin{aligned} \dot{V} = &-\boldsymbol{s}^T(\boldsymbol{J}_m^T(\boldsymbol{q})\boldsymbol{B}_t\boldsymbol{J}_m(\boldsymbol{q}))\boldsymbol{s} - (\Delta\dot{\hat{\boldsymbol{x}}} + \alpha\Delta\boldsymbol{x})^T\boldsymbol{K}(\Delta\dot{\hat{\boldsymbol{x}}} \\ &+ \alpha\Delta\boldsymbol{x}) + 2\alpha\Delta\boldsymbol{x}^T\boldsymbol{K}\Delta\dot{\boldsymbol{x}} - \Delta\boldsymbol{\theta}_k^T\boldsymbol{L}_k^{-1}\dot{\hat{\boldsymbol{\theta}}}_k - \beta\gamma\Delta F^2, \end{aligned} \tag{6.139}$$

where $\boldsymbol{R}^T(\boldsymbol{x})\boldsymbol{R}(\boldsymbol{x}) = \boldsymbol{I}_m$, and $\hat{\boldsymbol{d}}^T(\hat{\boldsymbol{r}})\hat{\boldsymbol{d}}(\hat{\boldsymbol{r}}) = 1$.

From Eq. (6.111), since $\dot{\hat{x}} = \dot{x} - Y_k(q, \dot{q})\Delta\theta_k$, one has

$$\Delta\dot{\hat{x}} = \Delta\dot{x} - Y_k(q, \dot{q})\Delta\theta_k. \tag{6.140}$$

Substituting Eqs. (6.130) and (6.140) into Eq. (6.139) yields

$$\begin{aligned}\dot{V} &= -s^T(J_m^T(q)B_t J_m(q))s - \Delta\dot{x}^T K\Delta\dot{x} - \alpha^2\Delta x^T K\Delta x\\ &\quad -\Delta\theta_k^T Y_k^T(q, \dot{q})KY_k(q, \dot{q})\Delta\theta_k - \beta\gamma\Delta F^2 \le 0.\end{aligned} \tag{6.141}$$

The following theorem summarizes these results:

Theorem 6.3 *The adaptive Jacobian control law* (6.126) *and the update laws* (6.127), (6.128), (6.129), (6.130) *for the robot system* (6.108) *result in the convergence of vision and force-tracking errors. That is,* $x(t) - x_d(t) \to 0$ *and* $\dot{x}(t) - \dot{x}_d(t) \to 0$ *as* $t \to \infty$. *In addition, the contact force* $\lambda(t)$ *also converges to* $\lambda_d(t)$ *as* $t \to \infty$.

Proof Since $M(q)$ is positive definite, V is positive definite in s, Δx, $\Delta\theta_d$, $\Delta\theta_f$, $\Delta\theta_k$, $\Delta\theta_{md}$, and ΔF. Since $\dot{V} \le 0$, V is bounded. Therefore, s, Δx, $\Delta\theta_d$, $\Delta\theta_f$, $\Delta\theta_k$, $\Delta\theta_{md}$, and ΔF are bounded vectors. This implies that $\hat{\theta}_d$, $\hat{\theta}_f$, $\hat{\theta}_k$, $\hat{\theta}_{md}$, and x are bounded, and $\hat{s}_x = \hat{J}(q, \hat{\theta}_k)s$ is also bounded.

As seen from Eq. (6.141), $\dot{V}$ is negative definite in s, $\Delta\dot{x}$, Δx, $Y_k(q, \dot{q})\Delta\theta_k$, and ΔF. To show the convergence of the errors, Barbalat's Lemma will be used to check the uniform continuity of $\dot{V}$. In the following development, we show uniform continuity by showing that $\ddot{V}$ is bounded.

First, note that $\dot{x}_r$, $\dot{\hat{x}}$ are bounded, as seen from Eqs. (6.114) and (6.116). From Eq. (6.118), we can conclude that $\dot{q}_r$ is bounded when $\hat{J}(q, \hat{\theta}_k)$ is nonsingular. Therefore, $\dot{q}$ is bounded, since s is bounded. The boundedness of $\dot{q}$ means that $\dot{x}$, $\dot{r}$ are bounded. Hence $\Delta\dot{x}$ is bounded and $\ddot{x}_r - \beta(\hat{J}_m(q, \hat{\theta}_f)\hat{J}^+(q, \hat{\theta}_k))^{-1}R(x)d(r)\Delta\lambda$ from Eq. (6.115) is also bounded if $\ddot{x}_d$ is bounded, $\dot{d}(r, \dot{r})$ is bounded because r, $\dot{r}$ are bounded.

As seen from Eq. (6.130), $\dot{\hat{\theta}}_k$ is bounded because $\dot{q}$, $\Delta\dot{x}$, and Δx are all bounded. Hence, $\dot{\hat{J}}(q, \hat{\theta}_k)$ is bounded. Using Eq. (6.119), one has

$$\begin{aligned}&\ddot{q}_r - \beta\hat{J}^+(q, \hat{\theta}_k)(\hat{J}_m(q, \hat{\theta}_f)\hat{J}^+(q, \hat{\theta}_k))^{-1}R(x)\hat{d}(\hat{r})\Delta\lambda\\ &= \hat{J}^+(q, \hat{\theta}_k))(\ddot{x}_r - \beta(\hat{J}_m(q, \hat{\theta}_f)\hat{J}^+(q, \hat{\theta}_k))^{-1}R(x)\hat{d}(\hat{r})\Delta\lambda)\\ &\quad + \dot{\hat{J}}^+(q, \hat{\theta}_k))\dot{x}_r + (I_n - \hat{J}^+(q, \hat{\theta}_k))\hat{J}(q, \hat{\theta}_k))\dot{\psi}\\ &\quad -(\dot{\hat{J}}^+(q, \hat{\theta}_k)\hat{J}(q, \hat{\theta}_k) + \hat{J}^+(q, \hat{\theta}_k)\dot{\hat{J}}(q, \hat{\theta}_k))\psi.\end{aligned} \tag{6.142}$$

Since all the terms on the right-hand side of the above equation are bounded, $\ddot{q}_r - \beta\hat{J}^+(q, \hat{\theta}_k)(\hat{J}_m(q, \hat{\theta}_f)\hat{J}^+(q, \hat{\theta}_k))^{-1}R(x)d(r)\Delta\lambda$ is therefore bounded. Next, let

$$\begin{aligned}
Y_d(q,\dot{q},\dot{q}_r,\ddot{q}_r)\Delta\theta_d &= (M(q)-\hat{M}(q))\ddot{q}_r + \left[\tfrac{1}{2}(\dot{M}(q)-\dot{\hat{M}}(q))\right.\\
&+S(q,\dot{q})-\hat{S}(q,\dot{q})+J_m^T(q)B_t J_m(q)-\hat{J}_m^T(q)B_t\hat{J}_m(q)\left.\right]\dot{q}_r+g(q)-\hat{g}(q)\\
&+\beta M(q)\hat{J}^+(q,\hat{\theta}_k)(\hat{J}_m(q,\hat{\theta}_f)\hat{J}^+(q,\hat{\theta}_k))^{-1}R(x)\\
&\times(\dot{d}(h(q),\,J_m(q)\dot{q})-\dot{\hat{d}}(\hat{h}(q),\,\hat{J}_m(q,\hat{\theta}_f)\dot{q}))\Delta F\\
&=(M(q)-\hat{M}(q))\ddot{q}_r+\bar{Z}_d(q,\dot{q},\dot{q}_r,x,\hat{\theta}_f,\hat{\theta}_k)\Delta\theta_d,
\end{aligned} \tag{6.143}$$

where

$$\begin{aligned}
\bar{Z}_d(q,\dot{q},\dot{q}_r,x,\hat{\theta}_f,\hat{\theta}_k)\Delta\theta_d &= \left(\tfrac{1}{2}(\dot{M}(q)-\dot{\hat{M}}(q))\right)+S(q,\dot{q})-\hat{S}(q,\dot{q})\\
&+J_m^T(q)B_t J_m(q)-\hat{J}_m^T(q)B_t\hat{J}_m(q)]\dot{q}_r+g(q)-\hat{g}(q)\\
&+\beta M(q)\hat{J}^+(q,\hat{\theta}_k)(\hat{J}_m(q,\hat{\theta}_f)\hat{J}^+(q,\hat{\theta}_k))^{-1}R(x)(\dot{d}(h(q),\,J_m(q)\dot{q})\\
&-\dot{\hat{d}}(\hat{h}(q),\,\hat{J}_m(q,\hat{\theta}_f)\dot{q})\Delta F.
\end{aligned} \tag{6.144}$$

Then from Eqs. (6.134) and (6.143), since $D(q)\ddot{q}=-\dot{D}(q)\dot{q}$, one has

$$\begin{aligned}
&-\dot{D}(q)\dot{q}+D(q)M^{-1}(q)\{-M(q)(\ddot{q}_r-\beta\hat{J}^+(q,\hat{\theta}_k)\\
&\quad(\hat{J}_m(q,\hat{\theta}_f)\hat{J}^+(q,\hat{\theta}_k))^{-1}R(x)\hat{d}(\hat{r})\Delta\lambda)+(M(q)\\
&-\hat{M}(q))(\ddot{q}_r-\beta\hat{J}^+(q,\hat{\theta}_k)(\hat{J}_m(q,\hat{\theta}_f)\hat{J}^+(q,\hat{\theta}_k))^{-1}R(x)\hat{d}(\hat{r})\Delta\lambda)+r(t)\}\\
&=D(q)M^{-1}(q)\{D_1\Delta\lambda-\hat{J}_m^T(q,\hat{\theta}_f)R(x)\hat{d}(\hat{r})\lambda_d\\
&+\beta\hat{M}(q)\hat{J}^+(q,\hat{\theta}_k)(\hat{J}_m(q,\hat{\theta}_f)\hat{J}^+(q,\hat{\theta}_k))^{-1}R(x)\hat{d}(\hat{r})\Delta\lambda+D_1\lambda_d\},
\end{aligned} \tag{6.145}$$

where

$$\begin{aligned}
r(t) &= \left(\tfrac{1}{2}\dot{M}(q)+S(q,\dot{q})\right)s+\bar{Z}_d(q,\dot{q},\dot{q}_r,x,\hat{\theta}_f,\hat{\theta}_k)\Delta\theta_d\\
&+\hat{J}^T(q,\theta_k)K(\Delta\dot{x}+\alpha\Delta x)-\gamma\hat{J}_m^T(q,\hat{\theta}_f)R(x)\hat{d}(\hat{r})\Delta F,\\
D_1 &= (Y_f(q)\Delta\theta_f+Y_{md}(q)\Delta\theta_{md}+\hat{J}_m^T(q,\hat{\theta}_f)R(x)\hat{d}(\hat{r})).
\end{aligned} \tag{6.146}$$

The above equation can be written as

$$\bar{r}(t)=k(t)\Delta\lambda, \tag{6.147}$$

where

$$\begin{aligned}
\bar{r}(t) &= -\dot{D}(q)\dot{q}+D(q)M^{-1}(q)\Big\{-M(q)(\ddot{q}_r-\beta\hat{J}^+(q,\hat{\theta}_k)(\hat{J}_m(q,\hat{\theta}_f)\hat{J}^+(q,\hat{\theta}_k))^{-1}\\
&\times R(x)\hat{d}(\hat{r})\Delta\lambda)+(M(q)-\hat{M}(q))(\ddot{q}_r-\beta\hat{J}^+(q,\hat{\theta}_k))(\hat{J}_m(q,\hat{\theta}_f)\hat{J}^+(q,\hat{\theta}_k))^{-1}\\
&\times R(x)\hat{d}(\hat{r})\Delta\lambda)+r(t)-(Y_f(q)\Delta\theta_f+Y_{md}(q)\Delta\theta_{md})\lambda_d\Big\},
\end{aligned} \tag{6.148}$$

and $k(t) = D(q)M^{-1}(q)\{D_1 + \beta\hat{M}(q)\hat{J}^+(q,\hat{\theta}_k)(\hat{J}_m(q,\hat{\theta}_f)\hat{J}^+(q,\hat{\theta}_k))^{-1}$ $R(x)\hat{d}(\hat{r}))\}$ are bounded scalars. Hence $\Delta\lambda$ is bounded, and the boundedness of $\Delta\lambda$ implies that $\ddot{x}_r$ and $\ddot{q}_r$ are bounded.

From the closed-loop Eq. (6.134), it can be concluded that $\dot{s}$ is bounded. The boundedness of $\dot{s}$ implies the boundedness of $\ddot{q}$, as seen from Eq. (6.121). Since $\ddot{x} = J(q)\ddot{q} + \dot{J}(q)\dot{q}$, $\ddot{x}$ is bounded, and hence $\Delta\ddot{x}$ is also bounded. To apply Barbalat's lemma, let us check the uniform continuity of $\dot{V}$. Differentiating Eq. (6.141) with respect to time gives

$$\ddot{V} = -2s^T(J_m^T(q)B_t J_m(q))\dot{s} + 2s^T J_m^T(q)B_t\dot{J}_m(q)s - 2\Delta\dot{x}^T K\Delta\ddot{x} - 2\alpha^2\Delta x^T K\Delta\dot{x}$$
$$+ 2\Delta\theta_k^T Y_k^T(q,\dot{q})KY_k(q,\dot{q})\dot{\hat{\theta}}_k - 2\Delta\theta_k^T Y_k^T(q,\dot{q})K\dot{Y}_k(q,\dot{q})\Delta\theta_k - 2\beta\gamma\Delta F\Delta\lambda. \tag{6.149}$$

This shows that $\ddot{V}$ is bounded, since Δx, $\Delta\dot{x}$, $\dot{q}$, $\Delta\ddot{x}$, $\dot{s}$, ΔF, $\Delta\lambda$ are all bounded. Hence, $\dot{V}$ is uniformly continuous. Using Barbalat's lemma, one has $\Delta x \to \mathbf{0}$, $\Delta\dot{x} \to \mathbf{0}$, $Y_k(q,\dot{q})\Delta\theta_k = \dot{x} - \dot{\hat{x}} \to \mathbf{0}$, $\Delta F \to 0$ as $t \to \infty$.

Next, the convergence of the force-tracking error will be shown. From Eqs. (6.127), (6.128), and (6.130), note that $\dot{\hat{\theta}}_d$, $\dot{\hat{\theta}}_f$, $\dot{\hat{\theta}}_k$ are bounded. Let $z_x = \ddot{x}_r - \beta\hat{J}_s R(x)d(r)$ $\Delta\lambda$ and $\hat{J}_s = (\hat{J}_m(q,\hat{\theta}_f)\hat{J}^+(q,\hat{\theta}_k))^{-1}$. From Eq. (6.115), one has

$$z_x = \ddot{x}_d - \alpha\Delta\dot{x} + \beta\hat{J}_s R(x)\dot{\hat{d}}(\hat{r},\dot{r})\Delta F$$
$$+ \beta\hat{J}_s\dot{R}(x)\hat{d}(\hat{r})\Delta F + \beta\dot{\hat{J}}_s R(x)\hat{d}(\hat{r})\Delta F. \tag{6.150}$$

Since $\Delta\ddot{x}, \dot{x}, \Delta\lambda, \ddot{q}$ are bounded, $\dot{z}_x$ is also bounded if the derivative of $\ddot{x}_d$ is bounded. Hence, the derivative of $\ddot{q}_r - \beta\hat{J}^+(q,\hat{\theta}_k)(\hat{J}_m(q,\hat{\theta}_f)\hat{J}^+(q,\hat{\theta}_k))^{-1}R(x)\hat{d}(\hat{r})\Delta\lambda$ is also bounded, as seen from Eq. (6.142). From Eq. (6.145), the derivative λ is therefore bounded. Finally, using Barbalat's lemma and the convergence of ΔF to 0, we have $\Delta\lambda \to 0$ as $t \to \infty$. $\triangle\triangle\triangle$

6.4 Multifingered Robot Hands

A human being is able to manipulate an object using fingers, without exact knowledge of the object and the locations of contact points between the fingers and the object. However, in many control methods of multifingered robot hands, such information is needed in the design of the controllers. In this section, the sensory task-space regulation problem of multifingered robot hands (see Fig. 6.3) is considered, and it is shown that sensory feedback can be used to overcome these uncertainties.

6.4.1 Dynamic Model of Multifingered Robot Hands

A set of k fingers holding an object as illustrated in Figs. 6.3 and 6.4 is considered. Let Σ denote the Cartesian coordinate, Σ_o the object coordinate frame fixed at the mass center of the object and moving with the object, Σ_{ci} the contact point frame located at the contact point of the ith finger, and Σ_{ei} the finger coordinate frame located at the ith finger, as shown in Fig. 6.3. The velocity vector $\boldsymbol{v}_o$ of the object in Σ_o is related to the velocity vector $\boldsymbol{v}_{ci}$ at the contact point of the ith finger in Σ_{ci} by

$$\boldsymbol{v}_{ci} = \boldsymbol{J}_{oi}\boldsymbol{v}_o, \tag{6.151}$$

where $\boldsymbol{J}_{oi} \in \Re^{n_o \times n_o}$ denotes the Jacobian matrix from Σ_o to Σ_{ci}. The velocity vector $\boldsymbol{v}_{ci}$ in Σ_{ci} is related to the velocity vector $\boldsymbol{v}_{ei}$ of the ith finger in Σ_{ei} by

$$\boldsymbol{v}_{ci} = \boldsymbol{J}_{fi}\boldsymbol{v}_{ei}, \tag{6.152}$$

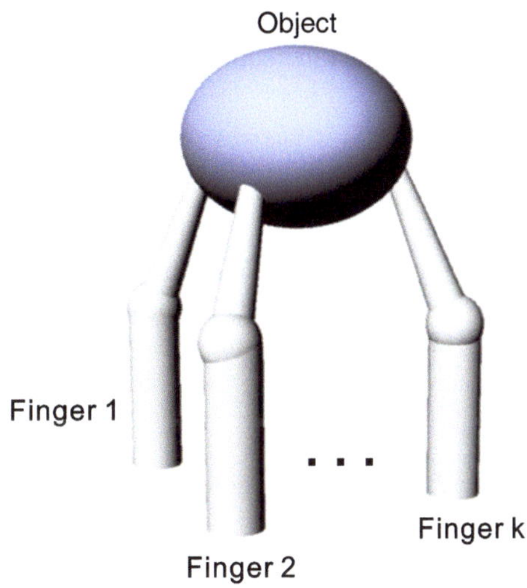

Fig. 6.3 Multifingered robot holding an object

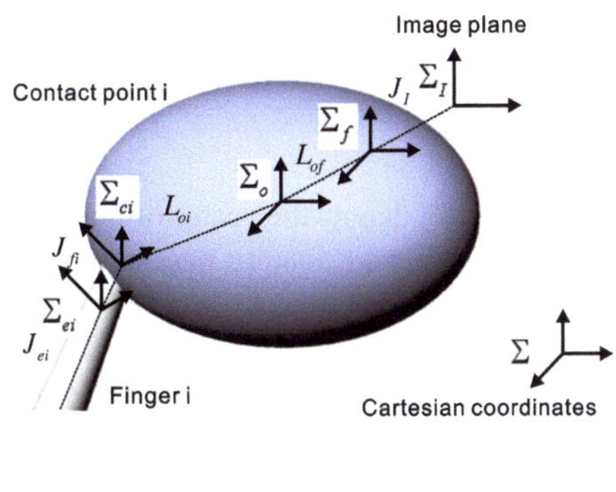

Fig. 6.4 Definition of coordinates

where $\boldsymbol{J}_{fi} \in \Re^{n_o \times n_o}$ denotes the Jacobian matrix from Σ_{ei} to Σ_{ci}. Since $\boldsymbol{v}_{ei}$ in Σ_{ei} is expressed as

$$\boldsymbol{v}_{ei} = \boldsymbol{J}_{ei}(\boldsymbol{q}_i)\dot{\boldsymbol{q}}_i, \tag{6.153}$$

where $\boldsymbol{q}_i \in \Re^{ni}(n_i > n_o)$ is the joint coordinates of the ith finger and $\boldsymbol{J}_{ei}(\boldsymbol{q}_i) \in \Re^{n_o \times n_i}$ is the Jacobian matrix of Σ_{ei} in $\boldsymbol{q}_i$. From Eqs. (6.151)–(6.153), the velocity of the joint variables $\boldsymbol{q} = [\boldsymbol{q}_1^T, \ldots, \boldsymbol{q}_k^T]^T$ and the velocity of the object $\boldsymbol{v}_o$ are constrained by the following equation:

$$\boldsymbol{J}_o\boldsymbol{v}_o = \boldsymbol{J}_e(\boldsymbol{q})\dot{\boldsymbol{q}},, \tag{6.154}$$

where $\boldsymbol{J}_o = [\boldsymbol{J}_{o1}^T, \ldots, \boldsymbol{J}_{ok}^T]^T$ and $\boldsymbol{J}_e(\boldsymbol{q}) = \text{diag}\{\boldsymbol{J}_{f1}\boldsymbol{J}_{e1}(\boldsymbol{q}_1), \ldots, \boldsymbol{J}_{fk}\boldsymbol{J}_{ek}(\boldsymbol{q}_k)\}$. In addition, let $\boldsymbol{r} \in \Re^{n_o}$ denote the position and orientation vector of Σ_o in Σ, Then the velocity vector $\dot{\boldsymbol{r}}$ is related to the velocity vector $\boldsymbol{v}_o$ by

$$\boldsymbol{v}_o = \boldsymbol{J}_v(\boldsymbol{r})\dot{\boldsymbol{r}}, \tag{6.155}$$

where $\boldsymbol{J}_v(\boldsymbol{r})$ is a nonsingular Jacobian mapping from Σ to Σ_o. From Eqs. (6.154) and (6.155), the kinematic constraint between $\dot{\boldsymbol{r}}$ and $\dot{\boldsymbol{q}}$ is

$$\boldsymbol{J}(\boldsymbol{r})\dot{\boldsymbol{r}} = \boldsymbol{J}_e(\boldsymbol{q})\dot{\boldsymbol{q}}, \tag{6.156}$$

where $\boldsymbol{J}(\boldsymbol{r}) = \boldsymbol{J}_o\boldsymbol{J}_v(\boldsymbol{r})$.

In this section, cameras that are fixed in the workspace are used to observe a feature on the object. We define a feature coordinate frame Σ_f fixed on the feature of the object, and $\boldsymbol{r}_f \in \Re^{n_o}$ is the position and orientation vector relative to the origin of Σ_f in Σ. Then $\dot{\boldsymbol{r}}$ has the following relation to $\dot{\boldsymbol{r}}_f$:

$$\dot{\boldsymbol{r}} = \boldsymbol{J}_{of}\dot{\boldsymbol{r}}_f, \tag{6.157}$$

where $\boldsymbol{J}_{of} \in \Re^{n_o \times n_o}$ denotes the Jacobian matrix from Σ_f to Σ_o. From Eqs. (6.156) and (6.157), one has

$$\boldsymbol{J}_f(\boldsymbol{r})\dot{\boldsymbol{r}}_f = \boldsymbol{J}_e(\boldsymbol{q})\dot{\boldsymbol{q}}, \tag{6.158}$$

where

$$\boldsymbol{J}_f(\boldsymbol{r}) = \boldsymbol{J}(\boldsymbol{r})\boldsymbol{J}_{of}. \tag{6.159}$$

Let $\boldsymbol{x} \in \Re^m$ denote the vector of image feature parameters, and $\dot{\boldsymbol{x}}$ the corresponding vector of image feature parameter rates of change. The relationship between Cartesian space and image space is represented by

$$\dot{\boldsymbol{x}} = \boldsymbol{J}_s(\boldsymbol{r}_f)\dot{\boldsymbol{r}}_f, \tag{6.160}$$

where $\boldsymbol{J}_s(\boldsymbol{r}_f) \in \Re^{m \times n_o}$ is the image Jacobian matrix.

The dynamics equations of the ith finger are described in the joint coordinates $\boldsymbol{q}_i$ by

$$\boldsymbol{M}_i(\boldsymbol{q}_i)\ddot{\boldsymbol{q}}_i + \left(\tfrac{1}{2}\dot{\boldsymbol{M}}_i(\boldsymbol{q}_i) + \boldsymbol{S}_i(\boldsymbol{q}_i, \dot{\boldsymbol{q}}_i)\right)\dot{\boldsymbol{q}}_i + \boldsymbol{g}_i(\boldsymbol{q}_i) = \boldsymbol{\tau}_i - \boldsymbol{J}_{ei}^T(\boldsymbol{q}_i)\boldsymbol{J}_{fi}^T\boldsymbol{f}_{ei}, \quad (6.161)$$

where $\boldsymbol{M}_i(\boldsymbol{q}_i) \in \Re^{ni\times ni}$ is the inertia matrix, which is symmetric and positive definite for all $\boldsymbol{q}_i$, $\boldsymbol{g}_i(\boldsymbol{q}_i) \in \Re^{ni}$ is the gravitational force, $\boldsymbol{\tau}_i \in \Re^{ni}$ is the control input, $\boldsymbol{f}_{ei} \in \Re^{ni}$ is the force exerted on the object by the ith finger.

Then the dynamic equation of the k fingers can be described in $\boldsymbol{q}$ as

$$\boldsymbol{M}(\boldsymbol{q})\ddot{\boldsymbol{q}} + \left(\tfrac{1}{2}\dot{\boldsymbol{M}}(\boldsymbol{q}) + \boldsymbol{S}(\boldsymbol{q}, \dot{\boldsymbol{q}})\right)\dot{\boldsymbol{q}} + \boldsymbol{g}(\boldsymbol{q}) = \boldsymbol{\tau} - \boldsymbol{J}_e^T(\boldsymbol{q})\boldsymbol{f}_e, \quad (6.162)$$

where $\boldsymbol{M}(\boldsymbol{q}) = \mathrm{diag}\{\boldsymbol{M}_1(\boldsymbol{q}_1), \ldots, \boldsymbol{M}_k(\boldsymbol{q}_k)\}$, $\boldsymbol{g}(\boldsymbol{q}) = [\boldsymbol{g}_1^T(\boldsymbol{q}), \ldots, \boldsymbol{g}_k^T(\boldsymbol{q})]^T$, $\boldsymbol{S}(\boldsymbol{q}, \dot{\boldsymbol{q}})) = \mathrm{diag}\{\boldsymbol{S}_1(\boldsymbol{q}_1, \dot{\boldsymbol{q}}_1), \ldots, \boldsymbol{S}_k(\boldsymbol{q}_k, \dot{\boldsymbol{q}}_k)\}$, $\boldsymbol{\tau} = [\boldsymbol{\tau}_1^T, \ldots, \boldsymbol{\tau}_k^T]^T$, and $\boldsymbol{f}_e = [\boldsymbol{f}_{e1}^T, \ldots, \boldsymbol{f}_{ek}^T]^T$. Note that $\boldsymbol{M}(\boldsymbol{q})$ is also symmetric and positive definite for all $\boldsymbol{q}$, and $\boldsymbol{S}(\boldsymbol{q}, \dot{\boldsymbol{q}})$ is skew-symmetric.

The equation of motion of the object can be written in Σ as

$$\boldsymbol{M}_o(\boldsymbol{r})\ddot{\boldsymbol{r}} + \left(\tfrac{1}{2}\dot{\boldsymbol{M}}_o(\boldsymbol{r}) + \boldsymbol{S}_o(\boldsymbol{r}, \dot{\boldsymbol{r}})\right)\dot{\boldsymbol{r}} + \boldsymbol{g}_o(\boldsymbol{r}) = \boldsymbol{F}, \quad (6.163)$$

where $\boldsymbol{F}$ is represented by $\boldsymbol{f}_e$ as

$$\boldsymbol{F} = \boldsymbol{J}^T(\boldsymbol{r})\boldsymbol{f}_e, \quad (6.164)$$

$\boldsymbol{M}_o(\boldsymbol{r})$ is a positive definite inertial matrix, $\boldsymbol{S}_o(\boldsymbol{r}, \dot{\boldsymbol{r}})$ is a skew-symmetric matrix came from Coriolis and $\boldsymbol{F}$ is the total force exerted on the object by the fingers. Substituting Eq. (6.157) into Eq. (6.163) yields

$$\boldsymbol{M}_o(\boldsymbol{r})\frac{\mathrm{d}}{\mathrm{d}t}(\boldsymbol{J}_{of}\dot{\boldsymbol{r}}_f) + \left(\tfrac{1}{2}\dot{\boldsymbol{M}}_o(\boldsymbol{r}) + \boldsymbol{S}_o(\boldsymbol{r}, \dot{\boldsymbol{r}})\right)\boldsymbol{J}_{of}\dot{\boldsymbol{r}}_f + \boldsymbol{g}_o(\boldsymbol{r}) = \boldsymbol{F}. \quad (6.165)$$

From Eqs. (6.164) and (6.158), it follows that

$$\boldsymbol{f}_e = (\boldsymbol{J}^T(\boldsymbol{r}))^+\boldsymbol{F} + \boldsymbol{Z}\boldsymbol{f}_{int}, \quad (6.166)$$

$$\dot{\boldsymbol{r}}_f = \boldsymbol{J}_f^+\boldsymbol{J}_e\dot{\boldsymbol{q}} = \boldsymbol{J}_r(\boldsymbol{q})\dot{\boldsymbol{q}}, \quad (6.167)$$

where $(\boldsymbol{J}^T(\boldsymbol{r}))^+\boldsymbol{F}$ is a particular solution to Eq. (6.164) that represents the force that causes the motion of the object, $(\boldsymbol{J}^T(\boldsymbol{r}))^+$ is the generalized inverse of $\boldsymbol{J}^T(\boldsymbol{r})$, $\boldsymbol{Z}\boldsymbol{f}_{int}$ represents the internal force, which does not effect motion of the object, $\boldsymbol{Z}$ is a matrix of orthonormal, which is obtained from a set of independent vectors of the null space of $\boldsymbol{J}^T$, $\boldsymbol{f}_{int} \in \Re^{n_I}$ is an arbitrary vector, which determines the magnitude of the internal force.

Substituting Eqs. (6.166) and (6.167) into Eq. (6.162) and using Eq. (6.165), the coordinated system can be rewritten as follows:

$$\begin{aligned} \boldsymbol{M}(\boldsymbol{q})\ddot{\boldsymbol{q}} + \boldsymbol{J}_e^T(\boldsymbol{q})(\boldsymbol{J}^+(\boldsymbol{r}))^T\boldsymbol{M}_o(\boldsymbol{r})\boldsymbol{J}_{of}\ddot{\boldsymbol{r}}_f + \left\{\tfrac{1}{2}\dot{\boldsymbol{M}}(\boldsymbol{q}) + \boldsymbol{S}(\boldsymbol{q},\dot{\boldsymbol{q}})\right\}\dot{\boldsymbol{q}} \\ +\boldsymbol{J}_e^T(\boldsymbol{q})(\boldsymbol{J}^+(\boldsymbol{r}))^T\left(\tfrac{1}{2}\dot{\boldsymbol{M}}_o(\boldsymbol{r}) + \boldsymbol{S}_o(\boldsymbol{r},\dot{\boldsymbol{r}})\right)\boldsymbol{J}_{of}\dot{\boldsymbol{r}}_f \\ + \boldsymbol{g}(\boldsymbol{q}) + \boldsymbol{J}_e^T(\boldsymbol{q})(\boldsymbol{J}^+(\boldsymbol{r}))^T\boldsymbol{g}_o(\boldsymbol{r}) = -\boldsymbol{J}_m^T(\boldsymbol{q})\boldsymbol{f}_{int} + \boldsymbol{\tau}, \end{aligned} \tag{6.168}$$

where

$$\boldsymbol{J}_m(\boldsymbol{q}) = \boldsymbol{Z}^T\boldsymbol{J}_e. \tag{6.169}$$

From Eqs. (6.169), (6.156), and (6.167), it follows that

$$\boldsymbol{J}_m(\boldsymbol{q})\dot{\boldsymbol{q}} = \boldsymbol{Z}^T\boldsymbol{J}_e\dot{\boldsymbol{q}} = \boldsymbol{Z}^T\boldsymbol{J}\dot{\boldsymbol{r}} = \boldsymbol{0}, \tag{6.170}$$

which indicates the geometric constraint on the joint angle velocity vector.

6.4.2 Adaptive Neural-Network Jacobian Control of Multifingered Robot Hands

When the fingertips are soft and deformable, the parameters and the structure of the Jacobian matrices are difficult to estimate exactly. In addition, the multifingered robot may be required to manipulate objects of various dimensions. To solve this problem, a neural-network Jacobian controller is used in this section for the design of the controller for multifingered robot hands. It is assumed that the fingers are in contact with the object and hold it tightly without rolling motions.

An important property of neural networks is the ability to approximate an arbitrary nonlinear function up to a small error. In order to represent the function, an approximating function is chosen first, and then the weights are updated based on the output errors. In this section, the radial basis function (RBF) network is used to approximate the uncertainties in the dynamic model of the robot, and the weights of neural network are adjusted online without a previous learning phase.

The structure of an RBF neural network is shown in Fig. 6.5. In the RBF neural network, $\boldsymbol{p} = [p_1, \ldots, p_{n_p}]^T \in \Re^{n_p}$ is the input of the network, $\boldsymbol{h}(\boldsymbol{p})$ is the output, and $\boldsymbol{\theta} = [\theta_1, \ldots, \theta_{n_a}]^T \in \Re^{n_a}$ is the activation function, where θ_i $(i = 1, \ldots, n_a)$ are specified as

$$\theta_i = \mathrm{e}^{\frac{-\|\boldsymbol{p}-\boldsymbol{\mu}_i\|^2}{\sigma_i^2}}, \tag{6.171}$$

where $\boldsymbol{\mu}_i = [\mu_{i1}, \ldots, \mu_{in_p}]^T \in \Re^{n_p}$ is called the center, $\boldsymbol{\sigma} = [\sigma_1, \ldots, \sigma_{n_a}]^T \in \Re^{n_a}$ is called the distance, and $\sigma_i > 0$. From Fig. 6.5, the function approximation using a

Fig. 6.5 An RBF neural network

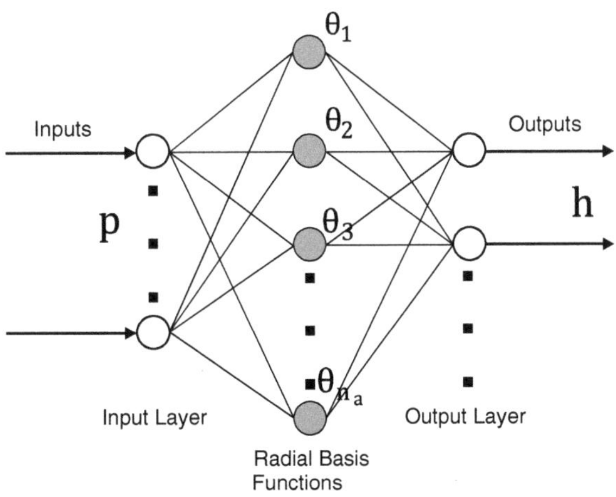

RBF neural network is seen to be

$$h(p) = W\theta(p) + E, \tag{6.172}$$

where W is the weight matrix, and E is called the neural-network functional approximation error, and the error generally decreases as the number of neurons increases.

By using a neural network, the gravity terms in Eq. (6.168) can be approximated by

$$g(q) + J_e^T(q)(J^+(r))^T g_o(r) = W_g\theta_g(q) + E_g, \tag{6.173}$$

where $\theta_g(q)$ is a vector of activation functions, W_g is a constant matrix of ideal network weights, which is unknown, and E_g is an approximation error of the gravity terms.

The Jacobian matrix $J_m(q)$ is also approximated by a neural network as

$$J_m^T(q) = (W_{f1}\theta_f(q), \ldots, W_{fn_l}\theta_f(q)) + E_f, \tag{6.174}$$

where $\theta_f(q)$ is a vector of activation functions, W_{fi} are constant matrices of ideal unknown network weights, and E_f is an approximation error of the force term.

The Jacobian matrix $J_x(q)$ can be approximated by neural networks as

$$J_x(q) = (W_{x1}\theta_x(q), \ldots, W_{xkni}\theta_x(q)) + E_x, \tag{6.175}$$

where $W_{xi}\theta_x(q)$ is the ith column vectors of $J_x(q)$, and E_x is the estimation error.

When the parameters and structure of the Jacobian matrix are uncertain, the Jacobian matrices are estimated as

$$\begin{aligned} \hat{J}_m^T(q, \hat{W}_f) &= (\hat{W}_{f1}\theta_f(q), \ldots, \hat{W}_{fn_l}\theta_f(q)) \\ \hat{J}_x(q, \hat{W}_x) &= (\hat{W}_{x1}\theta_x(q), \ldots, \hat{W}_{xkni}\theta_x(q)), \end{aligned} \tag{6.176}$$

where $\hat{\boldsymbol{W}}_f$ and $\hat{\boldsymbol{W}}_x$ are matrices of estimated network weights to be updated by update laws, and $\hat{\boldsymbol{W}}_{xi}\boldsymbol{\theta}_x(\boldsymbol{q})$ and $\hat{\boldsymbol{W}}_{fi}\boldsymbol{\theta}_f(\boldsymbol{q})$ are the ith column vectors of $\hat{\boldsymbol{J}}_x(\boldsymbol{q}, \hat{\boldsymbol{W}}_x)$ and $\hat{\boldsymbol{J}}_m^T(\boldsymbol{q}, \hat{\boldsymbol{W}}_f)$, respectively.

The adaptive Jacobian controller for the multifingered robot is designated as

$$\begin{aligned}\boldsymbol{\tau} = &-\hat{\boldsymbol{J}}_x^T(\boldsymbol{q}, \hat{\boldsymbol{W}}_x)\boldsymbol{K}_p\boldsymbol{s}(\Delta\boldsymbol{x}) - \hat{\boldsymbol{J}}_x^T(\boldsymbol{q}, \hat{\boldsymbol{W}}_x)\boldsymbol{K}_v\hat{\boldsymbol{J}}_x(\boldsymbol{q}, \hat{\boldsymbol{W}}_x)\dot{\boldsymbol{q}} + \hat{\boldsymbol{J}}_m(\boldsymbol{q}, \hat{\boldsymbol{W}}_f)^T\boldsymbol{f}_{int}\\ &+ \hat{\boldsymbol{J}}_m(\boldsymbol{q}, \hat{\boldsymbol{W}}_f)^T\boldsymbol{R}\boldsymbol{K}_m((\boldsymbol{f}_{int}-\boldsymbol{f}_d)+\gamma\int_o^t(\boldsymbol{f}_{int}-\boldsymbol{f}_d)\mathrm{d}\tau)+\hat{\boldsymbol{W}}_g\boldsymbol{\theta}_g(\boldsymbol{q})-\boldsymbol{K}_f\boldsymbol{f}_{int},\end{aligned} \tag{6.177}$$

where $\boldsymbol{s}(\Delta\boldsymbol{x}) = (s_1(\Delta x_1), \ldots, s_p(\Delta x_p))^T$ is a vector denoting the saturated position as defined in Sect. 2.3.2, $\Delta\boldsymbol{x} = \boldsymbol{x} - \boldsymbol{x}_d = (\Delta x_1, \ldots, \Delta x_p)^T$ is a positional deviation from a desired position of the image feature $\boldsymbol{x}_d$ in image space, $\boldsymbol{K}_p$ and $\boldsymbol{K}_v$ are diagonal feedback gain matrices for the position error and the velocity, respectively, γ is a positive constant, $\boldsymbol{f}_d$ is a desired internal force, $\boldsymbol{K}_m$ is a gain matrix, and $\boldsymbol{Y} = \dot{\boldsymbol{q}} + \alpha\hat{\boldsymbol{J}}_x^+(\boldsymbol{q})\boldsymbol{s}(\Delta\boldsymbol{x})$, $\boldsymbol{K}_f$ is a gain matrix defined as

$$\boldsymbol{K}_{fij} = \bar{k}_{ij}\mathrm{sgn}(y_i f_{intj}), \tag{6.178}$$

where $\bar{k}_{ij}$ is a positive constant, y_i and f_{intj} are the ith and jth elements of $\boldsymbol{Y}$ and $\boldsymbol{f}_{int}$. The estimated network weights $\hat{\boldsymbol{W}}_g$, $\hat{\boldsymbol{W}}_f$, and $\hat{\boldsymbol{W}}_x$ in Eqs. (6.177) and (6.176) are updated, respectively, by the following update laws:

$$\begin{aligned}\dot{\hat{\boldsymbol{W}}}_{gj}^T &= -k_1\hat{\boldsymbol{W}}_{gj}^T + \mathrm{proj}(\boldsymbol{\Omega}_{gj}),\\ \dot{\hat{\boldsymbol{W}}}_{fij}^T &= -k_2\hat{\boldsymbol{W}}_{fij}^T + \mathrm{proj}(\boldsymbol{\Omega}_{fij}),\\ \dot{\hat{\boldsymbol{W}}}_{xij}^T &= -k_3\hat{\boldsymbol{W}}_{xij}^T + \mathrm{proj}(\boldsymbol{\Omega}_{xij}),\end{aligned} \tag{6.179}$$

where

$$\begin{aligned}\boldsymbol{\Omega}_{gj} &= -\boldsymbol{L}_{gj}^{-1}\boldsymbol{\theta}_g(\boldsymbol{q})y_j,\\ \boldsymbol{\Omega}_{fij} &= -\boldsymbol{L}_{fij}^{-1}\boldsymbol{\theta}_f(\boldsymbol{q})y_j f_{inti},\\ \boldsymbol{\Omega}_{xij} &= \boldsymbol{L}_{xij}^{-1}\boldsymbol{\theta}_x(\boldsymbol{q})(k_{pi} + \alpha k_{vi})s(\Delta x_i)\dot{q}_j,\end{aligned} \tag{6.180}$$

$\hat{\boldsymbol{W}}_{gj}$ is the jth row vector of $\hat{\boldsymbol{W}}_g$, $\hat{\boldsymbol{W}}_{fij}$ is the ith row vector of $\hat{\boldsymbol{W}}_{fj}$, $\hat{\boldsymbol{W}}_{xij}$ is the ith row vector of $\hat{\boldsymbol{W}}_{xj}$, $\boldsymbol{L}_{gj}$, $\boldsymbol{L}_{fij}$, $\boldsymbol{L}_{xij}$ are positive definite matrices, and k_1, k_2, k_3 are positive constants. The function $\mathrm{proj}(\boldsymbol{\Omega}_{gj})$ is a projection algorithm defined as

$$\mathrm{proj}(\boldsymbol{\Omega}_{gj}) = \begin{cases}\boldsymbol{0}, & \text{if}\quad \hat{\boldsymbol{W}}_{gj} = \underline{\boldsymbol{W}}_{gj} \text{ and } \boldsymbol{\Omega}_{gj} < 0\\ & \text{or}\quad \hat{\boldsymbol{W}}_{gj} = \bar{\boldsymbol{W}}_{gj} \text{ and } \boldsymbol{\Omega}_{gj} > 0\\ \boldsymbol{\Omega}_{gj}, & \text{else}\end{cases} \tag{6.181}$$

where $\underline{W}_{gj}$ and $\bar{W}_{gj}$ are the lower and upper bounds of W_{gj}. The projection algorithms $\text{proj}(\boldsymbol{\Omega}_f)$ and $\text{proj}(\boldsymbol{\Omega}_x)$ can be similarly defined as above. The functions $\text{proj}(\cdot)$ are defined to ensure that $\hat{J}_x^T(q, \hat{W}_x)$ and $\hat{J}_m(q, \hat{W}_f)$ are bounded during adaptation.

Substituting Eqs. (6.177), (6.173), and (6.174) into Eq. (6.168), the following closed-loop equation is obtained:

$$
\begin{aligned}
&M(q)\ddot{q} + J_e^T(q)(J^+(r))^T M_o(r) J_{of}\ddot{r}_f + \left\{\tfrac{1}{2}\dot{M}(q) + S(q,\dot{q})\right\}\dot{q} \\
&+ J_e^T(q)(J^+(r))^T \left(\tfrac{1}{2}\dot{M}_o(r) + S_o(r,\dot{r})\right) J_{of}\dot{r}_f + \hat{J}_x^T(q, \hat{W}_x) K_p s(\Delta x) \\
&+ \hat{J}_x^T(q, \hat{W}_x) K_v \hat{J}_x(q, \hat{W}_x)\dot{q} + \Delta W_g \theta_g(q) + \Delta W_f \theta_f(q) f_{int} \\
&= \hat{J}_m^T(q, \hat{W}_f) R K_m\left((f_{int} - f_d) + \gamma \int_o^t (f_{int} - f_d)\mathrm{d}\tau\right) - E_g - E_f f_{int} - K_f f_{int},
\end{aligned}
\tag{6.182}
$$

where $\Delta W_g = W_g - \hat{W}_g$, and $\Delta W_f = W_f - \hat{W}_f$ are the weight estimation errors. To prove the stability of the system, a Lyapunov-like function candidate is defined as

$$
\begin{aligned}
V &= \tfrac{1}{2}\dot{q}^T M(q)\dot{q} + \tfrac{1}{2}\dot{r}_f^T J_{of}^T M_o(r) J_{of}\dot{r}_f + \alpha s^T(\Delta x)(\hat{J}_x^+(q, \hat{W}_x))^T M(q)\dot{q} \\
&+ \alpha s^T(\Delta x)(\hat{J}_x^+(q, \hat{W}_x))^T J_e^T(q)(J^+(r))^T M_o(r) J_{of}\dot{r}_f + \tfrac{1}{2}\sum_{j=1}^{n} \Delta W_{gj} L_{gj} \Delta W_{gj}^T \\
&+ \tfrac{1}{2}\sum_{i=1}^{n_I}\sum_{j=1}^{kni} \Delta W_{fij} L_{fij} \Delta W_{fij}^T + \tfrac{1}{2}\sum_{j=1}^{kni}\sum_{i=1}^{p} \Delta W_{xij} L_{xij} \Delta W_{xij}^T + \sum_{j=1}^{p}(k_{pj} + \alpha k_{vj}) S_j(\Delta x_j),
\end{aligned}
\tag{6.183}
$$

where $\Delta W_x = W_x - \hat{W}_x$ are the weight estimation errors.

To show the positive definiteness of the Lyapunov-like function candidate V in Eq. (6.183), note that

$$
\begin{aligned}
V &= \tfrac{1}{4}\dot{q}^T M(q)\dot{q} + \tfrac{1}{4}\dot{r}_f^T J_{of}^T M_o(r) J_{of}\dot{r}_f + \sum_{j=1}^{p}(k_{pj} + \alpha k_{vj}) S_j(\Delta x_j) \\
&+ \tfrac{1}{4}(\dot{q} + 2\alpha \hat{J}_x^+(q, \hat{W}_x)s(\Delta x))^T M(q)(\dot{q} + 2\alpha \hat{J}_x^+(q, \hat{W}_x)s(\Delta x)) \\
&+ \tfrac{1}{4}(J_{of}\dot{r}_f + 2\alpha J^+(r) J_e(q)\hat{J}_x^+(q, \hat{W}_x)s(\Delta x))^T M_o(r) \times \\
&(J_{of}\dot{r}_f + 2\alpha J^+(r) J_e(q)\hat{J}_x^+(q, \hat{W}_x)s(\Delta x)) - \alpha^2 s^T(\Delta x)((\hat{J}_x^+(q, \hat{W}_x))^T M(q)\hat{J}_x^+(q, \hat{W}_x) \\
&+ (\hat{J}_x^+(q, \hat{W}_x))^T J_e^T(q)(J^+(r))^T M_o(r) J^+(r) J_e(q)\hat{J}_x^+(q, \hat{W}_x))s(\Delta x) \\
&+ \tfrac{1}{2}\sum_{j=1}^{n} \Delta W_{gj} L_{gj} \Delta W_{gj}^T + \tfrac{1}{2}\sum_{i=1}^{n_I}\sum_{j=1}^{kni} \Delta W_{fij} L_{fij} \Delta W_{fij}^T + \tfrac{1}{2}\sum_{j=1}^{kni}\sum_{i=1}^{p} \Delta W_{xij} L_{xij} \Delta W_{xij}^T \\
&\geq \tfrac{1}{4}\dot{q}^T M(q)\dot{q} + \tfrac{1}{4}\dot{r}_f^T J_{of}^T M_o(r) J_{of}\dot{r}_f + \sum_{j=1}^{p}(k_{pj}\bar{c}_j + \alpha(k_{vj}\bar{c}_j - \alpha\lambda_m)) s_j^2(\Delta x_j) \\
&+ \tfrac{1}{2}\sum_{j=1}^{n} \Delta W_{gj} L_{gj} \Delta W_{gj}^T + \tfrac{1}{2}\sum_{i=1}^{n_I}\sum_{j=1}^{kni} \Delta W_{fij} L_{fij} \Delta W_{fij}^T + \tfrac{1}{2}\sum_{j=1}^{kni}\sum_{i=1}^{p} \Delta W_{xij} L_{xij} \Delta W_{xij}^T,
\end{aligned}
\tag{6.184}
$$

where λ_m denotes the maximum norm bound of the matrix $(\hat{\boldsymbol{J}}_x^+(\boldsymbol{q}, \hat{\boldsymbol{W}}_x))^T \boldsymbol{M}(\boldsymbol{q})\hat{\boldsymbol{J}}_x^+(\boldsymbol{q}, \hat{\boldsymbol{W}}_x) + (\hat{\boldsymbol{J}}_x^+(\boldsymbol{q}, \hat{\boldsymbol{W}}_x))^T \boldsymbol{J}_e^T(\boldsymbol{q})(\boldsymbol{J}^+(\boldsymbol{r}))^T \boldsymbol{M}_o(\boldsymbol{r})\boldsymbol{J}^+(\boldsymbol{r})\boldsymbol{J}_e(\boldsymbol{q})\hat{\boldsymbol{J}}_x^+(\boldsymbol{q}, \hat{\boldsymbol{W}}_x)$. Hence, V is positive definite in $\Delta\boldsymbol{x}$, $\dot{\boldsymbol{q}}$, $\Delta\boldsymbol{W}_g$, $\Delta\boldsymbol{W}_f$, $\Delta\boldsymbol{W}_x$, since α and k_v can be chosen such that

$$k_{vj}\bar{c}_j - \alpha\lambda_m > 0. \tag{6.185}$$

Next, differentiate V with respect to time. Then using Eqs. (6.168) and (6.177), one has

$$\begin{aligned}\dot{V} = &-\dot{\boldsymbol{q}}^T(\hat{\boldsymbol{J}}_x^T(\boldsymbol{q}, \hat{\boldsymbol{W}}_x)\boldsymbol{K}_v\hat{\boldsymbol{J}}_x(\boldsymbol{q}, \hat{\boldsymbol{W}}_x))\dot{\boldsymbol{q}} - \alpha\boldsymbol{s}^T(\Delta\boldsymbol{x})\boldsymbol{K}_p\boldsymbol{s}(\Delta\boldsymbol{x})\\ &+\boldsymbol{s}^T(\Delta\boldsymbol{x})(\boldsymbol{K}_p + \alpha\boldsymbol{K}_v)(\boldsymbol{J}_x(\boldsymbol{q}) - \hat{\boldsymbol{J}}_x(\boldsymbol{q}, \hat{\boldsymbol{W}}_x))\dot{\boldsymbol{q}} - \boldsymbol{Y}^T\Delta\boldsymbol{W}_g\boldsymbol{\theta}_g(\boldsymbol{q})\\ &-\boldsymbol{Y}^T\Delta\boldsymbol{W}_f\boldsymbol{\theta}_f(\boldsymbol{q})\boldsymbol{f}_{int} - \boldsymbol{Y}^T(\boldsymbol{E}_g + \boldsymbol{E}_f\boldsymbol{f}_{int} + \boldsymbol{K}_f\boldsymbol{f}_{int}) - h\\ &-\sum_{j=1}^{n}\Delta\boldsymbol{W}_{gj}\boldsymbol{L}_{gj}\dot{\hat{\boldsymbol{W}}}_{gj}^T - \sum_{i=1}^{n_I}\sum_{j=1}^{kni}\Delta\boldsymbol{W}_{fij}\boldsymbol{L}_{fij}\dot{\hat{\boldsymbol{W}}}_{fij}^T - \sum_{j=1}^{kni}\sum_{i=1}^{p}\Delta\boldsymbol{W}_{xij}\boldsymbol{L}_{xij}\dot{\hat{\boldsymbol{W}}}_{xij}^T,\end{aligned} \tag{6.186}$$

where

$$\begin{aligned}h = \alpha\{&\boldsymbol{s}^T(\Delta\boldsymbol{x})(\hat{\boldsymbol{J}}_x^+(\boldsymbol{q}, \hat{\boldsymbol{W}}_x))^T(\tfrac{1}{2}\dot{\boldsymbol{M}}(\boldsymbol{q}) + \boldsymbol{S}(\boldsymbol{q}, \dot{\boldsymbol{q}}))\dot{\boldsymbol{q}}\\ &+\boldsymbol{s}^T(\Delta\boldsymbol{x})(\hat{\boldsymbol{J}}_x^+(\boldsymbol{q}, \hat{\boldsymbol{W}}_x))^T\boldsymbol{J}_e^T(\boldsymbol{q})(\boldsymbol{J}^+(\boldsymbol{r}))^T\big(-\tfrac{1}{2}\dot{\boldsymbol{M}}_o(\boldsymbol{r}) + \boldsymbol{S}_o(\boldsymbol{r}, \dot{\boldsymbol{r}})\big)\boldsymbol{J}_{of}\dot{\boldsymbol{r}}_f\\ &-\dot{\boldsymbol{s}}^T(\Delta\boldsymbol{x})(\hat{\boldsymbol{J}}_x^+(\boldsymbol{q}, \hat{\boldsymbol{W}}_x))^T\boldsymbol{M}(\boldsymbol{q})\dot{\boldsymbol{q}} - \boldsymbol{s}^T(\Delta\boldsymbol{x})(\dot{\hat{\boldsymbol{J}}}_x^+(\boldsymbol{q}, \hat{\boldsymbol{W}}_x))^T\boldsymbol{M}(\boldsymbol{q})\dot{\boldsymbol{q}}\\ &-\tfrac{\mathrm{d}}{\mathrm{d}t}[\boldsymbol{s}^T(\Delta\boldsymbol{x})(\hat{\boldsymbol{J}}_x^+(\boldsymbol{q}, \hat{\boldsymbol{W}}_x))^T\boldsymbol{J}_e^T(\boldsymbol{q})(\boldsymbol{J}^+(\boldsymbol{r}))^T]\boldsymbol{M}_o(\boldsymbol{r})\boldsymbol{J}_{of}\dot{\boldsymbol{r}}_f\}.\end{aligned} \tag{6.187}$$

As seen from Eq. (6.187), since $\boldsymbol{s}(\Delta\boldsymbol{x})$ is a saturation function of the position error and the estimated network weights are bounded by the projection algorithms, there exists a constant c such that

$$|h| \leq \alpha c\|\dot{\boldsymbol{q}}\|^2. \tag{6.188}$$

Using Eqs. (6.175), (6.176), and (6.180), one has

$$\begin{aligned}\dot{V} \leq &-\dot{\boldsymbol{q}}^T(\hat{\boldsymbol{J}}_x^T(\boldsymbol{q}, \hat{\boldsymbol{W}}_x)\boldsymbol{K}_v\hat{\boldsymbol{J}}_x(\boldsymbol{q}, \hat{\boldsymbol{W}}_x))\dot{\boldsymbol{q}} - \alpha\boldsymbol{s}^T(\Delta\boldsymbol{x})\boldsymbol{K}_p\boldsymbol{s}(\Delta\boldsymbol{x})\\ &+\boldsymbol{s}^T(\Delta\boldsymbol{x})(\boldsymbol{K}_p + \alpha\boldsymbol{K}_v)\boldsymbol{E}_x\dot{\boldsymbol{q}} - \boldsymbol{Y}^T(\boldsymbol{E}_g + \boldsymbol{E}_f\boldsymbol{f}_{int} + \boldsymbol{K}_f\boldsymbol{f}_{int}) - h\\ &+k_1\sum_{j=1}^{n}\Delta\boldsymbol{W}_{gj}\boldsymbol{L}_{gj}\hat{\boldsymbol{W}}_{gj}^T + k_2\sum_{i=1}^{n_I}\sum_{j=1}^{kni}\Delta\boldsymbol{W}_{fij}\boldsymbol{L}_{fij}\hat{\boldsymbol{W}}_{fij}^T\\ &+k_3\sum_{j=1}^{kni}\sum_{i=1}^{p}\Delta\boldsymbol{W}_{xij}\boldsymbol{L}_{xij}\hat{\boldsymbol{W}}_{xij}^T.\end{aligned} \tag{6.189}$$

Since $\boldsymbol{E}_x$ is bounded, let b_{ex} be the upper bound of $\boldsymbol{E}_x$. Then

$$\boldsymbol{s}^T(\Delta\boldsymbol{x})(\boldsymbol{K}_p+\alpha\boldsymbol{K}_v)\boldsymbol{E}_x\dot{\boldsymbol{q}} \le \tfrac{1}{2}b_{ex}(k_{p\max}+\alpha k_{v\max})(\|\boldsymbol{s}(\Delta\boldsymbol{x})\|^2+\|\dot{\boldsymbol{q}}\|^2), \quad (6.190)$$

where $k_{p\max}=\lambda_{\max}[\boldsymbol{K}_p]$, and $k_{v\max}=\lambda_{\max}[\boldsymbol{K}_v]$. The notation $\lambda_{\max}[\cdot]$ denotes the maximum eigenvalue of the matrix. In addition,

$$\begin{aligned}
k_1\Delta\boldsymbol{W}_{gj}\boldsymbol{L}_{gj}^{-1}\hat{\boldsymbol{W}}_{gj}^T &\le \tfrac{k_1 l_{gj}}{2}(\|\boldsymbol{W}_{gj}\|^2-\|\Delta\boldsymbol{W}_{gj}\|^2),\\
k_2\Delta\boldsymbol{W}_{fij}\boldsymbol{L}_{fij}^{-1}\hat{\boldsymbol{W}}_{fij}^T &\le \tfrac{k_2 l_{fij}}{2}(\|\boldsymbol{W}_{fij}\|^2-\|\Delta\boldsymbol{W}_{fij}\|^2),\\
k_3\Delta\boldsymbol{W}_{xij}\boldsymbol{L}_{xij}^{-1}\hat{\boldsymbol{W}}_{xij}^T &\le \tfrac{k_3 l_{xij}}{2}(\|\boldsymbol{W}_{xij}\|^2-\|\Delta\boldsymbol{W}_{xij}\|^2).
\end{aligned} \quad (6.191)$$

Substituting inequality (6.188) into Eq. (6.189) and using Eq. (6.190) yields

$$\begin{aligned}
\dot{V} \le &-\big(\lambda_1-\alpha c-\tfrac{1}{2}b_{ex}(k_{p\max}+\alpha k_{v\max})\big)\|\dot{\boldsymbol{q}}\|^2\\
&-\big(\alpha k_{p\min}-\tfrac{1}{2}b_{ex}(k_{p\max}+\alpha k_{v\max})\big)\|\boldsymbol{s}(\Delta\boldsymbol{x})\|^2-\boldsymbol{Y}^T(\boldsymbol{E}_g+\boldsymbol{E}_f f_{int}+\boldsymbol{K}_f f_{int})\\
&-\tfrac{k_1 l_{gj}}{2}\sum_{i=1}^{n}\|\Delta\boldsymbol{W}_{gj}\|^2-\tfrac{k_3 l_{xij}}{2}\sum_{j=1}^{kni}\sum_{i=1}^{p}\|\Delta\boldsymbol{W}_{xij}\|^2-\tfrac{k_2 l_{fij}}{2}\sum_{i=1}^{n_I}\sum_{j=1}^{kni}\|\Delta\boldsymbol{W}_{fij}\|^2\\
&+\tfrac{k_1 l_{gj}}{2}\sum_{i=1}^{n}\|\boldsymbol{W}_{gj}\|^2+\tfrac{k_3 l_{xij}}{2}\sum_{j=1}^{kni}\sum_{i=1}^{p}\|\boldsymbol{W}_{xij}\|^2+\tfrac{k_2 l_{fij}}{2}\sum_{i=1}^{n_I}\sum_{j=1}^{kni}\|\boldsymbol{W}_{fij}\|^2,
\end{aligned} \quad (6.192)$$

where λ_1 denotes the lower norm bound of $\hat{\boldsymbol{J}}_x^T(\boldsymbol{q},\hat{\boldsymbol{W}}_x)\boldsymbol{K}_v\hat{\boldsymbol{J}}_x(\boldsymbol{q},\hat{\boldsymbol{W}}_x)$, $k_{p\min}=\lambda_{\min}[\boldsymbol{K}_p]$, $\lambda_{\min}[\cdot]$ denotes the minimum eigenvalue of the matrix. When $\bar{k}_{ij}$ is chosen sufficiently large that $\bar{k}_{ij}>|E_{fij}|$, then

$$\begin{aligned}
\dot{V} \le &-(\lambda_1-\alpha c-\tfrac{1}{2}b_{ex}(k_{p\max}+\alpha k_{v\max}))\|\dot{\boldsymbol{q}}\|^2\\
&-\big(\alpha k_{p\min}-\tfrac{1}{2}b_{ex}(k_{p\max}+\alpha k_{v\max})\big)\|\boldsymbol{s}(\Delta\boldsymbol{x})\|^2-\boldsymbol{Y}^T\boldsymbol{E}_g\\
&-\tfrac{k_1 l_{gj}}{2}\sum_{i=1}^{n}\|\Delta\boldsymbol{W}_{gj}\|^2-\tfrac{k_3 l_{xij}}{2}\sum_{j=1}^{kni}\sum_{i=1}^{p}\|\Delta\boldsymbol{W}_{xij}\|^2-\tfrac{k_2 l_{fij}}{2}\sum_{i=1}^{n_I}\sum_{j=1}^{kni}\|\Delta\boldsymbol{W}_{fij}\|^2\\
&+\tfrac{k_1 l_{gj}}{2}\sum_{i=1}^{n}\|\boldsymbol{W}_{gj}\|^2+\tfrac{k_3 l_{xij}}{2}\sum_{j=1}^{kni}\sum_{i=1}^{p}\|\boldsymbol{W}_{xij}\|^2+\tfrac{k_2 l_{fij}}{2}\sum_{i=1}^{n_I}\sum_{j=1}^{kni}\|\boldsymbol{W}_{fij}\|^2.
\end{aligned} \quad (6.193)$$

Next, note that

$$\begin{aligned}
\dot{\boldsymbol{q}}^T\boldsymbol{E}_g &\le \tfrac{1}{2}(\|\dot{\boldsymbol{q}}\|^2+\|\boldsymbol{E}_g\|^2),\\
\alpha\boldsymbol{s}^T(\Delta\boldsymbol{x})(\hat{\boldsymbol{J}}_x^+(\boldsymbol{q},\hat{\boldsymbol{W}}_x))^T\boldsymbol{E}_g &\le \tfrac{\alpha b_j}{2}(\|\boldsymbol{s}(\Delta\boldsymbol{x})\|^2+\|\boldsymbol{E}_g\|^2),
\end{aligned} \quad (6.194)$$

where b_j is the upper bound of $(\hat{\boldsymbol{J}}_x^+(\boldsymbol{q},\hat{\boldsymbol{W}}_x))^T$. Therefore,

$$\dot{V} \leq -\big(\lambda_1 - \alpha c - \tfrac{1}{2} b_{ex}(k_{p\max} + \alpha k_{v\max}) - \tfrac{1}{2}\big)\|\dot{\boldsymbol{q}}\|^2$$
$$-\big(\alpha k_{p\min} - \tfrac{1}{2} b_{ex}(k_{p\max} + \alpha k_{v\max}) - \tfrac{\alpha b_j}{2}\big)\|\boldsymbol{s}(\Delta \boldsymbol{x})\|^2 + \mu, \quad (6.195)$$

where

$$\mu = \tfrac{1+\alpha b_j}{2}\|\boldsymbol{E}_g\|^2 + \tfrac{k_1 l_{gj}}{2}\sum_{i=1}^{n}\|\boldsymbol{W}_{gj}\|^2 + \tfrac{k_3 l_{xij}}{2}\sum_{j=1}^{kni}\sum_{i=1}^{p}\|\boldsymbol{W}_{xij}\|^2 + \tfrac{k_2 l_{fij}}{2}\sum_{i=1}^{n_I}\sum_{j=1}^{kni}\|\boldsymbol{W}_{fij}\|^2. \quad (6.196)$$

From the upper bound of V, there exist $\boldsymbol{K}_v$, $\boldsymbol{K}_p$, and a positive constant $\bar{\gamma}_1$ such that

$$\big(\lambda_1 - \alpha c - \tfrac{1}{2} b_{ex}(k_{p\max} + \alpha k_{v\max}) - \tfrac{1}{2}\big)\|\dot{\boldsymbol{q}}\|^2$$
$$\geq \tfrac{\bar{\gamma}_1}{4}(\dot{\boldsymbol{q}}^T \boldsymbol{M}(\boldsymbol{q})\dot{\boldsymbol{q}} + \dot{\boldsymbol{r}}_f^T \boldsymbol{J}_{of}^T \boldsymbol{M}_o(\boldsymbol{r})\boldsymbol{J}_{of}\dot{\boldsymbol{r}}_f),$$
$$\big(\alpha k_{p\min} - \tfrac{1}{2} b_{ex}(k_{p\max} + \alpha k_{v\max}) - \tfrac{\alpha b_j}{2}\big)\|\boldsymbol{s}(\Delta \boldsymbol{x})\|^2$$
$$\geq \sum_{j=1}^{p}(k_{pj}\bar{c}_j + \alpha(k_{vj}\bar{c}_j - \alpha\lambda_m))s_j^2(\Delta x_j). \quad (6.197)$$

Let $\bar{\gamma} = \min\{\bar{\gamma}_1, k_1, k_2, k_3\}$. Then one has

$$\dot{V} \leq -\bar{\gamma} V + \mu. \quad (6.198)$$

The above inequality implies

$$V \leq \tfrac{\mu}{\bar{\gamma}} + \{V(0) - \tfrac{\mu}{\bar{\gamma}}\}\mathrm{e}^{-\bar{\gamma} t}. \quad (6.199)$$

Then it can be concluded that V is uniformly ultimately bounded.

Theorem 6.4 *The closed-loop system described by* (6.182) *with adaptive Jacobian matrices gives rise to the convergence of* $(\Delta \boldsymbol{x}, \dot{\boldsymbol{q}})$ *and boundedness of* $(\boldsymbol{f}_{int} - \boldsymbol{f}_d)$ *as* $t \to 0$ *if the feedback gains* $\boldsymbol{K}_p$ *and* $\boldsymbol{K}_v$ *are chosen to satisfy conditions* (6.185), (6.197).

Proof From inequality (6.199), it can be concluded that $\boldsymbol{s}(\Delta \boldsymbol{x}), \dot{\boldsymbol{q}}, \dot{\boldsymbol{r}}_f, \Delta \boldsymbol{W}_{gj}, \Delta \boldsymbol{W}_{fij}$, and $\Delta \boldsymbol{W}_{xij}$ are bounded.

Substituting Eqs. (6.157), (6.158), and (6.159) into (6.182), we have

$$(\boldsymbol{M}(\boldsymbol{q}) + \bar{\boldsymbol{M}}_o)\ddot{\boldsymbol{q}} + (\tfrac{1}{2}\dot{\boldsymbol{M}}(\boldsymbol{q}) + \boldsymbol{S}(\boldsymbol{q}, \dot{\boldsymbol{q}}) + \tfrac{1}{2}\dot{\bar{\boldsymbol{M}}}_o + \bar{\boldsymbol{S}}_o)\dot{\boldsymbol{q}}$$
$$+ \hat{\boldsymbol{J}}_x^T(\boldsymbol{q}, \hat{\boldsymbol{W}}_x)\boldsymbol{K}_p \boldsymbol{s}(\Delta \boldsymbol{x}) + \hat{\boldsymbol{J}}_x^T(\boldsymbol{q}, \hat{\boldsymbol{W}}_x)\boldsymbol{K}_v \hat{\boldsymbol{J}}_x(\boldsymbol{q}, \hat{\boldsymbol{W}}_x)\dot{\boldsymbol{q}}$$
$$+ \Delta \boldsymbol{W}_g \boldsymbol{\theta}_g(\boldsymbol{q}) + \Delta \boldsymbol{W}_f \boldsymbol{\theta}_f(\boldsymbol{q})\boldsymbol{f}_{int} + \boldsymbol{E}_g + \boldsymbol{E}_f \boldsymbol{f}_{int} + \boldsymbol{K}_f \boldsymbol{f}_{int}$$
$$= \hat{\boldsymbol{J}}_m^T(\boldsymbol{q}, \hat{\boldsymbol{W}}_f)\boldsymbol{R}\boldsymbol{K}_m((\boldsymbol{f}_{int} - \boldsymbol{f}_d) + \gamma \int_o^t (\boldsymbol{f}_{int} - \boldsymbol{f}_d)\mathrm{d}\tau). \quad (6.200)$$

Multiplying both sides of Eq. (6.200) by $J_m(q)\bar{M}^{-1}$ and using $J_m(q)\ddot{q} = -\dot{J}_m(q)\dot{q}$ shows that the force error is subject to

$$\Delta(t) = (f_{int} - f_d) + \gamma(K_m - \bar{E})^{-1}K_m \int_o^t (f_{int} - f_d)\mathrm{d}\tau, \qquad (6.201)$$

where

$$\begin{aligned}
\Delta(t) &= (K_m - \bar{E})^{-1}\kappa r(t),\\
\bar{E} &= J_m(q)\bar{M}^{-1}(\Delta W_f\theta_f(q) + E_f + K_f),\\
\kappa &= R^T(J_m(q)\bar{M}^{-1}\hat{J}_m^T(q, \hat{W}_f))^{-1},\\
r(t) &= -\dot{J}_m(q)\dot{q} + J_m(q)\bar{M}^{-1}\Big\{\big(\tfrac{1}{2}\dot{M}(q) + S(q, \dot{q}) + \tfrac{1}{2}\dot{\bar{M}}_o + \bar{S}_o\big)\dot{q} + \hat{J}_x^T(q, \hat{W}_x)K_p s(\Delta x)\\
&\quad + \hat{J}_x^T(q, \hat{W}_x)K_v\hat{J}_x(q, \hat{W}_x)\dot{q} + \Delta W_g\theta_g(q) + \Delta W_f\theta_f(q)f_d + E_g + E_f f_d + K_f f_d\Big\}.
\end{aligned} \qquad (6.202)$$

Since γ is a positive constant and all terms on the left-hand side of Eq. (6.201) are bounded, we can conclude that $f_{int} - f_d$ and its integration are bounded. The boundedness of the force errors implies that $\ddot{q}$ is bounded, because $s(\Delta x)$, $\dot{q}$, $\dot{r}_f$, ΔW_g, and ΔW_f are bounded. Hence, $s(\Delta x)$, $\dot{q}$ are uniformly continuous, and it follows that

$$\Delta x \to \mathbf{0}\,,\ \dot{q} \to 0 \text{ as } t \to \infty. \qquad (6.203)$$

△△△

6.5 Summary and Notes

Actuator dynamics represents an important component of a robotic system, and the actuator parameters are usually uncertain, due to calibration errors or parameter variation from overheating and changes in environmental temperature. Several control schemes with consideration of actuator dynamics have been developed, e.g. Dawson et al. (1992), Burg et al. (1996), Su and Stepanenko (1998). The task-space adaptive tracking control scheme with uncertain kinematics in Sect. 6.1 was proposed in Liu et al. (2006), which incorporates the actuator dynamics into the controller synthesis.

Due to gear elasticity, shaft wind-up and the use of harmonic drives, joint flexibility Spong (1987), De Luca and Lucibello (1998) is often found in robotic systems. Several joint-space controllers have been developed for robots with flexible joints, e.g., Lozano and Brogliato (1992), Kelly et al. (1994), De Luca et al. (2005). The study on task-space regulation problem in Sect. 6.2 for an RLFJ robot with uncertain kinematics was reported in Liu et al. (2008).

Position and force control problems with uncertain kinematics were considered in Cheah et al. (1998), Doulgeri and Arimoto (2002), Cheah et al. (2003). The control method specified in Sect. 6.3 was developed in Cheah et al. (2010), in which exact knowledge of the constraint surface, kinematics, dynamics, and camera model is not required. A review of the research in force control can be found in Yoshikawa (2000).

The results in Sect. 6.4 for multifingered robot hands were developed in Zhao and Cheah (2009). Many control laws have been developed for dynamic control of multifingered robot hands, e.g., Li et al. (1989), Yoshikawa and Nagai (1991), Cheah et al. (1998), Arimoto et al. (2000), Ozawa et al. (2005). Some good reviews can be found in Bicchi (2000), Yoshikawa (2010).

Appendix

Lyapunov Method

Consider the following nonlinear system:

$$\dot{\boldsymbol{x}}_s = \boldsymbol{f}_s(\boldsymbol{x}_s, t), \tag{A.1}$$

where $\boldsymbol{x}_s \in \Re^{n_s}$ denotes the state of the system, and $\boldsymbol{f}_s(\cdot)$ is a nonlinear function.

Definition A.1 The equilibrium points of the system (A.1) are defined as the state vector $\boldsymbol{x}_e$, for which if at specific time t_0, $\boldsymbol{x}_s = \boldsymbol{x}_e$, then $\boldsymbol{x}_s$ will remain unchanged for all $t > t_0$. In other words, at equilibrium points, the state of the system satisfies $\boldsymbol{f}_s(\boldsymbol{x}_e) = \boldsymbol{0}$. □

It can always be assumed that an equilibrium point is zero by introducing a change of variable $\boldsymbol{y} = \boldsymbol{x}_s - \boldsymbol{x}_e$. In addition, it is often important to know whether an equilibrium point is stable. We now give a definition of stable equilibrium point.

Definition A.2 The equilibrium point $\boldsymbol{x}_e = \boldsymbol{0}$ of the system (A.1) is said to be stable if for every $\varepsilon > 0$, there exists $\sigma > 0$ such that if $||\boldsymbol{x}_s(t_0) - \boldsymbol{0}|| < \sigma$, then $||\boldsymbol{x}_s(t) - \boldsymbol{0}|| < \varepsilon$ for all $t > t_0$. It can be mathematically represented as

$$\forall\, \varepsilon > 0\, \exists\, \sigma > 0, \text{ if } ||\boldsymbol{x}_s(t_0) - \boldsymbol{0}|| < \sigma \Rightarrow ||\boldsymbol{x}_s(t) - \boldsymbol{0}|| < \varepsilon \text{ for } t > t_0. \tag{A.2}$$

□

From the above definition, the system is stable if the trajectory of the state that starts around the equilibrium point can remain arbitrarily close to the equilibrium point. Moreover, the system is asymptotically stable if the following definition is satisfied.

Definition A.3 The equilibrium point $\boldsymbol{x}_e = \boldsymbol{0}$ of the system (A.1) is said to be asymptotically stable if $||\boldsymbol{x}_s(t_0) - \boldsymbol{0}|| < \sigma$, then the trajectory of the state that starts

C.C. Cheah and X. Li, *Task-Space Sensory Feedback Control of Robot Manipulators*,
Intelligent Systems, Control and Automation: Science and Engineering 73,
DOI 10.1007/978-981-287-062-9

around the equilibrium point converges to the equilibrium point at steady state. It can be mathematically represented as:

$$\exists\, \sigma > 0, \text{ if } ||\boldsymbol{x}_s(t_0) - \boldsymbol{0}|| < \sigma \Rightarrow ||\boldsymbol{x}_s(t)|| \to 0 \text{ as } t \to \infty. \tag{A.3}$$

□

Lyapunov theory can be used to examine the stability of the equilibrium point. The main advantage of Lyapunov theory is that the stability of the system can be determined without solving the differential equations of the system. Moreover, Lyapunov theory can be used to design controllers that stabilize nonlinear systems. Before presenting Lyapunov theory, a certain class of functions is introduced as follows:

Definition A.4 A continuous function $V\colon \Re^{n_s} \times \Re_+ \to \Re$ is a locally positive definite function if for some $\varepsilon > 0$ and some continuous, the following conditions hold:

$$\begin{cases} I : V(0,t) = 0, \\ II : V(\boldsymbol{x}_s, t) > 0 \,\forall\, \boldsymbol{x}_s \in B_\varepsilon, \forall\, t \geq 0. \end{cases} \tag{A.4}$$

where B_ε is a ball of size ε around the origin, which is mathematically expressed by $B_\varepsilon = \{\boldsymbol{x}_s \in \Re^{n_s} : ||\boldsymbol{x}_s|| < \varepsilon\}$. In addition, V is a positive definite function if condition (II) holds for all $\boldsymbol{x}_s \in \Re^{n_s}$. If in condition (II), $V(\boldsymbol{x}_s, t) \geq 0$, then $V(\boldsymbol{x}_s, t)$ is a (locally) positive semidefinite function. □

Theorem A.1 (Lyapunov Stability Theorem) *Suppose* $\boldsymbol{x}_e = \boldsymbol{0}$ *is an equilibrium point of the system* (A.1). *Let* $V(\boldsymbol{x}_s, t)$ *be a nonnegative function with derivative* $\dot{V}$ *along trajectories of the system dynamics:*

$$\frac{\mathrm{d}V}{\mathrm{d}t} = \frac{\partial V}{\partial t}\dot{\boldsymbol{x}}_s = \frac{\partial V}{\partial t} f(\boldsymbol{x}_s, t). \tag{A.5}$$

(i) *If* $V(\boldsymbol{x}_s, t)$ *is a locally positive definite function and* $\dot{V}(\boldsymbol{x}_s, t)$ *is a locally negative semidefinite function in* $\boldsymbol{x}_s$ *and for all* t*, then the origin of the system is locally stable.*
(ii) *If* $V(\boldsymbol{x}_s, t)$ *is a locally positive definite function and* $\dot{V}(\boldsymbol{x}_s, t)$ *is a locally negative definite function in* $\boldsymbol{x}_s$ *and for all* t*, then the origin of the system is locally asymptotically stable.*
(iii) *If* $V(\boldsymbol{x}_s, t)$ *is a positive definite function and* $\dot{V}(\boldsymbol{x}_s, t)$ *is a negative definite function in* $\boldsymbol{x}_s$ *and for all* t*, then the origin of the system is globally asymptotically stable. If such a function* $V(\boldsymbol{x}_s, t)$ *exists, then it is called a Lyapunov function.* △△△

The Lyapunov stability theorem provides only sufficient conditions for the stability of nonlinear systems; hence the failure to find a Lyapunov function whose determinant is negative definite does not prove the instability of the nonlinear system.

In the case that $\dot{V}(\boldsymbol{x}_s)$ is a negative semidefinite function, the Lyapunov stability theorem cannot provide any information on the asymptotic stability of the system. To show the asymptotic stability of nonlinear autonomous systems when $\dot{V}(\boldsymbol{x}_s)$ is a negative semidefinite function, LaSalle's invariance theorem can be used.

Lemma A.1 (LaSalle's Invariance Theorem) *Let* $V : \Re^{n_s} \to \Re$ *be a scalar function such that* $\dot{V}(\boldsymbol{x}_s) \le 0$ *in a compact set* Ω. *Let* D *be the set of all points in* Ω *where* $\dot{V}(\boldsymbol{x}_s) = 0$. *Therefore, every solution of the system* $\dot{\boldsymbol{x}}_s = \boldsymbol{f}(\boldsymbol{x}_s)$ *starting in* Ω *approaches the largest invariant set in* D. *In particular, if* D *contains no trajectories other than* $\boldsymbol{x}_s = \boldsymbol{0}$, *then the origin is locally asymptotically stable.* $\triangle\triangle\triangle$

LaSalle's invariance theorem enables one to conclude asymptotic stability only for autonomous systems. For nonautonomous systems, Barbalat's lemma and the associated Lyapunov-like lemma can be used.

Lemma A.2 (Barbalat's Lemma) *If a differentiable scalar function* $F(t)$ *defined over* $t \in (0, \infty)$ *has a finite limit as* $t \to \infty$ *and its derivative* $f(t)$ $(= \frac{\mathrm{d}F(t)}{\mathrm{d}t})$ *is uniformly continuous, then* $f(t) \to 0$ *as* $t \to \infty$. $\triangle\triangle\triangle$

The following lemma follows immediately from Barbalat's Lemma.

Lemma A.3 (Lyapunov-like Lemma) *If a scalar function* $V(\boldsymbol{x}_s, t)$ *satisfies the following conditions:*
(i) $V(\boldsymbol{x}_s, t)$ *is bounded from below;*
(ii) $\dot{V}(\boldsymbol{x}_s, t)$ *is negative semidefinite;*
(iii) $\dot{V}(\boldsymbol{x}_s, t)$ *is uniformly continuous in time;*
then $\dot{V}(\boldsymbol{x}_s, t) \to 0$ *as* $t \to \infty$. $\triangle\triangle\triangle$

The following lemma is also a variation of Barbalat's Lemma.

Lemma A.4 (Desoer and Vidyasagar 1975) *If an* n*-dimensional vector-valued function* $f(t)$ *defined over* $[0, \infty)$ *is bounded and uniformly continuous, and moreover* $f(t) \in L_1(0, \infty)$ *or* $L_2(0, \infty)$, *then* $f(t) \to 0$ *as* $t \to \infty$. $\triangle\triangle\triangle$

Force/Torque Relationship

The infinitesimal joint displacements $\delta \boldsymbol{q}$ are related to the infinitesimal displacements of the associated end effector in Cartesian space $\delta \boldsymbol{r}$ as follows:

$$\delta \boldsymbol{r} = \boldsymbol{J}_m(\boldsymbol{q}) \delta \boldsymbol{q}, \tag{A.6}$$

where $\boldsymbol{J}_m(\boldsymbol{q})$ is the Jacobian matrix.

The *virtual work* resulting from the application of joint torques and end effector forces is defined as

$$\delta W = \boldsymbol{\tau}^T \delta \boldsymbol{q} - \boldsymbol{F}^T \delta \boldsymbol{r}, \tag{A.7}$$

where $\boldsymbol{F}$ is the force on the robot end effector.

According to the principle of virtual work, δW reduces to zero for arbitrary virtual displacements, because of geometric constraints. Therefore, substituting Eq. (A.6) into Eq. (A.7) yields

$$\delta W = \boldsymbol{\tau}^T \delta \boldsymbol{q} - \boldsymbol{F}^T \delta \boldsymbol{r} = \boldsymbol{\tau}^T \delta \boldsymbol{q} - \boldsymbol{F}^T \boldsymbol{J}_m(\boldsymbol{q}) \delta \boldsymbol{q} = (\boldsymbol{\tau} - \boldsymbol{J}_m^T(\boldsymbol{q}) \boldsymbol{F})^T \delta \boldsymbol{q} = 0. \quad \text{(A.8)}$$

The above equation implies that

$$\boldsymbol{\tau} = \boldsymbol{J}_m^T(\boldsymbol{q}) \boldsymbol{F}, \quad \text{(A.9)}$$

which represents the relationship between force and torque.

Bibliography

Arimoto, S. (1996). *Control theory of nonlinear mechanical systems—a passivity-based and circuit-theoretic approach*. Oxford: Clarendon Press.
Arimoto, S., & Miyazaki, F. (1983). Stability and robustness of PID feedback control for robot manipulators of sensor capability. In M. Brady & R. P. Paul (Eds.), *Robotic Research: The First International Symposium* (pp. 783–799). Cambridge, MA: MIT Press.
Arimoto, S., & Miyazaki, F. (1985). Asymptotic stability of feedback control for robot manipulators. In *IFAC Symposium on Robot Control, Barcelona* (pp. 447–452).
Arimoto, S., Naniwa, T., Parra-Vega, V., & Whitcomb, L. (1994). A quasinatural potential and its role in design of hyper-stable PID servo-loop for robotic systems. In *CAI Pacific Symposium Control and Industrial Automation Application* (pp. 110–117).
Arimoto, S., Nguyen, P. T. A., Han, H., & Doulgeri, Z. (2000). Dynamics and control of a set of dual fingers with soft tips. *Robotica*, *18*(1), 71–80.
Berghuis, H., Ortega, R., & Nijmeijer, H. (1993). A robust adaptive robot controller. *IEEE Transactions on Robotics and Automation*, *9*(6), 825–830.
Bicchi, A. (2000). Hands for dexterous manipulation and robust grasping: A difficult road toward simplicity. *IEEE Transactions on Robotics and Automation*, *16*(6), 652–662.
Burg, T., Dawson, D., Hu, J., & de Queiroz, M. (1996). An adaptive partial state-feedback controller for RLED robot manipulators. *IEEE Transactions on Automatic Control*, *41*, 1024–1030.
Braganza, D., Dixon, W. E., Dawson, D. M., & Xian, B. (2005). Tracking control for robot manipulators with kinematic and dynamic uncertainty. In *IEEE Conference on Decision and Control* (pp. 5293–5297).
Cheah, C. C. (2003). Approximate Jacobian robot control with adaptive Jacobian matrix. In *IEEE Conference on Decision and Control* (pp. 5859–5864).
Cheah, C. C. (2008a). Task-space PD control of robot manipulators: Unified analysis and duality property. *International Journal of Robotics Research*, *27*(10), 1152–1170.
Cheah, C. C. (2008b). Region reaching control of robots with motion constraints. In *International Conference on Control, Automation, Robotics and Vision* (pp. 1752–1757).
Cheah, C. C., Han, H. Y., Kawamura, S., & Arimoto, S. (1998). Grasping and position control for multi-fingered robot handswith uncertain Jacobian matrices. In *IEEE International Conference on Robotics and Automation* (pp. 2403–2408).
Cheah, C. C., Hirano, M., Kawamura, S., & Arimoto, S. (2003). Approximate Jacobian control for robots with uncertain kinematics and dynamics. *IEEE Transactions on Robotics and Automation*, *19*(4), 692–702.

C.C. Cheah and X. Li, *Task-Space Sensory Feedback Control of Robot Manipulators*,
Intelligent Systems, Control and Automation: Science and Engineering 73,
DOI 10.1007/978-981-287-062-9

Cheah, C. C., Hou, S. P., & Slotine, J. J. E. (2009). Region-based shape control for a swam of robots. *Automatica*, *45*, 2406–2411.

Cheah, C. C., Hou, S. P., Zhao, Y., & Slotine, J. J. E. (2010). Adaptive vision and force tracking control for robots with constraint uncertainty. *IEEE/ASME Transactions on Mechatronics*, *15*(3), 389–399.

Cheah, C. C., Kawamura, S., & Arimoto, S. (1998). Hybrid position and force control for robotic manipulator with a class of constraint uncertainty. In *Proceedings of USA/Japan Symposium on Flexible Automation* (pp. 501–507).

Cheah, C. C., Kawamura, S., & Arimoto, S. (1999). Feedback control for robotic manipulator with an uncertain Jacobian matrix. *Journal of Robotic Systems*, *16*(2), 119–134.

Cheah, C. C., Kawamura, S., & Arimoto, S. (2003). Stability of hybrid position and force control for robotic manipulator with kinematics and dynamics uncertainties. *Automatica*, *39*, 847–855.

Cheah, C. C., Kawamura, S., Arimoto, S., & Lee, K. (1999). PID control for robotic manipulator with uncertain Jacobian matrix. In *IEEE International Conference on Robotics and Automation* (pp. 494–499).

Cheah, C. C., & Li, X. (2009). Human inspired task-space setpoint control of robot manipulator: See only when necessary. In *IEEE International Conference on Robotics and Biomimetics* (pp. 486–491).

Cheah, C. C., & Li, X. (2010). Reach then see: A new adaptive controller for robot manipulator based on dual task-space information. In *IEEE International Conference on Robotics and Automation* (pp. 5155–5160).

Cheah, C. C., & Li, X. (2011). Singularity-robust task-space tracking control of robot. In *International Conference on Robotics and Automation* (pp. 5819–5824).

Cheah, C. C., Liu, C., & Slotine, J. J. E. (2004). Approximate Jacobian adaptive control for robot manipulators. In *IEEE Conference on Robotics and Automation* (pp. 3075–3080).

Cheah, C. C., Liu, C., & Slotine, J. J. E. (2006). Adaptive tracking controls for robots with unknown kinematics and dynamic properties. *International Journal of Robotics Research*, *25*(3), 283–296.

Cheah, C. C., Liu, C., & Slotine, J. J. E. (2007). Adaptive vision based tracking control of robots with uncertainty in depth information. In *IEEE International Conference on Robotics and Automation* (pp. 2817–2822).

Cheah, C. C., Liu, C., & Slotine, J. J. E. (2010). Adaptive Jacobian vision based control for robots with uncertain depth information. *Automatica*, *46*(7), 1228–1233.

Cheah, C. C., & Wang, D. Q. (2005). Region reaching control of robots: Theory and experiments. In *IEEE International Conference on Robotics and Automation* (pp. 974–979)

Cheah, C. C., Wang, D. Q., & Sun, Y. C. (2007). Region-reaching control of robots. *IEEE Transactions on Robotics*, *23*(6), 1260–1264.

Chen, H., Wang, C., & Lou, Y. (2013). Flocking multiple microparticles with automatically controlled optical tweezers: Solutions and experiments. *IEEE Transactions on Biomedical Engineering*, *60*(6), 1518–1527.

Chesi, G., & Hashimoto, K. (2010). *Visual servoing via advanced numerical methods*. Springer, Heidelberg.

Chiaverini, S. (1997). Singularity-robust task-priority redundancy resolution for real-time kinematic control of robot manipulators. *IEEE Transactions on Robotics and Automation*, *13*(3), 398–410.

Cowan, N. J., Weingarten, J. D., & Koditschek, D. E. (2002). Visual servoing via navigation functions. *IEEE Transactions on Robotics and Automation*, *18*(4), 521–533.

Craig, J. J. (1988). *Adaptive control of mechanical manipulators*. New York: Addison-Wesley.

Craig, J. (1989). *Introduction to robotics: Mechanics and control*. Reading, MA: Addison-Wesley.

Craig, J. J., Hsu, P., & Sastry, S. S. (1987). Adaptive control of mechanical manipulators. *International Journal of Robotics Research*, *6*(2), 16–28.

Dawson, D. M., Qu, Z., & Carrol, J. J. (1992). Tracking control of rigid-link electrically-driven robot manipulators. *International Journal of Control*, *56*(5), 991–1006.

De Luca, A., & Lucibello, P. (1998). General algorithm for dynamic feedback linearization of robots with elastic joints. In *IEEE International Conference on Robotics and Automation* (pp. 504–510).

De Luca, A., Sicilianob, B., & Zolloc, L. (2005). PD control with on-line gravity compensation for robots with elastic joints: Theory and experiments. *Automatica*, *41*(10), 1809–1819.
Dixon, W. E. (2004). Adaptive regulation of amplitude limited robot manipulators with uncertain kinematics and dynamics. In *American Control Conference* (pp. 3839–3844).
Dixon, W. E. (2007). Adaptive regulation of amplitude limited robot manipulators with uncertain kinematics and dynamics. *IEEE Transactions on Automatic Control*, *52*(3), 488–493.
Doulgeri, Z., & Arimoto, S. (2002). A position/force control for a robot finger with soft tip and uncertain kinematics. *Journal of Robotic Systems*, *19*(3), 115–131.
Haghighi, R., & Cheah, C. C. (2012). Multi-group coordination control for robot swarms. *Automatica*, *48*, 2526–2534.
Gans, N. R., Hu, G., Nagarajan, K., & Dixon, W. E. (2011). Keeping multiple moving targets in the field of view of a mobile camera. *IEEE Transactions on Robotics*, *27*(4), 822–828.
Gans, N. R., & Hutchinson, S. A. (2007). Stable visual servoing through hybrid switched-system control. *IEEE Transactions on Robotics*, *23*(4), 530–540.
Garcia-Rodriguez, R., & Parra-Vega, V. (2012). Cartesian sliding PID control schemes for tracking robots with uncertain Jacobian. *Transactions of the Institute of Measurement and Control*, *34*(4), 448–462.
Kelly, R. (1993). Comments on adaptive PD contoller for robot manipulators. *IEEE Transactions on Robotics and Automation*, *9*, 117–119.
Kelly, R. (1998). Global positioning of robot manipulators via PD control plus a class of nonlinear integral actions. *IEEE Transactions Automatic Control*, *43*(7), 934–938.
Kelly, R., & Carelli, R. (1988). Unified approach to adaptive control of robotic manipulators. In *IEEE Conference on Decision and Control*.
Kelly, R., Ortega, R., Ailon, A., & Loria, A. (1994). Global regulation of flexible joint robots using approximate differentiation. *IEEE Transactions Automatic Control*, *39*(6), 1222–1224.
Kelly, R., Santibanez, V., & Loria, A. (2005). *Control of robot manipulators in joint space*. Berlin: Springer
Khalil, W., & Dombre, E. (2004). *Modeling, identification and control of robots*. Oxford: Butterworth-Heinemann
Koditschek, D. E. (1987). Adaptive techniques for mechanical systems. In *Fifth Yale Workshop on Applications of Adaptive Systems Theory* (pp. 259–265).
Lee, K. W., & Khalil, H. (1997). Adaptive output feedback control of robot manipulators using high gain observer. *International Journal of Control*, *67*(6), 869–886.
Lewis, F. L., Abdallah, C. T., & Dawson, D. M. (1993). *Control of robot manipulators*. New York: Macmillan Publishing Company.
Li, X., & Cheah, C. C. (2012a). Dynamic region control for robot-assisted cell manipulation using optical tweezers. In *IEEE International Conference on Robotics and Automation* (pp. 1057–1062).
Li, X., & Cheah, C. C. (2012b). Multiple task-space robot control: Sense locally, act globally. In *IEEE International Conference on Robotics and Automation* (pp. 265–270).
Li, X., & Cheah, C. C. (2012c). Adaptive regional feedback control of robot manipulator with uncertain kinematics and depth information. In *American Control Conference* (pp. 5472–5477).
Li, X., & Cheah, C. C. (2013). Global task-space adaptive control of robot. *Automatica*, *49*(1), 58–69.
Li, X., & Cheah, C. C. (2014). Adaptive neural network control of robot based on a unified objective bound. *IEEE Transactions on Control Systems Technology*, *22*(3), 1032–1043.
Li, Z., Hsu, P., & Sastry, S. S. (1989). Grasping and coordinated manipulation by a multifingered robot hand. *International Journal of Robotics Research*, *18*(4), 33–50.
Liang, X., Huang, X., Wang, M., & Zeng, X. (2010). Adaptive task-space tracking control of robots without task-space and joint-space velocity measurements. *IEEE Transactions on Robotics*, *26*(4), 733–742.
Liu, C., Cheah, C. C., & Slotine, J. J. E. (2006). Adaptive Jacobian tracking control of rigid-link electrically driven robots based on visual task-space information. *Automatica*, *42*(9), 1491–1501.

Liu, C., Cheah, C. C., & Slotine, J. J. E. (2008). Adaptive task-space regulation of rigid-link flexible-joint robots with uncertain kinematics. *Automatica*, *44*(7), 1806–1814.

Lozano, R., & Brogliato, B. (1992). Adaptive control of robot manipulators with flexible joints. *IEEE Transactions on Automatic Control*, *37*(2), 174–181.

Malis, E. (2004). Visual servoing invariant to changes in camera-intrinsic parameters. *IEEE Transactions on Robotics and Automation*, *20*(1), 72–81.

Mezouar, Y., & Chaumette, F. (2002). Path planning for robust image-based control. *IEEE Transactions on Robotics and Automation*, *18*, 534–549.

Middleton, R. H., & Goodwin, G. C. (1988). Adaptive computed torque control for rigid link manipulators. *Systems and Control Letter*, *10*, 9–16.

Miyazaki, F., & Masutani, Y. (1990). Robustness of sensory feedback control based on imperfect Jacobian. In *Robotic Research: The Fifth International Symposium* (pp. 201–201).

Nakamura, Y. (1991). Advanced robotics: Redundancy and optimization. Reading, MA: Addison-Wesley.

Nakamura, Y., & Hanafusa, H. (1986). Inverse kinematic solutions with singularity robustness for robot manipulator control. *Journal of Dynamic Systems, Measurement and Control, Transactions of the ASME*, *108*(3), 163–171.

Nakamura, Y., Hanafusa, H., & Yoshikawa, T. (1987). Task-priority based redundncy control of robot manipulators. *International Journal of Robotics Research*, *6*(2), 3–15.

Niemeyer, G. & Slotine, J. J. E. (1991). Performance in adaptive manipulator control. *International Journal of Robotics Research*, *10*(2), 149–161.

Ortega, R., Loria, A., & Kelly, R. (1995). A semi-globally stable output feedback PI2D regulator for robot manipulators. *IEEE Transactions on Automatic Control*, *40*(8), 1432–1436.

Ortega, R., & Spong, M. W. (1989). Adaptive motion control of rigid robots: A tutorial. *Automatica*, *25*(6), 877–888.

Ozawa, R., Arimoto, S., Nakamura, S., & Bae, J. H. (2005). Control of an object with parallel surfaces by a pair of finger robots without object sensing. *IEEE Transactions on Robotics*, *21*(5), 965–976.

Paden, B., & Panja, R. (1988). A globally asymptotically stable 'PD+' controller for robot manipulator. *International Journal of Control*, *47*(6), 1697–1712.

Sadegh, N., & Horowitz, R. (1990). Stability and robustness analysis of a class of adaptive controllers for robotic manipulators. *International Journal of Robotics Research*, *9*(3), 74–92.

Schilling, R. J. (1990). *Fundamentals of robotics: Analysis and control*. Englewood Cliffs, NJ: Prentice Hall.

Sciavicco, L., & Siciliano, B. (1988). A solution algorithm to the inverse kinematic problem for redundant manipulators. *IEEE Journal of Robotics and Automation*, *4*, 403–410.

Slotine, J. J. E., & Li, W. (1987). On the adaptive control of robot manipulators. *International Journal of Robotics Research*, *6*, 49–59.

Slotine, J. J. E., & Li, W. (1987). Adaptive strategies in constrained manipulation. In *Proceedings of the IEEE International Conference on Robotics and Automation* (pp. 595–601).

Slotine, J. J. E., & Li, W. (1988). Adaptive manipulator control: A case study. *IEEE Transactions on Automatic Control*, *33*(11), 995–1003.

Slotine, J. J. E., & Li, W. (1991). *Applied nonlinear control*. Englewood Cliffs, NJ: Prentice Hall.

Spong, M. W. (1987). Modeling and control of elastic joint robots. *Journal of Dynamic Systems, Measurement, and Control*, *109*(4), 310–319.

Spong, M. W., Hutchinson, S., & Vidyasagar, M. (2006). *Robot modeling and control*. New York: Wiley.

Su, C. Y., & Stepanenko, Y. (1998). Redesign of hybrid adaptive/robust motion control of rigid-link electrically-driven robot manipulators. *IEEE Transactions on Robotics and Automation*, *14*, 651–655.

Takegaki, M., & Arimoto, S. (1981). A new feedback method for dynamic control of manipulators. *Journal of Dynamic Systems, Measurement, and Control*, *103*, 119–125.

Tomei, P. (1991). Adaptive PD contoller for robot manipulators. *IEEE Transactions on Robotics and Automation*, *7*, 565–570.

Tomei, P. (2000). Robust adaptive friction compensation for tracking control of robot manipulators. *IEEE Transactions on Automatic Control*, *45*(11), 2164–2169.

Wang, H., Liu, Y. H., & Zhou, D. (2007). Dynamic visual tracking for manipulators using an uncalibrated fixed camera. *IEEE Transactions on Robotics*, *23*(3), 610–617.

Wang, H., & Xie, Y. (2009). Prediction error based adaptive Jacobian tracking of robots with uncertain kinematics and dynamics. *IEEE Transactions on Automatic Control*, *54*(12), 2889–2894.

Wen, J. T., & Bayard, D. (1988). New class of control laws for robotic manipulators-part 2. Adaptive case. *International Journal of Control*, *47*(5), 1387–1406.

Yoshikawa, T. (2000). Force control of robot manipulators. In *IEEE International Conference on Robotics and Automation* (pp. 220–226).

Yoshikawa, T. (2010). Multifingered robot hands: Control for grasping and manipulation. *Annual Reviews in Control*, *34*(2), 199–208.

Yoshikawa, T., & Nagai, K. (1991). Manipulating and grasping forces in manipulation by multifingered robot hands. *IEEE Transactions on Robotics and Automation*, *7*(1), 67–77.

Zhao, Y., & Cheah, C. C. (2009). Neural network control of multifingered robot hands using visual feedback. *IEEE Transactions on Neural Networks*, *20*(5), 758–767.

MIX
Papier aus verantwortungsvollen Quellen
Paper from responsible sources
FSC® C105338

If you have any concerns about our products,
you can contact us on
ProductSafety@springernature.com

In case Publisher is established outside the EU,
the EU authorized representative is:
Springer Nature Customer Service Center GmbH
Europaplatz 3, 69115 Heidelberg, Germany

Printed by Libri Plureos GmbH
in Hamburg, Germany